Fractional-Order System: Control Theory and Applications

Fractional-Order System: Control Theory and Applications

Editors

Thach Ngoc Dinh
Shyam Kamal
Rajesh Kumar Pandey

MDPI • Basel • Beijing • Wuhan • Barcelona • Belgrade • Manchester • Tokyo • Cluj • Tianjin

Editors
Thach Ngoc Dinh
Conservatoire National des Arts et Métiers
(CNAM)
France

Shyam Kamal
Banaras Hindu University
India

Rajesh Kumar Pandey
Indian Institute of Technology (BHU) Varanasi
India

Editorial Office
MDPI
St. Alban-Anlage 66
4052 Basel, Switzerland

This is a reprint of articles from the Special Issue published online in the open access journal *Fractal and Fractional* (ISSN 2504-3110) (available at: https://www.mdpi.com/journal/fractalfract/special_issues/foscta).

For citation purposes, cite each article independently as indicated on the article page online and as indicated below:

LastName, A.A.; LastName, B.B.; LastName, C.C. Article Title. *Journal Name* **Year**, *Volume Number*, Page Range.

ISBN 978-3-0365-6422-7 (Hbk)
ISBN 978-3-0365-6423-4 (PDF)

Contents

About the Editors

Thach Ngoc Dinh

Thach Ngoc Dinh is a Vietnamese French control theorist. He was born in Vung Tau, Vietnam in 1988. He is a tenure track Associate Professor at CNAM Paris (French "grand établissement" and "grande école" in engineering) and at the Cedric-Lab EA4629, France. From 2016 to 2017, he held a Temporary Position of Assistant Professor at INSA Hauts-de-France of Université Polytechnique Hauts-de-France and at the LAMIH-Lab UMR CNRS 8201, France. From 2015 to 2016, he was a JSPS Postdoctoral Fellow at Kyushu Institute of Technology, Japan. He received the Ph.D. degree from Université Paris-Saclay funded by INRIA in the laboratories CAOR of Mines ParisTech-PSL and L2S UMR CNRS 8506, France in 2014. He obtained the MScRes in Automated Systems Engineering and the "Diplôme d'Ingénieur" (Master's Degree) in Electrical Engineering, both from INSA de Lyon, France in 2011. He has held a visiting position at Kyushu Institute of Technology, Japan (February–March 2019). He was recipient of the JSPS Postdoctoral Fellowship for North American and European Researchers in March of 2015. His research interests include robust control and interval estimation, fault detection with a special emphasis on autonomous systems.

Shyam Kamal

Shyam Kamal received his Bachelor's degree in Electronics and Communication Engineering from the Gurukula Kangri Vishwavidyalaya Haridwar, Uttrakhand, India in 2009, and Ph.D. in Systems and Control Engineering from the Indian Institute of Technology Bombay, India in 2014. From 2015 to 2016, he was with the Department of Systems Design and Informatics, Kyushu Institute of Technology, Japan as a Project Assistant Professor. Currently, he is an Associate Professor in the Department of Electrical Engineering, Indian Institute of Technology (BHU) Varanasi, India. He has published one monograph, four book chapters and 110 journal articles and conference papers. His research interests include the areas of nonlinear control and its applications. Particularly he is working on fractional-order systems, contraction analysis, discrete and continuous higher-order sliding mode control and multi-agent systems. He has received excellence awards in Ph.D. thesis from IIT Bombay in 2015 for his thesis "Sliding Mode Control of Fractional-order Systems" and INAE Young Engineer Award in 2019 for his research work. He is now an INAE Young Associate. He has also received honorary appointment as a Visiting professor from RMIT Melbourne Australia in July 2017 and in the duration of November and December 2019 and Visiting Professor from Harbin Institute of Technology China in August 2019.

Rajesh Kumar Pandey

Rajesh Kumar Pandey is currently working as Associate Professor in the Department of Mathematical Sciences, Indian Institute of Technology (BHU), Varanasi, India. He is recipient of Indo-US fellowship and INSA Visiting Fellowship for the year 2012. He had been visiting faculty in Southern Illinois University Carbondale Illinois, IL, USA during 2012-13, the University of Tokyo, Japan during Nov–Dec 2010, Central South University China and Shanghai University Shanghai, China. He has also worked as Assistant Professor in Mathematics at BITS Pilani and Indian Institute of Information Technology Design and Manufacturing Jabalpur during July 2009–May 2014. He received his Ph.D. degree from the Department of Applied Mathematics, Institute of Technology (BHU), Varanasi, India in 2009. He has published 70 research papers in refereed journals and guided 8 Ph D students. He had been a member of the advisory committees in various national and international conferences. His current areas of research include Fractional derivatives, Image processing, and Numerical methods for fractional integro-differential equations and fractional partial differential equations.

Preface to "Fractional-Order System: Control Theory and Applications"

Thach Ngoc Dinh would like to dedicate this book to the 70th birthday of his father Dinh Ngoc Thuan and to all his beloved family members for their constant support. Shyam Kamal would like to dedicate this book to his daughter Adhita Kamal and his wife Aprajita Kamal for their constant support. Rajesh Kumar Pandey would like to dedicate this book to his parents and his beloved family members for their constant support.

Thach Ngoc Dinh, Shyam Kamal, and Rajesh Kumar Pandey
Editors

fractal and fractional

Editorial

Fractional-Order System: Control Theory and Applications

Thach Ngoc Dinh [1,*], Shyam Kamal [2] and Rajesh Kumar Pandey [3]

[1] Conservatoire National des Arts et Métiers (CNAM), Cedric-Laetitia, Rue St-Martin, CEDEX 03, 75141 Paris, France
[2] Department of Electrical Engineering, Indian Institute of Technology, Banaras Hindu University, Varanasi 221005, Uttar Pradesh, India
[3] Department of Mathematical Sciences, Indian Institute of Technology, Banaras Hindu University, Varanasi 221005, Uttar Pradesh, India
* Correspondence: ngoc-thach.dinh@lecnam.net

Citation: Dinh, T.N.; Kamal, S.; Pandey, R.K. Fractional-Order System: Control Theory and Applications. *Fractal Fract.* **2023**, *7*, 48. https://doi.org/10.3390/fractalfract7010048

Received: 21 December 2022
Accepted: 27 December 2022
Published: 31 December 2022

(Fractional) differential equations have seen increasing use in physics, signal processing, fluid mechanics, viscoelasticity, mathematical biology, electrochemistry, and many other fields over the last two decades, providing a new and more realistic way to capture memory-dependent phenomena and irregularities inside systems using more sophisticated mathematical analysis (see, for example, [1] and the references therein).

The study of the stability of (fractional) differential equations has attracted a lot of attention as a result of its growing applications. Furthermore, fractional- and integer-order controllers have received increased attention in recent years. Among these are optimal control, CRONE controllers, fractional PID controllers, lead-lag compensators, and sliding mode control.

The purpose of this Special Issue is to carry out studies on fractional/integer-order control theory and its applications to practical systems modeled using fractional/integer-order differential equations such as design, implementation, and application of fractional/integer-order control to electrical circuits and systems, mechanical systems, chemical systems, biological systems, finance systems, etc.

Ten high-quality papers were accepted for publication in this Special Issue. The papers were written by different authors (note that no author published more than one paper, which proves the wide scope of the Special Issue). The published papers are briefly summarized as follows.

According to [2], the discrete fractional Fourier transform (DFRFT) has several definitions, the most common of which is the multiweighted fractional Fourier transform (M-WFRFT). It is difficult to demonstrate its unitarity. The weighted-type fractional Fourier transform, fractional-order matrix, and eigendecomposition-type fractional Fourier transform are used as basic functions to demonstrate and describe unitarity. They observed that the M-WFRFT has just four effective weighting terms, none of which are extended to M terms, as stated by the definition. Furthermore, the program code is examined, and the results demonstrate that the prior work (Digit Signal Process 2020: 104: 18) for unitary verification based on MATLAB is incorrect.

According to [3], there has been a recent surge in the number of papers addressing the overall issue of fractional-order controllers, with a concentration on fractional-order PID. This controller has been offered in several versions, each with its own set of tweaking techniques and implementation possibilities. A number of recent studies have discussed the practical application of such controllers. However, industrial acceptance of these controllers is still a long way off. Auto-tuning approaches for fractional-order PIDs may increase their desirability in relation to industrial applications. The existing auto-tuning approaches for fractional-order PIDs are reviewed in this work. The emphasis is on the most recent discoveries. For various processes, a comparison of many auto-tuning algorithms is addressed. Numerical examples are provided to demonstrate the applicability of the methodologies, which might be applied to simple industrial operations.

The study [4] proposes an interval estimator for a fourth-order nonlinear susceptible–exposed–infected–recovered (SEIR) model with disturbances using noisy counts of susceptible patients given by Public Health Services (PHS). According to the World Health Organization, infectious diseases are the leading cause of mortality among the top 10 causes of death worldwide (WHO). As a result, tracking and assessing the progression of these diseases is critical for developing intervention methods. The authors investigate a real-world situation in which some uncertain variables, such as model disturbances and uncertain input and output measurement noise, are not precisely available but fall within an interval. Furthermore, the unclear transmission bound rate from the susceptible to the exposed stage cannot be measured. They created an interval estimator based on an observability matrix that yields a tight interval vector for the SEIR model's actual states in a guaranteed manner without computing the observer gain. The developed approach provides additional freedom because it is not dependent on observer gain. For the estimated state vector, the convergence of the width to a known value in a finite period is explored to demonstrate the stability of the estimation error. Finally, simulation results show that the suggested approach performs well.

Ref. [5] discusses a novel finite time stability (FTS) of neutral fractional-order systems with a time delay (NFOTSs). In light of this, the Gronwall inequality is used to demonstrate the FTSs of NFOTSs in the literature. The application of fixed-point theory to show the FTS of NFOTSs is a novel component of our proposed study. Finally, two instances are used to validate and substantiate the theoretical contributions.

The authors of [6] introduce a framework of distributed interval observers for fractional-order multiagent systems with nonlinearity. First, a frame was created to construct the system's upper and lower boundaries. The positivity of the error dynamics might be ensured by applying monotone system theory, implying that the constraints could trap the initial state. Second, a sufficient condition was used to ensure that distributed interval observers are bounded. The adequate condition was then based on an expansion of the Lyapunov function in the realm of fractional calculus. An algorithm related to the observer design technique was also provided. Finally, a numerical simulation was utilized to demonstrate the distributed interval observer's usefulness.

The paper [7] investigates an approximate method for solving the generalized fractional diffusion equation that combines the finite difference and collocation methods (GFDE). The presented method's convergence and stability analyses are also thoroughly established. To ensure the proposed method's effectiveness and accuracy, test examples with different scale and weight functions are taken into account, and the numerical results obtained are compared to the existing methods in the literature. The suggested method works particularly well with generalized fractional derivatives (GFDs), as the existence of scale and weight functions in a GFD makes discretization and further analysis problematic.

According to [8], autonomous underwater vehicles (AUVs) have a wide range of uses due to their capacity to travel great distances, their ability conceal themselves well, their high level of intelligence, and their ability to replace humans in dangerous missions. AUV motion control systems, which can assure steady operation in the complicated ocean environment, have piqued the interest of researchers. The authors suggest a single-input fractional-order fuzzy logic controller (SIFOFLC) as an AUV motion-control system in this research. First, a single-input fuzzy logic controller (SIFLC) based on the signed distance approach was presented, with its control input being a linear combination of the error signal and its derivative. The SIFLC reduces the controller design and calculation procedure significantly. Then, a SIFOFLC with the error signal's derivative extending to a fractional order was produced, providing additional flexibility and adaptability. Finally, comparative numerical simulations of spiral dive motion control were performed to validate the superiority of the suggested control algorithm. Meanwhile, the hybrid particle swarm optimization (HPSO) technique was used to optimize the parameters of several controllers. The simulation results demonstrate the suggested control algorithm's enhanced stability and transient performance.

The traditional approach to the integration of fractional-order starting value issues, according to [9], is based on the Caputo derivative, whose beginning conditions are employed to build the classical integral equation. The authors show, using a simple counter example, that this technique results in incorrect free-response transients. The frequency-distributed model of the fractional integrator and its distributed beginning conditions are used to solve this fundamental problem. They answer the preceding counter-example using this model and provide a methodology that is a generalization of the integer-order approach. Finally, in the linear situation, this technique is used to model Fractional Differential Systems (FDS) and for the formulation of their transients. Two expressions are constructed, one based on the Mittag–Leffler function and the other on the notion of a distributed exponential function.

According to [10], fractional-order differential equations are effective tools for modeling dynamic systems with long-term memory effects. The verified simulation of such system models using interval tools enables the computation of assured enclosures of attainable pseudo-state regions over a finite time horizon. In prior work, the author published an iteration method based on Picard iteration that uses Mittag–Leffler functions to determine guaranteed pseudo-state enclosures. In this study, the corresponding iteration is generalized to use exponential functions during the iteration scheme evaluation. A validated solution of integer-order sets of differential equations yields such exponential functions. The goal of this work is to show that using exponential functions for Mittag–Leffler functions instead of pure box-type interval enclosures not only improves the tightness of the calculated pseudo-state enclosures, but also minimizes the required computational cost. These claims are supported by a realistic simulation model of the charging/discharging kinetics of Lithium-ion batteries.

Finally, Ref. [11] investigates the synchronization of fractional-order uncertain delayed neural networks with an event-triggered communication strategy. By developing an appropriate Lyapunov–Krasovskii functional (LKF) and inequality approaches, sufficient criteria for the stability of delayed neural networks are obtained. The criteria are expressed as linear matrix inequalities (LMIs). To accomplish synchronization, a controller is derived using the drive-response idea, the LMI technique, and the Lyapunov stability theorem. Finally, numerical examples are provided to validate the effectiveness of the major findings.

Acknowledgments: The Guest Editors of this Special Issue would like to thank the anonymous reviewers and the editorial office for their hard work during the review and publication process.

References

1. Kamal, S.; Sharma, R.K.; Dinh, T.N.; Harikrishnan, M S.; Bandyopadhyay, B. Sliding Mode Control of Uncertain Fractional Order Systems: A Reaching Phase Free Approach. *Asian J. Control* **2021**, *23*, 199–208. [CrossRef]
2. Zhao, T.; Chi, Y. Multiweighted-Type Fractional Fourier Transform: Unitarity. *Fractal Fract.* **2021**, *5*, 205. [CrossRef]
3. Muresan, C.I.; Birs, I.; Ionescu, C.; Dulf, E.H.; De Keyser, R. A Review of Recent Developments in Autotuning Methods for Fractional-Order Controllers. *Fractal Fract.* **2022**, *6*, 37. [CrossRef]
4. Khan, A.; Bai, X.; Ilyas, M.; Rauf, A.; Xie, W.; Yan, P.; Zhang, B. Design and Application of an Interval Estimator for Nonlinear Discrete-Time SEIR Epidemic Models. *Fractal Fract.* **2022**, *6*, 213. [CrossRef]
5. Ben Makhlouf, A.; Baleanu, D. Finite Time Stability of Fractional Order Systems of Neutral Type. *Fractal Fract.* **2022**, *6*, 289. [CrossRef]
6. Zhang, H.; Huang, J.; He, S. Fractional-Order Interval Observer for Multiagent Nonlinear Systems. *Fractal Fract.* **2022**, *6*, 355. [CrossRef]
7. Kumar, S.; Pandey, R.K.; Kumar, K.; Kamal, S.; Dinh, T.N. Finite Difference–Collocation Method for the Generalized Fractional Diffusion Equation. *Fractal Fract.* **2022**, *6*, 387. [CrossRef]
8. Cui, Z.; Liu, L.; Zhu, B.; Zhang, L.; Yu, Y.; Zhao, Z.; Li, S.; Liu, M. Spiral Dive Control of Underactuated AUV Based on a Single-Input Fractional-Order Fuzzy Logic Controller. *Fractal Fract.* **2022**, *6*, 519. [CrossRef]
9. Maamri, N.; Trigeassou, J.-C. A Plea for the Integration of Fractional Differential Systems: The Initial Value Problem. *Fractal Fract.* **2022**, *6*, 550. [CrossRef]

10. Rauh, A. Exponential Enclosures for the Verified Simulation of Fractional-Order Differential Equations. *Fractal Fract.* **2022**, *6*, 567. [CrossRef]
11. Hymavathi, M.; Syed Ali, M.; Ibrahim, T.F.; Younis, B.A.; Osman, K.I.; Mukdasai, K. Synchronization of Fractional-Order Uncertain Delayed Neural Networks with an Event-Triggered Communication Scheme. *Fractal Fract.* **2022**, *6*, 641. [CrossRef]

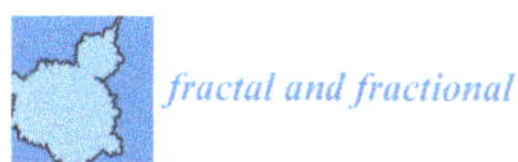

fractal and fractional

Article

Finite Time Stability of Fractional Order Systems of Neutral Type

Abdellatif Ben Makhlouf [1,*] **and Dumitru Baleanu** [2,3,4]

[1] Mathematics Department, College of Science, Jouf University, P.O. Box 2014, Sakaka 72388, Saudi Arabia
[2] Department of Mathematics, Cankaya University, 06790 Ankara, Turkey; dumitru@cankaya.edu.tr
[3] Institute of Space Sciences, Magurele, 077125 Bucharest, Romania
[4] Department of Medical Research, China Medical University Hospital, China Medical University, Taichung 40447, Taiwan
* Correspondence: abmakhlouf@ju.edu.sa

Abstract: This work deals with a new finite time stability (FTS) of neutral fractional order systems with time delay (NFOTSs). In light of this, FTSs of NFOTSs are demonstrated in the literature using the Gronwall inequality. The innovative aspect of our proposed study is the application of fixed point theory to show the FTS of NFOTSs. Finally, using two examples, the theoretical contributions are confirmed and substantiated.

Keywords: fractional calculus; neutral systems; fixed-point theory

Citation: Ben Makhlouf, A.; Baleanu, D. Finite Time Stability of Fractional Order Systems of Neutral Type. *Fractal Fract.* **2022**, *6*, 289. https://doi.org/10.3390/fractalfract6060289

Academic Editors: Ricardo Almeida and Yongguang Yu

Received: 23 March 2022
Accepted: 24 May 2022
Published: 26 May 2022

1. Introduction

The Fractional Order System (FOS) is a nonlinear system presented with a non-integer derivative. It is well established that mathematical models can be used to describe physical systems. These mathematical models are used to operate such systems in a variety of ways, including controlling, observing, and detecting. The faults and errors of modelization may affect the system quality and performance. Therefore, the use of Fractional derivatives can approach such a mathematical model to physical reality. This fact is proved in many real physical systems, see for example [1]. Recently, the fractional calculus has attracted the attention of many researchers and numerous works have been published in this context [2–11]. In fact, by using quantum calculus, the work in [6] deals with the extension of a hybrid fractional differential operator. Utilizing the local fractional Laplace variational iteration methods and the local fractional reduced differential transform, authors in [7] have obtained an approximation of the solutions for coupled Korteweg De Vries Equations. The application of these FOSs is numerous in different domain applications, whether in electricity [10], thermal [5], chemistry [11], signal processing [12], biology [13,14] or control theory, such as fault estimation [15], stabilization [16], observer design [16,17], optimal control [18], and asymptotic stability [19,20].

The study of FTS for the Fractional Order Time Delay Systems (FOTDSs) has been largely studied in the literature in the case of continuous and discrete time [21–30]. In [30], H. Ye et al., have shown a Generalized Gronwall Inequality (GGI). After that, authors in [25] have used the GGI to study the FTS for FOTDSs. The stability of neutral fractional order time delay systems with Lipschitz nonlinearities in finite time has been investigated by F. Du et al. in [23]. The finite-time stability of a class of fractional delayed neural networks with commensurate order between 0 and 1 was studied by the authors in [28]. Additionally, the authors in [26] have provided an analytical method based on the Laplace transform and the 'inf-sup' approach for evaluating the finite-time stability of singular fractional-order switching systems with delay. The authors have proposed a constructive geometric design for switching laws based on the partitioning of the stability state regions in convex cones. The suggested technique allows for the development of novel delay-dependent adequate conditions for the system's regularity, impulse-free, and finite-time stability in terms of tractable matrix inequalities and Mittag–Leffler functions. A case study is offered to demonstrate the proposed method's efficacy. Using the Lyapunov method, Thanh et al. in [27] have investigated a novel FTS analysis of FOTDSs. By using Banach fixed point method, author in [21] has studied the FTS for FOTDSs. In the discrete case, one has the following references [22,24,29]. Indeed, authors in [24] have proposed a sufficient condition for ensuring the FTS for Nabla uncertain FOS. Furthermore, authors in [22] have established a new Gronwall Inequality and they have used it to study the FTS of a class of nonlinear fractional delay difference systems. Furthermore, in [29], the FTS of Caputo delta fractional difference equations is investigated. On a finite time domain, a generalized Gronwall inequality is given. For fractional differential equations, a finite-time stability condition is suggested. The concept is then generalized to discrete fractional cases. There are finite-time stable conditions for a linear fractional difference equation with constant delays. To support the theoretical result, one example is numerically shown.

Motivated by the above study, this article treats the FTS for FOS of neutral type by using a version of the Banach fixed point theorem and some properties of the Mittag–Leffler Function (MLF). The contribution of this work is summarized as follows:

- Knowing that, FTS of NFOTSs are proved in the literature based on the Gronwall inequality, see [23]. The novelty of our suggested work comes from the use of the fixed point theory to demonstrate the FTS of NFOTSs;
- A novel FTS result of FOS of neutral type is given;
- The theoretical contributions are confirmed and validated by two examples.

The rest of the paper is organized as follows. The second section deals with some preliminaries. Some basic results related to fractional calculus, fixed point theory, as well as finite time stability are shown. In regards to the third section, the stability analysis of the suggested system (2), in the case of $(\lambda_1 < \lambda_2)$ and $(\lambda_1 = \lambda_2)$, is investigated and described. Note that the fixed point approach is used to demonstrate the main results. The fourth section is concentrated to show the validity of the proposed results. Two examples are suggested to demonstrate the efficiency of the main results. Finally, to end the work, a conclusion is presented in the fifth section showing the principle fundamentals of the work.

2. Basic Results

Definition 1 ([31]). *Given $0 < \chi < 1$. The CFD is given by,*

$$^{C}D_a^{\chi}g(s) = \frac{1}{\Gamma(1-\chi)}\frac{d}{ds}\int_a^s (s-\omega)^{-\chi}\big(g(\omega) - g(a)\big)d\omega. \tag{1}$$

Definition 2 ([31]). *The MLF is defined by :*

$$E_\chi(s) = \sum_{q=0}^{+\infty} \frac{s^q}{\Gamma(q\chi + 1)},$$

with $\chi > 0, s \in \mathbb{C}$.

Lemma 1 ([21]). *We have for $s \geq 0$*

$$\frac{s^\chi}{E_\chi(\lambda s^\chi)} \leq \frac{\Gamma(\chi+1)}{\lambda},$$

where $0 < \chi < 1$ and $\lambda > 0$.

Remark 1. *The function $d(t) = E_\chi\big(b(t-\tau)^\chi\big)$ satisfies $^{C}D_a^{\chi}d(t) = bd(t)$, where $b \in \mathbb{R}^*$.*

Definition 3. *A mapping $\beta : B \times B \to [0, \infty]$ is called a generalized metric on a nonempty set B if:*

S1 $\beta(\omega_1, \omega_2) = 0$ *if, and only if, $\omega_1 = \omega_2$;*

S2 $\beta(\omega_1, \omega_2) = \beta(\omega_2, \omega_1)$ *for all $\omega_1, \omega_2 \in B$;*

S3 $\beta(\omega_1, \omega_3) \leq \beta(\omega_1, \omega_2) + \beta(\omega_2, \omega_3)$ *for all $\omega_1, \omega_2, \omega_3 \in B$.*

Theorem 1. *Let (B, β) be a generalized complete metric space. Suppose that $K : B \to B$ is contractive with $k < 1$. If there is an integer $k_0 \geq 0$, such that $\beta(K^{k_0+1}b_0, K^{k_0}b_0) < \infty$ for some $b_0 \in B$, so:*

(a) $\displaystyle\lim_{n \longrightarrow +\infty} K^n b_0 = b_1$ *with $K(b_1) = b_1$;*

(b) *b_1 is the unique fixed point of K in $B^* := \{b_2 \in B : \beta(K^{k_0}b_0, b_2) < \infty\}$;*

(c) *If $b_2 \in B^*$, then $\beta(b_1, b_2) \leq \frac{1}{1-k}\beta(Kb_2, b_2)$.*

We consider the following system:

$$
\begin{aligned}
{}^C D_0^{\lambda_2} x(t) - C \, {}^C D_0^{\lambda_1} x(t - \varsigma(t)) \;=\; & B_0 x(t) + B_1 x(t - \varsigma(t)) \\
& + B_2 v(t) + F(t, x(t), x(t - \varsigma(t)), v(t)), \; t \geq 0, \quad (2)
\end{aligned}
$$

with the initial condition $x(s) = \zeta(s)$ for $-\varsigma \leq s \leq 0$, with $0 < \lambda_1 \leq \lambda_2 < 1$, $\varsigma(t)$ is continuous, $0 \leq \varsigma(t) \leq \varsigma$, $v(t) \in \mathbb{R}^p$ is the disturbance, $\zeta \in C^1([-\varsigma, 0], \mathbb{R}^q)$, $C \in \mathbb{R}^{q \times q}$, $B_0 \in \mathbb{R}^{q \times q}$, $B_1 \in \mathbb{R}^{q \times q}$, $B_2 \in \mathbb{R}^{q \times p}$.

The function F is continuous and satisfies:

$$
\|F(\tau, \sigma_1, \sigma_2, \sigma_3) - F(\tau, \psi_1, \psi_2, \psi_3)\| \leq f(\tau)\big(\|\sigma_1 - \psi_1\| + \|\sigma_2 - \psi_2\| + \|\sigma_3 - \psi_3\|\big), \quad (3)
$$

and $F(\tau, 0, 0, 0) = 0$, for all $(\tau, \sigma_1, \sigma_2, \sigma_3, \psi_1, \psi_2, \psi_3) \in \mathbb{R}_+ \times \mathbb{R}^q \times \mathbb{R}^q \times \mathbb{R}^p \times \mathbb{R}^q \times \mathbb{R}^q \times \mathbb{R}^p$ where f is a continuous function.

The function v is continuous and satisfies:

$$
\exists \varrho > 0 : \quad v^T(t)v(t) \leq \varrho^2. \tag{4}
$$

Definition 4. *The FOS (2) possesses FTS w.r.t. $\{\gamma_1, \gamma_2, \varrho, T\}$, $\gamma_1 < \gamma_2$ if*

$$
\|\zeta\| \leq \gamma_1,
$$

implies:

$$
\|x(t)\| \leq \gamma_2, \; \forall t \in [0, T],
$$

for all v satisfying (4), where $\|\zeta\| = \displaystyle\sup_{\tau \in [-\varsigma, 0]} \|\zeta(\tau)\|$.

3. Stability Analysis

This section is used to show our main results.

First, let us denote $b_i = \displaystyle\max_{r \in [0, T]} \big(f(r) + \|B_i\|\big)$ for $i = 0, 1, 2$ and $c = \|C\|$.

In the next subsections, we study the FTS of (2) when $\lambda_1 < \lambda_2$ and when $\lambda_1 = \lambda_2$.

3.1. The Case $\lambda_1 < \lambda_2$

From Theorem 1 in [23], we have the solution of the FOS (2) is the solution of the following system

$$
\begin{aligned}
x(t) \;=\;& \zeta(0) - C\zeta(-\varsigma(0))\frac{t^{\lambda_2-\lambda_1}}{\Gamma(\lambda_2-\lambda_1+1)} + \frac{1}{\Gamma(\lambda_2-\lambda_1)}\int_0^t (t-s)^{\lambda_2-\lambda_1-1}Cx(s-\varsigma(s))ds \\
& + \frac{1}{\Gamma(\lambda_2)}\int_0^t (t-s)^{\lambda_2-1}\Big[B_0 x(s) + B_1 x(s-\varsigma(s)) \\
& + B_2 v(s) + F(s,x(s),x(s-\varsigma(s)),v(s))\Big]ds,\; 0 \le t \le T,
\end{aligned}
$$

$$
x(t) = \zeta(t),\; -\varsigma \le t \le 0.
$$

Theorem 2. *The FOS (2) is FTS w.r.t. $\{\gamma_1,\gamma_2,\varrho,T\}$, $\gamma_1 < \gamma_2$ if there exist $\eta_1,\eta_2 > 0$, such that*

$$
G(\gamma_1,\varrho) \le \gamma_2,
\tag{5}
$$

where

$$
\begin{aligned}
G(\gamma_1,\varrho) \;=\;& \Big(\delta + c_1 E_{\lambda_2-\lambda_1}\big((c+\eta_1)T^{\lambda_2-\lambda_1}\big)E_{\lambda_2}\big((b_0+b_1+\eta_2)T^{\lambda_2}\big)\Big)\gamma_1 \\
& + c_2 E_{\lambda_2-\lambda_1}\big((c+\eta_1)T^{\lambda_2-\lambda_1}\big)E_{\lambda_2}\big((b_0+b_1+\eta_2)T^{\lambda_2}\big)\varrho,
\end{aligned}
\tag{6}
$$

$$
\delta = 1 + c\frac{T^{\lambda_2-\lambda_1}}{\Gamma(\lambda_2-\lambda_1+1)},\; c_1 = \frac{1}{(1-\eta)}\left(\frac{c\delta M_1}{\Gamma(\lambda_2-\lambda_1+1)} + \frac{b_0\delta M_2}{\Gamma(\lambda_2+1)} + \frac{b_1\delta M_2}{\Gamma(\lambda_2+1)}\right),
$$

$$
c_2 = \frac{b_2 M_2}{(1-\eta)\Gamma(\lambda_2+1)},\; M_1 = \sup_{\tau\in[0,T]}\left(\frac{\tau^{\lambda_2-\lambda_1}}{E_{\lambda_2-\lambda_1}\big((c+\eta_1)\tau^{\lambda_2-\lambda_1}\big)}\right),
$$

$$
M_2 = \sup_{\tau\in[0,T]}\left(\frac{\tau^{\lambda_2}}{E_{\lambda_2}\big((b_0+b_1+\eta_2)\tau^{\lambda_2}\big)}\right)\; and\; \eta = \left(\frac{c}{c+\eta_1} + \frac{b_0+b_1}{b_0+b_1+\eta_2}\right).
$$

Proof. Let $\zeta \in C^1([-\varsigma,0],\mathbb{R}^q)$, such that $\|\zeta\| \le \gamma_1$.

Let $\mathcal{F} = C([-\varsigma,T],\mathbb{R}^q)$ and consider the metric β on $\mathcal{F}$ by

$$
\beta(y_1,y_2) = \inf\left\{r \in [0,\infty] : \|y_1(t) - y_2(t)\| \le rg(t), \forall t \in [-\varsigma,T]\right\},
$$

where g is given by $g(\tau) = E_{\lambda_2-\lambda_1}\big((c+\eta_1)\tau^{\lambda_2-\lambda_1}\big)E_{\lambda_2}\big((b_0+b_1+\eta_2)\tau^{\lambda_2}\big)$ for $\tau \in [0,T]$ and $g(\tau) = 1$, for $\tau \in [-\varsigma,0]$.

We consider the operator: $\mathcal{D} : \mathcal{F} \to \mathcal{F}$, such that

$$
\begin{aligned}
(\mathcal{D}X)(w) \;=\;& \zeta(0) - C\zeta(-\varsigma(0))\frac{w^{\lambda_2-\lambda_1}}{\Gamma(\lambda_2-\lambda_1+1)} \\
& + \frac{1}{\Gamma(\lambda_2-\lambda_1)}\int_0^w (w-s)^{\lambda_2-\lambda_1-1}CX(s-\varsigma(s))ds \\
& + \frac{1}{\Gamma(\lambda_2)}\int_0^w (w-s)^{\lambda_2-1}\Big[B_0 X(s) + B_1 X(s-\varsigma(s)) \\
& + B_2 v(s) + F(s,X(s),X(s-\varsigma(s)),v(s))\Big]ds,
\end{aligned}
\tag{7}
$$

for $w \in [0,T]$ and $(\mathcal{D}X)(w) = \zeta(w)$, for $w \in [-\varsigma,0]$.

Note that, $\mathcal{D}$ is well defined, $(\mathcal{F},\beta)$ is a generalized complete metric space, $\beta(\mathcal{D}X_0,X_0) < \infty$, and $\{X_1 \in \mathcal{F} : \beta(X_0,X_1) < \infty\} = \mathcal{F}$, $\forall X_0 \in \mathcal{F}$.

Let X_1, $X_2 \in \mathcal{F}$, for $w \in [-\varsigma, 0]$, we get $(\mathcal{D}X_1)(w) - (\mathcal{D}X_2)(w) = 0$. For $w \in [0, T]$, we have

$$
\left\| (\mathcal{D}X_1)(w) - (\mathcal{D}X_2)(w) \right\|
$$
$$
\leq \int_0^w \frac{(w-r)^{\lambda_2 - \lambda_1 - 1}}{\Gamma(\lambda_2 - \lambda_1)} c \| X_1(r - \varsigma(r)) - X_2(r - \varsigma(r)) \| dr
$$
$$
+ \int_0^w \frac{(w-r)^{\lambda_2 - 1}}{\Gamma(\lambda_2)} \Big[\Big(f(r) + \|B_0\| \Big) \| X_1(r) - X_2(r) \|
$$
$$
+ \Big(f(r) + \|B_1\| \Big) \| X_1(r - \varsigma(r)) - X_2(r - \varsigma(r)) \| \Big] dr
$$
$$
\leq c \int_0^w \frac{(w-r)^{\lambda_2 - \lambda_1 - 1}}{\Gamma(\lambda_2 - \lambda_1)} \| X_1(r - \varsigma(r)) - X_2(r - \varsigma(r)) \| dr
$$
$$
+ b_0 \int_0^w (w-r)^{\lambda_2 - 1} \frac{\| X_1(r) - X_2(r) \|}{\Gamma(\lambda_2)} dr
$$
$$
+ b_1 \int_0^w (w-r)^{\lambda_2 - 1} \frac{\| X_1(r - \varsigma(r)) - X_2(r - \varsigma(r)) \|}{\Gamma(\lambda_2)} dr. \tag{8}
$$

Then,

$$
\left\| (\mathcal{D}X_1)(w) - (\mathcal{D}X_2)(w) \right\|
$$
$$
\leq c \int_0^w \frac{(w-r)^{\lambda_2 - \lambda_1 - 1}}{\Gamma(\lambda_2 - \lambda_1)} \frac{\| X_1(r - \varsigma(r)) - X_2(r - \varsigma(r)) \|}{g(r - \varsigma(r))} g(r - \varsigma(r)) dr
$$
$$
+ \frac{b_0}{\Gamma(\lambda_2)} \int_0^w (w-r)^{\lambda_2 - 1} \frac{\| X_1(r) - X_2(r) \|}{g(r)} g(r) dr
$$
$$
+ \frac{b_1}{\Gamma(\lambda_2)} \int_0^w (w-r)^{\lambda_2 - 1} \frac{\| X_1(r - \varsigma(r)) - X_2(r - \varsigma(r)) \|}{g(r - \varsigma(r))} g(r - \varsigma(r)) dr
$$
$$
\leq c\beta(X_1, X_2) \int_0^w \frac{(w-r)^{\lambda_2 - \lambda_1 - 1}}{\Gamma(\lambda_2 - \lambda_1)} g(r - \varsigma(r)) dr
$$
$$
+ \frac{b_0 \beta(X_1, X_2)}{\Gamma(\lambda_2)} \int_0^w (w-r)^{\lambda_2 - 1} g(r) dr
$$
$$
+ \frac{b_1 \beta(X_1, X_2)}{\Gamma(\lambda_2)} \int_0^w (w-r)^{\lambda_2 - 1} g(r - \varsigma(r)) dr.
$$

Therefore,

$$
\begin{aligned}
\left\| (\mathcal{D}X_1)(w) - (\mathcal{D}X_2)(w) \right\| \leq{} & c\beta(X_1, X_2) \int_0^w \frac{(w-\tau)^{\lambda_2 - \lambda_1 - 1}}{\Gamma(\lambda_2 - \lambda_1)} g(\tau) d\tau \\
& + \frac{(b_0 + b_1)\beta(X_1, X_2)}{\Gamma(\lambda_2)} \int_0^w (w-\tau)^{\lambda_2 - 1} g(\tau) d\tau \\
\leq{} & c\beta(X_1, X_2) E_{\lambda_2}\big((b_0 + b_1 + \eta_2) w^{\lambda_2} \big) \\
& \times \int_0^w \frac{(w-\tau)^{\lambda_2 - \lambda_1 - 1}}{\Gamma(\lambda_2 - \lambda_1)} E_{\lambda_2 - \lambda_1}\big((c + \eta_1)\tau^{\lambda_2 - \lambda_1} \big) d\tau \\
& + (b_0 + b_1)\beta(X_1, X_2) E_{\lambda_2 - \lambda_1}\big((c + \eta_1) w^{\lambda_2 - \lambda_1} \big) \\
& \times \int_0^w \frac{(w-\tau)^{\lambda_2 - 1}}{\Gamma(\lambda_2)} E_{\lambda_2}\big((b_0 + b_1 + \eta_2)\tau^{\lambda_2} \big) d\tau.
\end{aligned}
$$

Using Remark 1, we get

$$\left\|(\mathcal{D}X_1)(w) - (\mathcal{D}X_2)(w)\right\| \leq \frac{c}{c+\eta_1}\beta(X_1,X_2)g(w) + \frac{b_0}{b_0+b_1+\eta_2}\beta(X_1,X_2)g(w)$$
$$+ \frac{b_1}{b_0+b_1+\eta_2}\beta(X_1,X_2)g(w)$$
$$\leq \left(\frac{c}{c+\eta_1} + \frac{b_0+b_1}{b_0+b_1+\eta_2}\right)\beta(X_1,X_2)g(w). \tag{9}$$

Then,

$$\frac{\left\|(\mathcal{D}X_1)(w) - (\mathcal{D}X_2)(w)\right\|}{g(w)} \leq \left(\frac{c}{c+\eta_1} + \frac{b_0+b_1}{b_0+b_1+\eta_2}\right)\beta(X_1,X_2).$$

Thus,

$$\beta(\mathcal{D}X_1,\mathcal{D}X_2) \leq \left(\frac{c}{c+\eta_1} + \frac{b_0+b_1}{b_0+b_1+\eta_2}\right)\beta(X_1,X_2).$$

Therefore, $\mathcal{D}$ is contractive.

Let x_0 be the function given by $x_0(\tau) = \zeta(\tau)$, for $\tau \in [-\varsigma,0]$ and $x_0(\tau) = \zeta(0) - C\zeta\big(-\varsigma(0)\big)\frac{\tau^{\lambda_2-\lambda_1}}{\Gamma(\lambda_2-\lambda_1+1)}$ for $\tau \in [0,T]$.

Then, we have

$$\|x_0(\tau)\| \leq \left(\|\zeta\| + c\|\zeta\|\frac{T^{\lambda_2-\lambda_1}}{\Gamma(\lambda_2-\lambda_1+1)}\right),$$

for all $\tau \in [-\varsigma,T]$.

For $\tau \in [-\varsigma,0]$, we get $(\mathcal{D}x_0)(\tau) - x_0(\tau) = 0$.

For $w \in [0,T]$, we have

$$\left\|(\mathcal{D}x_0)(w) - x_0(w)\right\| \leq \int_0^w \frac{(w-s)^{\lambda_2-\lambda_1-1}}{\Gamma(\lambda_2-\lambda_1)} c\|x_0(s-\varsigma(s))\| ds$$
$$+ \frac{1}{\Gamma(\lambda_2)}\int_0^w (w-s)^{\lambda_2-1}\big[b_0\|x_0(s)\| + b_1\|x_0(s-\varsigma(s))\| + b_2\varrho\big] ds$$
$$\leq c\left(\|\zeta\| + c\|\zeta\|\frac{T^{\lambda_2-\lambda_1}}{\Gamma(\lambda_2-\lambda_1+1)}\right)\frac{w^{\lambda_2-\lambda_1}}{\Gamma(\lambda_2-\lambda_1+1)}$$
$$+ \left(b_0\left(\|\zeta\| + c\|\zeta\|\frac{T^{\lambda_2-\lambda_1}}{\Gamma(\lambda_2-\lambda_1+1)}\right) + b_1\big(\|\zeta\|\right.$$
$$+ \left.c\|\zeta\|\frac{T^{\lambda_2-\lambda_1}}{\Gamma(\lambda_2-\lambda_1+1)}\big) + b_2\varrho\right)\frac{w^{\lambda_2}}{\Gamma(\lambda_2+1)}$$
$$\leq c\|\zeta\|\delta\frac{w^{\lambda_2-\lambda_1}}{\Gamma(\lambda_2-\lambda_1+1)}$$
$$+ \left(b_0\|\zeta\|\delta + b_1\|\zeta\|\delta + b_2\varrho\right)\frac{w^{\lambda_2}}{\Gamma(\lambda_2+1)}. \tag{10}$$

Then

$$\frac{\left\|(\mathcal{D}x_0)(w) - x_0(w)\right\|}{g(w)} \leq \frac{c\|\zeta\|\delta M_1}{\Gamma(\lambda_2-\lambda_1+1)}$$
$$+ \left(b_0\|\zeta\|\delta + b_1\|\zeta\|\delta + b_2\varrho\right)\frac{M_2}{\Gamma(\lambda_2+1)}, \tag{11}$$

for all $w \in [0,T]$.

Therefore,

$$\beta(\mathcal{D}x_0, x_0) \leq \frac{c\|\zeta\|\delta M_1}{\Gamma(\lambda_2 - \lambda_1 + 1)}$$
$$+ \ (b_0\|\zeta\|\delta + b_1\|\zeta\|\delta + b_2\varrho)\frac{M_2}{\Gamma(\lambda_2 + 1)}. \tag{12}$$

It follows from Theorem 1 that there is a unique solution x of (2) with initial conditions of ζ, such that

$$\beta(x_0, x) \leq \frac{1}{1 - \eta}\Bigg[\frac{c\|\zeta\|\delta M_1}{\Gamma(\lambda_2 - \lambda_1 + 1)}$$
$$+ \ (b_0\|\zeta\|\delta + b_1\|\zeta\|\delta + b_2\varrho)\frac{M_2}{\Gamma(\lambda_2 + 1)}\Bigg]$$
$$\leq \ c_1\gamma_1 + c_2\varrho. \tag{13}$$

Therefore,

$$\|x_0(t) - x(t)\| \leq (c_1\gamma_1 + c_2\varrho)E_{\lambda_2 - \lambda_1}\big((c + \eta_1)T^{\lambda_2 - \lambda_1}\big)E_{\lambda_2}\big((b_0 + b_1 + \eta_2)T^{\lambda_2}\big),$$

for every $t \in [0, T]$.

Then,

$$\|x(t)\| \leq \|x_0(t)\| + \|x(t) - x_0(t)\|$$
$$\leq \Big(\delta + c_1 E_{\lambda_2 - \lambda_1}\big((c + \eta_1)T^{\lambda_2 - \lambda_1}\big)E_{\lambda_2}\big((b_0 + b_1 + \eta_2)T^{\lambda_2}\big)\Big)\gamma_1$$
$$+ \ c_2 E_{\lambda_2 - \lambda_1}\big((c + \eta_1)T^{\lambda_2 - \lambda_1}\big)E_{\lambda_2}\big((b_0 + b_1 + \eta_2)T^{\lambda_2}\big)\varrho, \tag{14}$$

for every $t \in [0, T]$.

Thus, $\|x(t)\| \leq \gamma_2$, for all $t \in [0, T]$, if (5) is satisfied. $\square$

Remark 2. *Using Lemma 1, we get*

$$c_1 \leq \frac{1}{(1 - \eta)}\left(\frac{c\delta}{c + \eta_1} + \frac{b_0\delta}{b_0 + b_1 + \eta_2} + \frac{b_1\delta}{b_0 + b_1 + \eta_2}\right)$$

and

$$c_2 \leq \frac{1}{(1 - \eta)}\frac{b_2}{b_0 + b_1 + \eta_2}.$$

Let

$$\tilde{c}_1 = \frac{1}{(1 - \eta)}\left(\frac{c\delta}{c + \eta_1} + \frac{b_0\delta}{b_0 + b_1 + \eta_2} + \frac{b_1\delta}{b_0 + b_1 + \eta_2}\right)$$

and

$$\tilde{c}_2 = \frac{1}{(1 - \eta)}\frac{b_2}{b_0 + b_1 + \eta_2}.$$

Therefore, the condition (5) can be relaxed by:

$$\tilde{G}(\gamma_1, \varrho) \leq \gamma_2, \tag{15}$$

where

$$\tilde{G}(\gamma_1, \varrho) = \Big(\delta + \tilde{c}_1 E_{\lambda_2 - \lambda_1}\big((c + \eta_1)T^{\lambda_2 - \lambda_1}\big)E_{\lambda_2}\big((b_0 + b_1 + \eta_2)T^{\lambda_2}\big)\Big)\gamma_1$$
$$+ \ \tilde{c}_2 E_{\lambda_2 - \lambda_1}\big((c + \eta_1)T^{\lambda_2 - \lambda_1}\big)E_{\lambda_2}\big((b_0 + b_1 + \eta_2)T^{\lambda_2}\big)\varrho. \tag{16}$$

3.2. The Case $\lambda_1 = \lambda_2$

The solution of the FOS (2) is the solution of

$$
\begin{aligned}
x(t) &= \zeta(0) + C\left(x(t - \varsigma(t)) - \zeta(-\varsigma(0))\right) + \frac{1}{\Gamma(\lambda_2)} \int_0^t (t - s)^{\lambda_2 - 1}\Big[B_0 x(s) + B_1 x(s - \varsigma(s)) \\
&+ B_2 v(s) + F(s, x(s), x(s - \varsigma(s)), v(s))\Big] ds,\ 0 \le t \le T,
\end{aligned}
$$

$$
x(t) = \zeta(t),\ -\varsigma \le t \le 0.
$$

Theorem 3. *The FOS (2) is FTS w.r.t. $\{\gamma_1, \gamma_2, \varrho, T\}$, $\gamma_1 < \gamma_2$ if there exist $\theta > 0$, such that*

$$
\eta < 1,
$$

and

$$
K(\gamma_1, \varrho) \le \gamma_2, \tag{17}
$$

where

$$
\eta = \left(c + \frac{b_0 + b_1}{b_0 + b_1 + \theta}\right),
$$

$$
\begin{aligned}
K(\gamma_1, \varrho) &= \left(1 + c_1 E_{\lambda_2}\big((b_0 + b_1 + \theta) T^{\lambda_2}\big)\right)\gamma_1 \\
&+ c_2 E_{\lambda_2}\big((b_0 + b_1 + \theta) T^{\lambda_2}\big)\varrho, \tag{18}
\end{aligned}
$$

$$
c_1 = \frac{1}{(1 - \eta)}\left(2c + \frac{b_0 M}{\Gamma(\lambda_2 + 1)} + \frac{b_1 M}{\Gamma(\lambda_2 + 1)}\right),\ c_2 = \frac{b_2 M}{(1 - \eta)\Gamma(\lambda_2 + 1)}\ and
$$

$$
M = \sup_{\tau \in [0,T]}\left(\frac{\tau^{\lambda_2}}{E_{\lambda_2}\big((b_0 + b_1 + \theta)\tau^{\lambda_2}\big)}\right).
$$

Proof. Let $\zeta \in C^1\big([-\varsigma, 0], \mathbb{R}^q\big)$, such that $\|\zeta\| \le \gamma_1$.

Let $\mathcal{F} = C\big([-\varsigma, T], \mathbb{R}^q\big)$ and consider the metric β on $\mathcal{F}$ by

$$
\beta(y_1, y_2) = \inf\left\{r \in [0, \infty] : \frac{\|y_1(l) - y_2(l)\|}{g(l)} \le r, \forall l \in [-\varsigma, T]\right\},
$$

where g is given by $g(l) = 1$, for $l \in [-\varsigma, 0]$ and $g(l) = E_{\lambda_2}\big((b_0 + b_1 + \theta)l^{\lambda_2}\big)$ for $l \in [0, T]$.

We consider the operator: $\mathcal{D} : \mathcal{F} \to \mathcal{F}$, such that

$$
\begin{aligned}
(\mathcal{D}X)(w) &= \zeta(0) + C\left(X(w - \varsigma(w)) - \zeta(-\varsigma(0))\right) \\
&+ \frac{1}{\Gamma(\lambda_2)} \int_0^w (w - s)^{\lambda_2 - 1}\Big[B_0 X(s) + B_1 X(s - \varsigma(s)) \\
&+ B_2 v(s) + F(s, X(s), X(s - \varsigma(s)), v(s))\Big] ds, \tag{19}
\end{aligned}
$$

for $w \in [0, T]$ and $(\mathcal{D}X)(w) = \zeta(w)$, for $w \in [-\varsigma, 0]$.

Note that, $\mathcal{D}$ is well defined, $(\mathcal{F}, \beta)$ is a generalized complete metric space, $\beta(\mathcal{D}X_0, X_0) < \infty$, and $\{X_1 \in \mathcal{F} : \beta(X_0, X_1) < \infty\} = \mathcal{F}$, $\forall X_0 \in \mathcal{F}$.

Let $X_1, X_2 \in \mathcal{F}$, for $w \in [-\varsigma, 0]$, we get $(\mathcal{D}X_1)(w) - (\mathcal{D}X_2)(w) = 0$.

For $w \in [0, T]$, we have

$$\left\| (\mathcal{D}X_1)(w) - (\mathcal{D}X_2)(w) \right\|$$
$$\leq c\|X_1(w - \varsigma(w)) - X_2(w - \varsigma(w))\|$$
$$+ \int_0^w \frac{(w - r)^{\lambda_2 - 1}}{\Gamma(\lambda_2)} \Big[\big(f(r) + \|B_0\| \big) \|X_1(r) - X_2(r)\|$$
$$+ \big(f(r) + \|B_1\| \big) \|X_1(r - \varsigma(r)) - X_2(r - \varsigma(r))\| \Big] dr$$
$$\leq c \frac{\|X_1(w - \varsigma(w)) - X_2(w - \varsigma(w))\|}{g(w - \varsigma(w))} g(w - \varsigma(w))$$
$$+ b_0 \int_0^w \frac{(w - u)^{\lambda_2 - 1}}{\Gamma(\lambda_2)} \frac{\|X_1(u) - X_2(u)\|}{g(u)} g(u) du$$
$$+ b_1 \int_0^w \frac{(w - u)^{\lambda_2 - 1}}{\Gamma(\lambda_2)} \frac{\|X_1(u - \varsigma(u)) - X_2(u - \varsigma(u))\|}{g(u - \varsigma(u))} g(u - \varsigma(u)) du$$
$$\leq c\beta(X_1, X_2) g(w - \varsigma(w)) + \frac{b_0 \beta(X_1, X_2)}{\Gamma(\lambda_2)} \int_0^w (w - u)^{\lambda_2 - 1} g(u) du$$
$$+ \frac{b_1 \beta(X_1, X_2)}{\Gamma(\lambda_2)} \int_0^w (w - u)^{\lambda_2 - 1} g(u) du. \tag{20}$$

Using Remark 1, we get

$$\left\| (\mathcal{D}X_1)(w) - (\mathcal{D}X_2)(w) \right\| \leq c\beta(X_1, X_2) g(w) + \frac{b_0}{b_0 + b_1 + \theta} \beta(X_1, X_2) g(w)$$
$$+ \frac{b_1}{b_0 + b_1 + \theta} \beta(X_1, X_2) g(w)$$
$$\leq \left(c + \frac{b_0 + b_1}{b_0 + b_1 + \theta} \right) \beta(X_1, X_2) g(w). \tag{21}$$

Then,

$$\frac{\left\| (\mathcal{D}X_1)(w) - (\mathcal{D}X_2)(w) \right\|}{g(w)} \leq \left(c + \frac{b_0 + b_1}{b_0 + b_1 + \theta} \right) \beta(X_1, X_2),$$

Thus,

$$\beta(\mathcal{D}X_1, \mathcal{D}X_2) \leq \left(c + \frac{b_0 + b_1}{b_0 + b_1 + \theta} \right) \beta(X_1, X_2).$$

Therefore, $\mathcal{D}$ is contractive.

Let x_0 be the function given by $x_0(\tau) = \zeta(\tau)$, for $\tau \in [-\varsigma, 0]$ and $x_0(\tau) = \zeta(0)$ for $\tau \in [0, T]$.

Then, we have

$$\|x_0(\tau)\| \leq \|\zeta\|,$$

for all $t \in [-\varsigma, T]$.

For $\tau \in [-\varsigma, 0]$, we get $(\mathcal{D}x_0)(\tau) - x_0(\tau) = 0$.

For $w \in [0, T]$, we have

$$\left\| (\mathcal{D}x_0)(w) - x_0(w) \right\| \leq 2c\|\zeta\|$$
$$+ \frac{1}{\Gamma(\lambda_2)} \int_0^w (w - s)^{\lambda_2 - 1} \big[b_0 \|x_0(s)\| + b_1 \|x_0(s - \varsigma(s))\| + b_2 \varrho \big] ds$$
$$\leq 2c\|\zeta\| + \frac{w^{\lambda_2}}{\Gamma(\lambda_2 + 1)} \Big(b_0 \|\zeta\| + b_1 \|\zeta\| + b_2 \varrho \Big). \tag{22}$$

Then,

$$\frac{\left\|(\mathcal{D}x_0)(w) - x_0(w)\right\|}{g(w)} \leq 2c\|\zeta\|$$
$$+ \; \left(b_0\|\zeta\| + b_1\|\zeta\| + b_2\varrho\right)\frac{M}{\Gamma(\lambda_2 + 1)}, \tag{23}$$

for all $w \in [0, T]$.

Therefore,

$$\beta(\mathcal{D}x_0, x_0) \leq 2c\|\zeta\|$$
$$+ \; \left(b_0\|\zeta\| + b_1\|\zeta\| + b_2\varrho\right)\frac{M}{\Gamma(\lambda_2 + 1)}. \tag{24}$$

Theorem 1 implies that (2) has a unique solution x with initial conditions of ζ, such that

$$\beta(x_0, x) \leq \frac{1}{1 - \eta}\Big[2c\|\zeta\|$$
$$+ \; \left(b_0\|\zeta\| + b_1\|\zeta\| + b_2\varrho\right)\frac{M}{\Gamma(\lambda_2 + 1)}\Big]$$
$$\leq \; c_1\gamma_1 + c_2\varrho. \tag{25}$$

Therefore,

$$\|x_0(t) - x(t)\| \leq \left(c_1\gamma_1 + c_2\varrho\right)E_{\lambda_2}\left((b_0 + b_1 + \theta)T^{\lambda_2}\right),$$

for all $t \in [0, T]$.

Then,

$$\|x(t)\| \leq \|(x - x_0)(t)\| + \|x_0(t)\|$$
$$\leq \left(1 + c_1 E_{\lambda_2}\left((b_0 + b_1 + \theta)T^{\lambda_2}\right)\right)\gamma_1$$
$$+ \; c_2 E_{\lambda_2}\left((b_0 + b_1 + \theta)T^{\lambda_2}\right)\varrho. \tag{26}$$

Thus, $\|x(t)\| \leq \gamma_2$, for all $t \in [0, T]$, if (17) is satisfied. $\square$

Remark 3. *Using Lemma 1, we get*

$$c_1 \leq \frac{1}{(1 - \eta)}\left(2c + \frac{b_0}{b_0 + b_1 + \theta} + \frac{b_1}{\theta + b_1 + b_0}\right)$$

and

$$c_2 \leq \frac{1}{(1 - \eta)}\frac{b_2}{b_0 + b_1 + \theta}.$$

Let us consider

$$\tilde{c}_1 = \frac{1}{(1 - \eta)}\left(2c + \frac{b_0}{\theta + b_1 + b_0} + \frac{b_1}{b_0 + b_1 + \theta}\right)$$

and

$$\tilde{c}_2 = \frac{1}{(1 - \eta)}\frac{b_2}{\theta + b_1 + b_0}.$$

Therefore, the condition (17) can be relaxed by:

$$\tilde{K}(\gamma_1, \varrho) \leq \gamma_2, \tag{27}$$

where

$$\tilde{K}(\gamma_1, \varrho) \;=\; \left(1 + \tilde{c}_1 E_{\lambda_2}\big((b_0 + b_1 + \theta)T^{\lambda_2}\big)\right)\gamma_1$$
$$+\quad \tilde{c}_2 E_{\lambda_2}\big((b_0 + b_1 + \theta)T^{\lambda_2}\big)\varrho. \tag{28}$$

Remark 4. *In the Theorem 3, $c < 1$ it is a necessary condition.*

Remark 5. *In the case when $C = 0$, we get the results in [21].*

4. Examples

Two examples are studied to prove the applicability of Theorems 2 and 3.

Example 1. *Consider the NFOTDSs (2), with $\lambda_2 = 0.7$, $\lambda_1 = 0.2$, $\varsigma(s) = 0.1$,*

$$v(\tau) = (0.5, 0)^T, \quad \zeta(\tau) = (0.05, 0)^T, \text{ for } \tau \in [-0.1, 0],$$

$$F(s, x(s), x(s - \varsigma(s)), v(s)) = 0.01\Big(\sin\big(x_2(s - \varsigma(s))\big), \sin\big(x_1(s)\big)\Big)^T,$$

and

$$B_0 = \begin{pmatrix} 0 & 0.4 \\ 0.1 & 0 \end{pmatrix}, \; B_1 = \begin{pmatrix} -0.6 & 0 \\ -0.2 & 0 \end{pmatrix}, \; B_2 = \begin{pmatrix} 0.3 & 0 \\ 0.4 & 0 \end{pmatrix}, \; C = \begin{pmatrix} 0.2 & 0 \\ -0.1 & 0 \end{pmatrix}.$$

We get $b_0 = 0.41$, $b_1 = 0.64$, $b_2 = 0.51$ and $c = 0.2236$.
For $\eta_1 = \eta_2 = 1$, $\varrho = 1$, $\gamma_1 = 0.3$ and $\gamma_2 = 60$. Moreover, if we calculate δ, $\tilde{c}_1$ and $\tilde{c}_2$, then $\tilde{G}(\gamma_1, \varrho) \simeq 59 < \gamma_2$, for $T = 0.61$. Based on theorem 2 it is clear that the NFOTDSs is FTS w.r.t $(0.3, 60, 1, 0.61)$.

Example 2. *Consider the NFOTDSs (2), with $\lambda_2 = \lambda_1 = 0.6$, $\varsigma(s) = 0.1$,*

$$v(\tau) = (0, 0.5, 0)^T, \quad \zeta(\tau) = (0.04, 0, 0.02)^T, \text{ for } \tau \in [-0.1, 0],$$

$$F(s, x(s), x(s - \varsigma(s)), v(s)) = 0.01\Big(\sin\big(x_2(s - \varsigma(s))\big), \sin\big(x_3(s - \varsigma(s))\big), \sin\big(x_1(s)\big)\Big)^T,$$

and

$$B_0 = \begin{pmatrix} 0.01 & -0.2 & 0.25 \\ -0.02 & 0.05 & 0.1 \\ 0.2 & -0.01 & 0.15 \end{pmatrix}, \; B_1 = \begin{pmatrix} 0.01 & -0.15 & 0.31 \\ 0.25 & 0.12 & -0.14 \\ 0.13 & -0.12 & 0.22 \end{pmatrix},$$

$$B_2 = \begin{pmatrix} 0.08 & 0.07 & 0.2 \\ 0.08 & -0.07 & -0.06 \\ -0.12 & -0.03 & -0.14 \end{pmatrix}, \; C = \begin{pmatrix} 0.1 & 0.2 & 0.03 \\ 0.12 & 0.22 & 0.05 \\ -0.17 & 0.05 & -0.21 \end{pmatrix}.$$

We get $b_0 = 0.37$, $b_1 = 0.47$, $b_2 = 0.30$, and $c = 0.35$.
For $\varrho = 1$, $\theta = 1$, $\gamma_1 = 0.4$, $\gamma_2 = 100$, and $T = 1.05$, we get $\tilde{K}(\gamma_1, \varrho) \simeq 97 < \gamma_2$.
Theorem 3 implies that the NFOTDSs is FTS w.r.t $(0.4, 100, 1, 1.05)$.

5. Conclusions

In this paper, a new robust FTS for NFOTDSs was described. By suggesting an approach based on the fixed point theory, novel sufficient conditions for the robust FTS of such systems are obtained. Finally, two examples were described to show the validity and the useless of the suggested result.

Author Contributions: Formal analysis, A.B.M.; writing—original draft preparation, A.B.M.; Supervision, D.B.; Visualization, D.B. All authors have read and agreed to the published version of the manuscript.

Funding: This research received no external funding.

Institutional Review Board Statement: Not applicable.

Informed Consent Statement: Not applicable.

Data Availability Statement: Not applicable.

Conflicts of Interest: The authors declare no conflict of interest.

References

1. Warrier, P.; Shah, P. Fractional Order Control of Power Electronic Converters in Industrial Drives and Renewable Energy Systems: A Review. *IEEE Access* **2021**, *9*, 58982–59009. [CrossRef]
2. Afshari, A. Solution of fractional differential equations in quasi-b-metric and bmetric- like spaces. *Adv. Differ. Equ.* **2019**, *2019*, 285 [CrossRef]
3. Afshari, A.; Gholamyan, H.; Zhai, C.B. New applications of concave operators to existence and uniqueness of solutions for fractional differential equations. *Math. Commun.* **2020**, *25*, 157–169.
4. Afshari, A.; Sajjadmanesh, M.; Baleanu, D. Existence and uniqueness of positive solutions for a new class of coupled system via fractional derivatives. *Adv. Differ. Equ.* **2020**, *2020*, 111. [CrossRef]
5. Feng, Y.Y.; Yang, X.J.; Liu, J.G.; Chen, Z.Q. A new fractional Nishihara-type model with creep damage considering thermal effect. *Eng. Fract. Mech.* **2021**, *242*, 107451. [CrossRef]
6. Ibrahim, R.W.; Baleanu, D. On quantum hybrid fractional conformable differential and integral operators in a complex domain. *Rev. R. Acad. Cienc. Exactas Fis. Nat. Ser. A Mat.* **2021**, *31*, 514–531. [CrossRef]
7. Jafari, H.; Jassim, H.K.; Baleanu, D.; Chu, Y. On the Approximate Solutions for a System of Coupled Korteweg De Vries Equations with Local Fractional Derivative. *Fractals* **2021**, *29*, 2140012. [CrossRef]
8. Sakar, M.G. Numerical solution of neutral functional-differential equations with proportional delays. *Int. J. Optim. Control. Theor. Appl. (IJOCTA)* **2017**, *7*, 186–194. [CrossRef]
9. Veeresha, P.; Yavuz, M.; Baishya, C. A computational approach for shallow water forced Korteweg-De Vries equation on critical flow over a hole with three fractional operators. *Int. J. Optim. Control. Theor. Appl. (IJOCTA)* **2021**, *11*, 52–67. [CrossRef]
10. Vigya; Mahto, T.; Malik, H.; Mukherjee, V.; Alotaibi, M.A.; Almutairi, A. Renewable generation based hybrid power system control using fractional order-fuzzy controller. *Energy Rep.* **2021**, *7*, 641–653. [CrossRef]
11. Zhang, K.; Wu, L. Using a fractional order grey seasonal model to predict the dissolved oxygen and pH in the Huaihe River. *Water Sci. Technol.* **2021**, *83*, 475–486. [CrossRef] [PubMed]
12. Daoui, A.; Yamni, M.; Karmouni, H.; Sayyouri, M.; Qjidaa, H. Biomedical signals reconstruction and zero-watermarking using separable fractional order Charlier-Krawtchouk transformation and Sine Cosine Algorithm. *Signal Process.* **2021**, *180*, 107854. [CrossRef]
13. Higazy, M.; Allehiany, F.M.; Mahmoud, E.E. Numerical study of fractional order COVID-19 pandemic transmission model in context of ABO blood group. *Results Phys.* **2021**, *22*, 103852. [CrossRef] [PubMed]
14. Liu, D.; Zhao, S.; Luo, X.; Yuan, Y. Synchronization for fractional-order extended Hindmarsh-Rose neuronal models with magneto-acoustical stimulation input. *Chaos Solitons Fractals* **2021**, *144*, 110635. [CrossRef]
15. Zhang, C.; Yang, H.; Jiang, B. Fault Estimation and Accommodation of Fractional-Order Nonlinear, Switched, and Interconnected Systems. *IEEE Trans. Cybern.* **2020**, *52*, 1443–1453. [CrossRef]
16. Amiri, S.; Keyanpour, M.; Asaraii, A. Observer-based output feedback control design for a coupled system of fractional ordinary and reaction-diffusion equations. *IMA J. Math. Control. Inf.* **2021**, *38*, 90–124. [CrossRef]
17. Feng, T.; Wang, Y.E.; Liu, L.; Wu, B. Observer-based event-triggered control for uncertain fractional-order systems. *J. Frankl. Inst.* **2020**, *357*, 9423–9441. [CrossRef]
18. Edrisi-Tabriz, Y.; Lakestani, M.; Razzaghi, M. Study of B-spline collocation method for solving fractional optimal control problems. *Trans. Inst. Meas. Control* **2021**, *43*, 2425–2437. [CrossRef]
19. Brandibur, O.; Kaslik, E. Stability analysis of multi-term fractional-differential equations with three fractional derivatives. *J. Math. Anal. Appl.* **2021**, *495*, 124751. [CrossRef]
20. Ivanescu, M.; Popescu, N.; Popescu, D. Physical Significance Variable Control for a Class of Fractional-Order Systems. *Circuits Syst. Signal Process.* **2021**, *40*, 1525–1541. [CrossRef]
21. Ben Makhlouf, A. A Novel Finite Time Stability Analysis of Nonlinear Fractional-Order Time Delay Systems: A Fixed Point Approach. *Asian J. Control* **2021**. [CrossRef]
22. Du, F.; Jia, B. Finite-time stability of a class of nonlinear fractional delay difference systems. *Appl. Math. Lett.* **2019**, *98*, 233–239. [CrossRef]
23. Du, F.; Lu, J.G. Finite-time stability of neutral fractional order time delay systems with Lipschitz nonlinearities. *Appl. Math. Comput.* **2020**, *375*, 125079. [CrossRef]
24. Lu, Q.; Zhu, Y.; Li, B. Finite-time stability in mean for Nabla Uncertain Fractional Order Linear Difference Systems. *Chaos Solitons Fractals* **2021**, *29*, 2150097. [CrossRef]

25. Phat, V.N.; Thanh, N.T. New criteria for finite-time stability of nonlinear fractional-order delay systems: A Gronwall inequality approach. *Appl. Math. Lett.* **2018**, *83*, 169–175. [CrossRef]
26. Thanh, N.T.; Phat, V.N. Switching law design for finite-time stability of singular fractional-order systems with delay. *IET Control Theory Appl.* **2019**, *13*, 1367–1373. [CrossRef]
27. Thanh, N.T.; Phat, V.N.; Niamsup, T. New finite-time stability analysis of singular fractional differential equations with time-varying delay. *Fract. Calc. Appl. Anal.* **2020**, *23*, 504–519. [CrossRef]
28. Wu, R.; Lu, Y.; Chen, L. Finite-time stability of fractional delayed neural networks. *Neurocomputing* **2015**, *149*, 700–707. [CrossRef]
29. Wu, G.C.; Baleanu, D.; Zeng, S.D. Finite-time stability of discrete fractional delay systems: Gronwall inequality and stability criterion. *Commun. Nonlinear Sci. Numer. Simul.* **2018**, *57*, 299–308. [CrossRef]
30. Ye, H.; Gao, J.; Ding, Y. A generalized Gronwall inequality and its application to a fractional differential equation. *J. Math. Anal. Appl.* **2007**, *328*, 1075–1081. [CrossRef]
31. Podlubny, I. *Fractional Differential Equations*; Academic Press: New York, NY, USA, 1999.

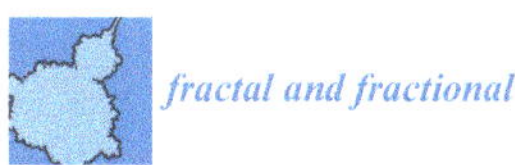

fractal and fractional

Article

A Plea for the Integration of Fractional Differential Systems: The Initial Value Problem

Nezha Maamri [1],* and Jean-Claude Trigeassou [2]

1   LIAS Laboratory, Poitiers University, 2 rue Pierre Brousse, CEDEX 9, 86073 Poitiers, France
2   IMS Laboratory, Bordeaux University, 351 Cours de la libération, CEDEX, 33403 Talence, France
*   Correspondence: nezha.maamri@univ-poitiers.fr

Abstract: The usual approach to the integration of fractional order initial value problems is based on the Caputo derivative, whose initial conditions are used to formulate the classical integral equation. Thanks to an elementary counter example, we demonstrate that this technique leads to wrong free-response transients. The solution of this fundamental problem is to use the frequency-distributed model of the fractional integrator and its distributed initial conditions. Using this model, we solve the previous counter example and propose a methodology which is the generalization of the integer order approach. Finally, this technique is applied to the modeling of Fractional Differential Systems (FDS) and the formulation of their transients in the linear case. Two expressions are derived, one using the Mittag–Leffler function and a new one based on the definition of a distributed exponential function.

Keywords: initial value problem; fractional differential systems; fractional integral equation; infinite-state approach; Riemann–Liouville integral; frequency distributed exponential function

Citation: Maamri, N.; Trigeassou, J.-C. A Plea for the Integration of Fractional Differential Systems: The Initial Value Problem. *Fractal Fract.* **2022**, *6*, 550. https://doi.org/10.3390/fractalfract6100550

Academic Editors: Thach Ngoc Dinh, Shyam Kamal, Rajesh Kumar Pandey and Norbert Herencsar

Received: 14 July 2022
Accepted: 19 September 2022
Published: 28 September 2022

Publisher's Note: MDPI stays neutral with regard to jurisdictional claims in published maps and institutional affiliations.

1. Introduction

The integration of Fractional Differential Equations (FDE) and Systems (FDS) is considered to be a well-founded and approved topic for most fractional calculus researchers. Therefore, the title of the paper appears as an ingenuous and unrealistic objective to revisit an established mathematical result. Nevertheless, our purpose is to provide an objective analysis of this fundamental problem and to formulate a satisfactory solution to fractional-order initial-value problems.

In fact, the initial-value problem, or Cauchy problem, is obviously trivial in the integer order case [1,2]. On the other hand, the solution of the fractional-order case appears as a generalization of the integer-order one. However, due to the multiplicity of fractional-order derivative definitions, researchers have considered it necessary to adapt the classical approach by referring to a particular derivative and its corresponding initial conditions [3–5]. Practically, most of the time, the Caputo derivative [3,6] is used because its "initial conditions" can be physically interpreted. Many critics have already addressed this choice, based on initialization considerations [7–15]. In those papers, the authors emphasize the inability of the Caputo derivative technique to solve the initialization problem, but, contrary to the history function technique [9,10,16–21] and the infinite-state approach [13,22–24], they do not provide a solution to this problem. Recently, some solutions based on new fractional derivatives (see, for example, [25–28]), which are in fact local derivatives [29,30], have been proposed. Practically, the direct consequence of these multiple choices is that different theoretical free responses are possible for the same FDE/FDS problem, which is of course physically inconsistent.

Our objective in this paper is to prove (in fact to recall) theoretically, using an elementary initial value problem, that the solution predicted by the Caputo derivative approach leads to a false free response. Then, we treat the same example with the frequency-distributed model of the fractional integrator [31–34]. We demonstrate that using a distributed initial condition, in fact that of the fractional integrator, provides the good solution to the considered problem. The conclusion of this analysis is that any fractional-order initial-value problem has to be treated such as in the integer-order case, using essentially the fractional integrator and its distributed initial state or its initialization function. Then, this technique is applied to the modeling of FDE/FDS and the formulation of their transients in the linear case. Two expressions are derived, one using the classic Mittag–Leffler function and a new one based on the definition of a distributed exponential function.

The theory developed in the paper is not a new one, since the first paper [35] related to the fractional integrator was published in 1999. Since that original publication, this research has been applied to the modeling and identification of real-world diffusive processes: electrochemical [36], thermal [37], and rotor skin effect, see chapter 5, volume 1 of [34]. The modeling of fractional systems based on the fractional integrator, known as the infinite-state approach, has been presented in several articles, see, for example, [22,33,38], with a particular focus on system initialization [23,24]. Moreover, it has been applied to the stability analysis of linear and nonlinear systems with a distributed formulation of the Lyapunov function [39]. The theory of the infinite state approach and its applications to various domains of control theory are presented in a two-volume monograph [34]. However, in spite of its contributions to initialization and Lyapunov system stability, this theory is ignored or considered as an exotic contribution to fractional calculus, although it has been adopted by researchers for initialization purposes [21,40–42] and Lyapunov stability analysis [43–48]. Moreover, although the pseudo initial conditions of the Caputo derivative are frequently criticized [8,9,16,20], mainly for their use in system initialization [7,14,15,21], they are still used because they provide apparently simple solutions. Consequently, there is an important challenge to provide a general and satisfactory solution to the initialization problem, using the same approach as in the integer-order case, where the initial conditions are those of the fractional integrator.

Thus, this paper intends to treat the FDE initial-value problem with a new and theoretical presentation of the infinite state approach, demonstrating that we do not have to refer to any fractional derivative and, on the contrary, can focus on the Riemann–Liouville integral and its distributed initial conditions. It is important to note that the authors have privileged a theoretical formalism contrary to their previous publications, where numerical simulations were abundantly used. So, the reader can refer to these previous papers to find numerical illustrations related to initialization. A restricted version of the paper has already been published in a recent conference [49].

The paper is composed of six sections and a conclusion. Section 1 is the introduction. Sections 2–4 present the materials and methods related to initial-value problems and the infinite state approach. In Section 5, an elementary counter-example permits us to invalidate the usual Caputo derivative initial value approach. In Section 6, the frequency-distributed integrator model is used to solve the previous counter-example and to formulate a new approach to the FDE initial-value problem. This methodology is used in Section 6 to express the dynamics of the general FDS initial-value problem.

2. Materials and Methods

2.1. The ODE Initial-Value Problem (or Cauchy Problem [1])

Let us consider the following Ordinary Differential System:

$$\frac{dx(t)}{dt} = f(x(t), u(t)) \tag{1}$$

where $x(t) = x(0)$ at t $= 0$ is the initial value.

The Picard-Lindelöf theorem [50] guarantees the existence of a unique solution to (1). The principle of this theorem consists in reformulating the problem as an equivalent integral equation:

$$x(t) = I_t^1\left[\frac{dx(t)}{dt}\right] + x(0) = \int_0^t f(x(\tau), u(\tau))d\tau + x(0) \tag{2}$$

and to construct a sequence of functions

$$\phi_{k+1}(t) = \int_0^t f(\phi_k(\tau), u(\tau))d\tau + x(0) \text{ with } \phi_1(t) = x(0), \tag{3}$$

which converges to the solution of (1) and thus to the solution of the initial-value problem. Such a construction is called Picard's method [51] or the method of successive approximations. In the linear and multidimensional case, Equation (1) can be expressed as:

$$\frac{d\underline{x}(t)}{dt} = A\underline{x}(t) + Bu(t) \; x(t) = x(0) \text{ at } t = 0, \tag{4}$$

where $x(t) \in R^N$, and A and B are matrices of appropriate dimensions.

It is well known that its solution, based on the exponential matrix function or transition matrix

$$\Phi(t) = e^{At} \text{ with } \Phi(t) \in R^{N \times N}$$

is given by [52]:

$$x(t) = \Phi(t)x(0) + \int_0^t \Phi(t - \tau)Bu(\tau)d\tau. \tag{5}$$

2.2. The FDE/FDS Initial Value Problem

Let us consider the elementary FDE

$$D_t^n(x(t)) = f(x(t), u(t)) \; 0 < n < 1, \tag{6}$$

where n is the fractional order and $x(t) = x(0)$ at $t = 0$.

Contrary to the integer-order case, several approaches are derived from the fractional derivative definitions of $D_t^n(x(t))$. The main popular ones are the Caputo and Riemann–Liouville derivatives [3].

Practically, Equation (6) is integrated with the Caputo derivative definition, since its initial condition is considered equal to $x(0)$.

Then, in order to prove the existence and the uniqueness of the solution $x(t)$ of (6), Picard's method [3–5] is frequently used.

In the linear multidimensional commensurate order case, Equation (6) becomes:

$$\begin{cases} D_t^n(\underline{x}(t)) = A\underline{x}(t) + Bu(t) \\ \qquad y(t) = \underline{C}x(t) \end{cases} \quad 0 < n < 1, \tag{7}$$

where $x(t) \in R^N$, and A and B are matrices of appropriate dimensions.

The general solution of (7), expressed in terms of the Mittag–Leffler matrix function [53]

$$\Phi(t) = E_{n,1}(At^n), \; \Phi(t) \in R^{N \times N}$$

is

$$\underline{x}(t) = \Phi(t)\underline{x}(0) + \int_0^t \Phi(t - \tau)B\tilde{u}(\tau)d\tau \quad \text{with} \quad \tilde{u}(\tau) = D^{1-n}(u(\tau)). \tag{8}$$

As mentioned in the introduction, the main objective of this paper is to revisit the integration of the FDE/FDS initial-value problem, using the infinite-state approach, which is directly related to the integer order ODE case and does not need any derivative definition, as it is exhibited in Section 4.

3. Integration of FDE/FDS Based on Derivative Definitions

3.1. Riemann–Liouville Integral

The fractional integral of a function $v(t)$, also called the Riemann–Liouville integral is defined by

$$x(t) = {}_0I_t^n(f(t)) = \int_0^t \frac{(t-\tau)^{n-1}}{\Gamma(n)}v(\tau)d\tau \quad 0 < n < 1, \tag{9}$$

where (n) is the gamma function.

The fractional integral is in fact a convolution integral, characterized by the impulse response or Kernel, $h_n(t)$, such that:

$$h_n(t) = \frac{t^{n-1}}{\Gamma(n)} \quad and \quad x(t) = h_n(t) * v(t). \tag{10}$$

Using the Laplace transform, we obtain

$$L\{h_n(t)\} = \frac{1}{s^n}, \tag{11}$$

where $\frac{1}{s^n}$ corresponds to the fractional order integration operator.

3.2. Fractional Derivatives Definitions

Contrary to the fractional integral, the fractional derivative is not uniquely defined. Usually, two main derivatives are considered; since they are used for the integration of FDE/FDS, we focus on the case $0 < n < 1$ [34].

3.2.1. Caputo Derivative Definition

This definition corresponds to first differentiate $x(t)$ and then calculates a fractional integral with order $(1-n)$. Since $0 < n < 1$, then $0 < 1-n < 1$.

$$^C D_t^n(x(t)) = I_t^{1-n}\left(\frac{dx}{dt}\right) = h_{1-n}(t) * \frac{dx}{dt} \tag{12}$$

Definition (12) clearly shows that the Caputo derivative corresponds to the Riemann–Liouville integral of the derivative of $x(t)$.

In the Laplace domain, the Caputo derivative definition leads to

$$L\left\{\left(^C D_t^n(x)\right)\right\} = \frac{1}{s^{1-n}}L\left\{\frac{dx(t)}{dt}\right\} = \frac{1}{s^{1-n}}[sX(s) - x(0)]. \tag{13}$$

$$\text{So, } X(s) = \frac{1}{s^n}L\left\{^C D_t^n(x)\right\} + \frac{x(0)}{s}, \tag{14}$$

Or in the time domain,

$$x(t) = I_t^n\left(^C D_t^n(x)\right) + x(0). \tag{15}$$

Thus, the solution of FDE/FDS (6) according to the Caputo approach is

$$x(t) = {}_0I_t^n(f(x(t), u(t))) + x(0) \quad \text{for } t \geq 0 \quad 0 < n < 1, \tag{16}$$

where $x(0)$ is interpreted as the initial condition of the FDE/FDS and also as the initial value of the Riemann–Liouville integral.

This simple integral Equation (16), apparently equivalent to the integer order case (2), has made the success of the Caputo derivative approach.

3.2.2. Riemann–Liouville Derivative Definition

This definition shows that the Riemann–Liouville derivative corresponds to the integer-order derivative of the Riemann–Liouville integral of $x(t)$.

$$^{RL}D_t^n(x(t)) = \frac{d}{dt}\left[I_t^{1-n}(x)\right] = \frac{d}{dt}[h_{1-n}(t) * x(t)] \tag{17}$$

Using the Laplace transform, we obtain

$$L\{^{RL}D_t^n(x)\} = s\left(L\{I_t^{1-n}(x)\}\right) - g(0) = s\left(\frac{1}{s^{1-n}}X(s)\right) - g(0) = s^n X(s) - g(0)$$
$$\text{with } g(0) = \left\{_{-\infty}I_t^{1-n}(x)\right\}_{t=0}.$$

Since $g(0)$ does not have a physical and direct interpretation, the Riemann–Liouville derivative is generally not used to integrate system (6) or (7).

It is important to note that the two derivative definitions only require integer-order differentiation $\left(\frac{d}{dt}\right)$ and fractional-order integration $\left(I^{1-n}\right)$. So, contrary to a common belief, the basic operation of fractional calculus is not fractional differentiation but fractional integration.

3.2.3. The Grünwald–Letnikov Derivative

Instead of the two previous fractional derivatives, it is possible to use the Grünwald-Letnikov (G.L.) derivative, with appropriate initial conditions. In fact, it is preferable to consider the G.L. integrator that corresponds to the discretization of the Riemann–Liouville integral.

The Nth integer order Euler derivative of $x(t)$ is defined as

$$\left(D^N(x(t))\right)_{t=kT_e} = \lim_{T_e \to 0} \frac{(1-q^{-1})^N}{T_e^N} x_k, \tag{18}$$

where T_e is the sample time, $x_k = x(kT_e)$, and q^{-1} is the delay operator.

The generalization to the fractional order case provides the Grünwald–Letnikov derivative

$$\left(^{GL}D^n(x(t))\right)_{t=kT_e} = \lim_{T_e \to 0} \frac{(1-q^{-1})^n}{T_e^n} x_k \quad 0 < n < 1. \tag{19}$$

Since $L\{q^{-1}\} = e^{-T_e s}$, we obtain

$$L\{^{GL}D^n(x(t))\} = \lim_{T_e \to 0} \frac{(1-e^{-T_e s})^n}{T_e^n} L\{x(t)\} = s^n X(s). \tag{20}$$

Notice that

$$\frac{(1-q^{-1})^n}{T_e^n} = \frac{1}{T_e^n}\left[1 + \sum_{i=0}^{\infty} \alpha_{i,GL} q^{-i}\right], \tag{21}$$

with $\alpha_{i,GL} = (-1)^i \frac{n}{1} \frac{n-1}{2} \frac{n-2}{3} \ldots \ldots \frac{n-(i+1)}{i}$,

so $\left(^{GL}D^n(x(t))\right)_{t=kT_e} = \dfrac{1 + \sum\limits_{i=0}^{\infty} \alpha_{i,GL} q^{-i}}{T_e^n} x_k, \tag{22}$

which is the Moving Average formulation of the Grünwald–Letnikov derivative.

Reciprocally, we can define the Grünwald–Letnikov integral operator [34] as

$$^{GL}I^n(f_k) = \frac{T_e{}^n q^{-1}}{1 + \sum\limits_{i=0}^{\infty} \alpha_{i,GL} q^{-i}} f_k, \tag{23}$$

which is the Auto-Regressive formulation of the Grünwald–Letnikov integrator.

Notice that

$$L\left\{{}^{GL}I^n(f(t))\right\} = \lim_{T_e \to 0} \frac{T_e{}^n e^{-T_e s}}{(1 - e^{-T_e s})^n} L\{f(t)\} = \frac{1}{s^n} F(s), \tag{24}$$

which means that the Grünwald–Letnikov integrator is the time discretization of the Riemann–Liouville integral.

Consider now the elementary FDE initial value problem (6):

$$D_t^n(x(t)) = f(x(t), u(t)) \quad 0 < n < 1.$$

Using the Grünwald–Letnikov integrator, we can express $\{x(k)\}$ as:

$$\{x(k+1)\} = {}^{GL}I^n(f(\{xk\}, \{uk\})) + g\{x_{init}\}, \tag{25}$$

where $\{x_{init}\} = \{x(0), x(-1), \ldots\ldots, x(-i), \ldots\ldots, x(-\infty)\}$.

This means that the initial conditions are composed of all the past values of $x(-i)$, since $k = -\infty$.

Practically, this technique is used for the numerical simulation of the FDE/FDS problem.

The interested reader can refer to chapter 3 volume 1 of [34], where different initializations of the G.L. integral and the short memory principle [3] are analyzed.

4. The Infinite State Approach

4.1. Introduction

The infinite-state approach (do not confuse it with diffusive representation, see chapter 7 of [34]) is a modeling technique based on the fractional-integration operator, which is at the heart of any modeling and simulation system, either integer or fractional order, linear or non-linear [22,52].

4.2. The Frequency-Distributed Model of the Fractional Integrator

As shown in Section 3, the Riemann–Liouville integral of a function $v(t)$ is defined as the convolution of $v(t)$ with the impulse response $h_n(t)$ of the fractional integrator.

Another expression of $h_n(t)$ can be derived from the inverse Laplace transform of $\frac{1}{s^n}$ [31,32,34], for $0 < n < 1$, i.e., $h_n(t) = L^{-1}\left\{\frac{1}{s^n}\right\}$.

Using a Bromwich contour, we can write (see [34] and the references therein):

$$h_n(t) = \begin{cases} \dfrac{1}{2\pi j} \displaystyle\int_{\gamma - j\omega}^{\gamma + j\omega} \dfrac{1}{s^n} e^{st}\, ds & \text{for } t > 0 \\[2ex] 0 & \text{for } t < 0 \end{cases}.$$

Thus, we obtain

$$h_n(t) = \int_0^{\infty} \mu_n(\omega) e^{-\omega t} d\omega = \frac{t^{n-1}}{\Gamma(n)} \quad \text{for } 0 < n < 1$$

$$\text{with } \mu_n(\omega) = \frac{\sin(n\pi)}{\pi} \omega^{-n} \tag{26}$$

Note that, in the particular case where $v(t) = (t)$ is an impulse function, the output $x(t)$ corresponds to the impulse response $h_n(t)$ and is provided by the following distributed integer-order differential system

$$\begin{cases} \frac{\partial z(\omega,t)}{\partial t} = -\omega z(\omega,t) + \delta(t) \; \omega \in [0,+\infty) \\ h_n(t) = \int\limits_0^\infty \mu(\omega)z(\omega,t)d\omega \end{cases} \tag{27}$$

Thus $z(\omega,t) = e^{-\omega t}$, which leads to $h_n(t) = \int\limits_0^\infty \mu(\omega)e^{-\omega t}d\omega$.

More generally, for any input $v(t)$, the corresponding output $x(t)$ of the fractional integrator is provided by the following distributed frequency system:

$$\begin{cases} \frac{\partial z(\omega,t)}{\partial t} = -\omega z(\omega,t) + v(t) \quad \omega \in [0,\infty) \\ x(t) = \int\limits_0^\infty \mu_n(\omega)z(\omega,t)d\omega \\ \text{with } \mu_n(\omega) = \frac{\sin(n\pi)}{\pi}\omega^{-n} \end{cases} \tag{28}$$

It is fundamental to notice that the original model of the fractional integrator has been transformed into an infinite-dimension integer-order differential system (28), where integer-order differentiation $\frac{\partial}{\partial t}$ has been substituted to fractional-order differentiation, and where the fractional order n appears in the weighting function $\mu_n(\omega)$. This means that the fractional integrator $\frac{1}{s^n}$ is an infinite-dimension linear system [54]. These models are the two "faces" of the fractional integrator.

In fact, according to linear system theory [52], the fractional integrator has two types of models, as does any other linear system:

- the Equation $X(s) = \frac{1}{s^n}V(s)$ is the input/output representation of the fractional integrator, characterized by its impulse response $h_n(t)$ and its frequency response $\frac{1}{(j\omega)^n}$.
- the distributed differential system (28) is the infinite-dimension state-space model of the integrator, where the internal state $z(\omega,t)$ permits a complete representation of system dynamics and particularly its free response from an initial condition $z(\omega,0)$.

Remark 1: let us consider the Laplace transform of (27):

$$L\{h_n(t)\} = L\left\{\int\limits_0^\infty \mu(\omega)e^{-\omega t}d\omega\right\} = \int\limits_0^\infty \mu_n(\omega)L\{e^{-\omega t}\}d\omega = \int\limits_0^\infty \mu_n(\omega)\frac{1}{s+\omega}d\omega$$

The previous equality and Equation (11) lead to

$$\frac{1}{s^n} = \int\limits_0^\infty \mu_n(\omega)\frac{1}{s+\omega}d\omega \quad 0 < n < 1. \tag{29}$$

This relation exhibits that the fractional integrator is composed of an infinity of modes ω, ranging from 0 to $+\infty$, whereas the integer-order integrator corresponds to only one mode situated at $\omega = 0$. Figure 1 displays the graphic representation of Equation (28). Note that the distributed differential Equations of (28) correspond to the first-order systems displayed in Figure 1. Due to the distributed nature of the differential system, the graph of Figure 1 is composed of an infinity of first-order systems.

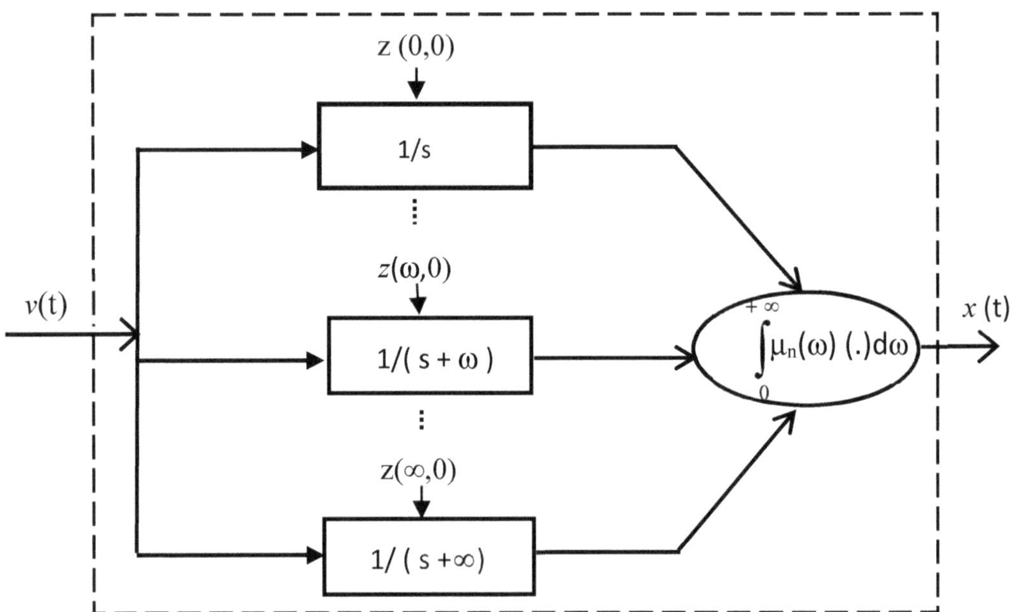

Figure 1. The frequency-distributed model of the fractional integrator.

Remark 2: The classical Laplace transform of the integer order derivative is known as $L\left\{\frac{dx(t)}{dt}\right\} = sX(s) - x(0)$, which corresponds in fact to the relation $X(s) = \frac{1}{s}L\left\{\frac{dx(t)}{dt}\right\} + \frac{x(0)}{s}$ i.e., in the time domain, to $x(t) = \int_0^t \frac{dx(\tau)}{d\tau}d\tau + x(0)$, which means that $x(0)$ is not the initial condition of the derivative but is, in fact, the initial condition of the integrator which has memorizing capability.

4.3. Transients of the Fractional Integrator

Consider the Laplace transform of (28):

$$\begin{cases} sZ(\omega,s) - z(\omega,0) = -\omega Z(\omega,s) + V(s)\ \omega \in [0,\infty) \\ X(s) = \int_0^\infty \mu_n(\omega)Z(\omega,s)d\omega \end{cases} \tag{30}$$

where $z(\omega,0)$ is the initial value of $z(\omega,t)$ at $t = 0$.

This means that $Z(\omega,s) = \frac{z(\omega,0)}{s+\omega} + \frac{V(s)}{s+\omega}$

$$\text{and } X(s) = \int_0^\infty \mu_n(\omega)\frac{z(\omega,0)}{s+\omega}d\omega + \int_0^\infty \mu_n(\omega)\frac{V(s)}{s+\omega}d\omega \tag{31}$$

Since $\frac{1}{s^n} = \int_0^\infty \mu_n(\omega)\frac{1}{s+\omega}d\omega\quad 0 < n < 1$

we can write in the time domain:

$$x(t) = \int_0^\infty \mu_n(\omega)z(\omega,0)e^{-\omega t}d\omega + {}_0I_t^n(v(t)) \tag{32}$$

where:

- $x(t) = \int_0^\infty \mu_n(\omega)z(\omega,0)e^{-\omega t}d\omega$ is the free response of the fractional integrator initialized by the distributed initial conditions $z(\omega,0)$ $\forall \omega \in [0,\infty)$.
- $I_t^n(v(t))$ is the forced response of the fractional integrator caused by the input $v(t)$.

Previously, using the definition of the Caputo derivative "initial condition", we wrote
(20) $x(t) = I_t^n(v(t)) + x(0)$.

The conclusion is that this expression of the free response is wrong, since $\int_0^\infty \mu_n(\omega)z(\omega,0)e^{-\omega t}d\omega$ is the initialization function of the integrator. In fact, $x(0) = \int_0^\infty \mu_n(\omega)z(\omega,0)d\omega$, and Equation (15) is correct only at $t = 0$ and is wrong for $t > 0$.

The conclusion is that the fractional-integrator transients require to refer to its distributed model.

Basically, the initial condition of the differential system $D_t^n(x(t)) = f(x(t),u(t))$ is related to the initial condition of the fractional integrator $\frac{1}{s^n}$ used for the integration of the FDE/FDS, not to the pseudo-initial condition of any fractional derivative. Notice that if this FDE/FDS is related to a real system, its dynamics must not depend on the fractional derivative definition choice of the user.

5. A Counter Example

In previous papers related to the infinite-state representation, we have already demonstrated that the so-called initial conditions of the Caputo derivative are unable to correctly express the dynamics of FDE/FDS free responses. However, attracted by the apparent simplicity of the Caputo initial conditions, most fractional calculus researchers ignore the more complex (in fact not too complex) infinite-state approach.

Consequently, this paper intends to prove the fundamental errors of the usual Caputo derivative approach using an elementary theoretical counter example. Then, with the frequency-distributed integrator allowing the theoretical computation of the true free response, we prove the necessity to use frequency-distributed initial conditions to solve any FDE/FDS initial-condition problem.

5.1. Problem Formulation

Consider the simplest FDE initial value problem

$$\begin{cases} D^n(x(t)) = u(t) & 0 < n < 1 \\ \quad\quad x(t) = x(0) & at\ t = 0 \end{cases} \tag{33}$$

Consider the special function $u(t)$, composed of two delayed Heaviside functions, $UH(t)$ and $-UH(t-T)$, with $u(t) = UH(t) - UH(t-T)$.

Consequently, see Figure 2

$$u(t) = \begin{cases} U & for\ 0 \le t < T \\ 0 & for\ t \ge T \end{cases} \tag{34}$$

Moreover, assume that the system is at rest at $t = 0$, i.e., $x(0) = 0$.

Remark 3: The interest of this example is to create a realistic initial condition at $t = T$, where the free response can be calculated with two approaches, the first one from $t = 0$ with no ambiguity using usual fractional calculus theory and the second one from $t = T$, using Equation (15) at $t_0 = T$.

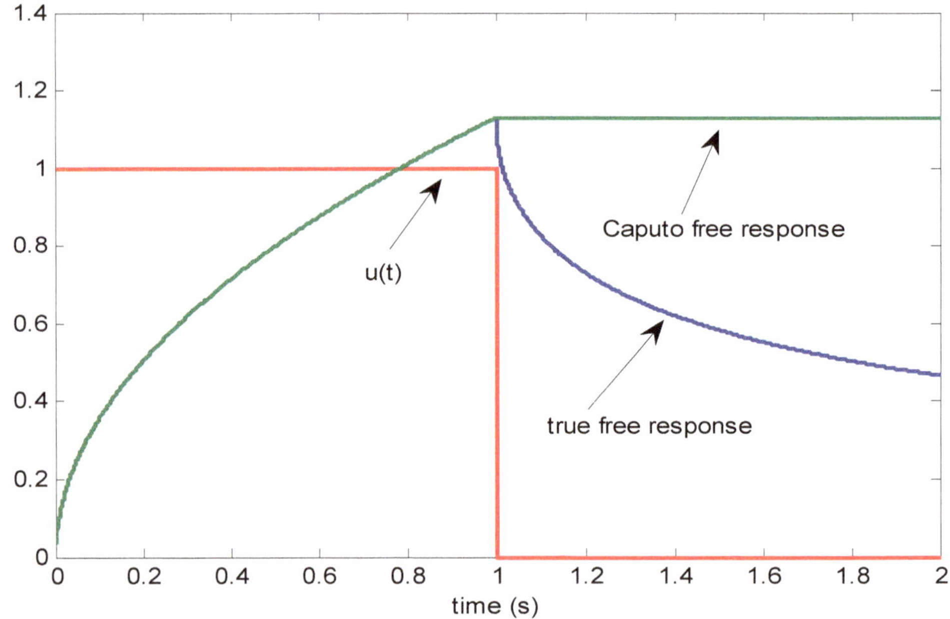

Figure 2. True free response and Caputo derivative initialization for n = 0.5.

5.2. The Exact Solution

Since $x(0) = 0$, we obtain $x(t) = I_t^n(UH(t) - UH(t - T)) = x^+(t) + x^-(t)$.

So $x^+(t) = h_n(t) * UH(t) = \int_0^t \frac{(t-\tau)^{n-1}}{\Gamma(n)} UH(\tau)d\tau = \frac{t^n}{\Gamma(n+1)} UH(t)$ and

$x^-(t) = -h_n(t) * UH(t - T) = -\frac{(t-T)^n}{\Gamma(n+1)} UH(t - T)$.

Consequently, see Figure 2 (for n = 0.5):

$$x(t) = \begin{cases} \frac{t^n U}{\Gamma(n+1)} & for\ 0 \leq t \leq T \\ \frac{U}{\Gamma(n+1)}\left[t^n - (t - T)^n\right] & for\ t \geq T \end{cases} \tag{35}$$

where $x(t)\ for\ t \geq T$ represents the free response of (33) at $t_0 = T$.

5.3. Solution Derived from the Caputo Derivative Definition

This free response can also be expressed using Equation (15) at $t_0 = T$ with $x(t_0) = x(T)$, i.e., $x(t) = I_t^n(u(t)) + x(T)\ for\ t \geq T$.

Thus (see Figure 2),

$$x(t) = x(T)\ for\ t \geq T \tag{36}$$

This result is obviously in complete contradiction with Equation (35), i.e., $x(t) = \frac{U}{\Gamma(n+1)}\left[t^n - (t - T)^n\right]\ for\ t \geq T$.

Notice that, for $n = 1$, we obtain:

$x(T) = UT\ for\ t = T$ and $x(t) = U[t - (t - T)] = UT\ for\ t \geq T$.

Thus, we verify that $x(t) = x(T)\ for\ t \geq T$, i.e., Equation (36) is only correct in the integer-order case.

With this very simple example, we have demonstrated that Equation (15) is wrong in the fractional-order case. So, what is the reason of this basic error?

5.4. Solution Derived from the Distributed Frequency Model of the Fractional Integrator

Consider again the elementary example (33):

$$D^n(x(t)) = u(t) \quad 0 < n < 1$$

This system is supposed at rest at $t = 0$, i.e., $z(\omega, 0) \quad \forall \omega \in [0, \infty)$.
So, using (30) we obtain

$$Z^+(\omega, s) = \frac{U}{s(s + \omega)} \text{ for } u(t) = UH(t)$$

Thus, according to (29), $X(s) = \int\limits_0^\infty \mu_n(\omega) \frac{1}{(s+\omega)} \frac{U}{s} d\omega = \frac{U}{s^{n+1}}$

and
$$\begin{cases} x^+(t) = \frac{t^n U}{\Gamma(n+1)} H(t) \\ x^-(t) = -\frac{U(t-T)^n}{\Gamma(n+1)} H(t - T) \end{cases}.$$

Obviously, we recover the same result as (35) using the distributed model. Moreover, this model allows us to express $z(\omega, t)$, i.e.,

$$z^+(\omega, t) = \frac{U}{\omega}\left(1 - e^{-\omega t}\right) H(t)$$

So
$$\begin{cases} x^+(t) = UH(t)\int\limits_0^\infty \frac{\mu_n(\omega)}{\omega}\left(1 - e^{-\omega t}\right) d\omega \\ x^-(t) = -UH(t - T)\int\limits_0^\infty \frac{\mu_n(\omega)}{\omega}\left(1 - e^{-\omega(t-T)}\right) d\omega \end{cases} \tag{37}$$

Consequently, the free response is expressed as:

$$x(t) = U\int\limits_0^\infty \frac{\mu(\omega)}{\omega}\left(e^{-\omega(t-T)} - e^{-\omega t}\right) d\omega \text{ for } t \geq T, \tag{38}$$

which is the distributed equivalent of Equation (35).

Moreover, we can verify that it is now possible to calculate the response of the integrator for $t \geq T$ using the expression (32).

So, consider the response initialized at $t = T$ with $u(t) = 0$ for $t \geq T$.
Since $z(\omega, T)$ is the initial condition at $t = T$, with

$$z(\omega, T) = z^+(\omega, T) = \frac{U}{\omega}\left(1 - e^{-\omega T}\right), \tag{39}$$

and since $I_t^n(0) = 0$, we obtain

$$x(t) = U\int\limits_0^\infty \mu_n(\omega)z(\omega, T)e^{-\omega(t-T)}d\omega \quad \text{for } t \geq T$$

So

$$x(t) = U\int\limits_0^\infty \frac{\mu_n(\omega)}{\omega}\left(1 - e^{-\omega T}\right)e^{-\omega(t-T)}d\omega \quad \text{for } t \geq T$$

And we obtain the same result as previously (38), i.e.,

$$x(t) = U\int\limits_0^\infty \frac{\mu_n(\omega)}{\omega}\left(e^{-\omega(t-T)} - e^{-\omega t}\right)d\omega \quad \text{for } t \geq T$$

We can conclude that the distributed state-space model provides the exact expression of the free response using the usual tools of linear system theory. Consequently, this distributed model is the necessary tool to express transients of the fractional integrator.

Notice that numerical simulations corresponding to this counter example are available in [33].

5.5. Conclusions

Two main conclusions can be stated from this counter example:

- The integration of FDE/FDS based on the Caputo derivative definition (or on the Riemann–Liouville derivative) are wrong approaches leading to erroneous free responses.
- The frequency-distributed state-space model provides the exact expression of the free response using the usual tools of linear system theory. Consequently, this distributed model is the necessary tool to express transients of the fractional integrator and thus those of FDE/FDS.

Notice that the Grünwald–Letnikov approach based on relation (25) provides a correct solution to the integration of FDE/FDS. However, its initial conditions, composed of past values of $x(-i)$ since $k = -\infty$, are not easy to use, particularly for an initialization objective.

5.6. The Caputo Derivative Definition Revisited

We have demonstrated with the previous elementary counter example that the integration technique based on the Caputo derivative definition is unable to provide a correct expression of the free response of an elementary initial-value problem. Of course, we have pointed out the reason of this failure, i.e., $x(0)$ does not represent the initial condition of the fractional integrator. Basically, what is the origin of this error?

In order to understand why so many researchers have been misled by the so-called "initial condition" $x(0)$, we have to once again consider the definition (12) of the Caputo derivative:

$$^{C}D_t^n(x(t)) = I_t^{1-n}\left(\frac{dx(t)}{dt}\right)$$

This derivative relies on a fractional integrator $\frac{1}{s^{1-n}}$, so we have to take into account its internal state variables $z_C(\omega, t)$ at $t = 0$ (notice that $z_C(\omega, t) \neq z(\omega, t)$) [34].

Thus, the distributed-frequency model of the Caputo derivative corresponds to that of the fractional integrator $I_t^{1-n}(.)$, where, in this case, the input and the output are $\frac{dx(t)}{dt}$ and $^{C}D_t^n(x(t))$, respectively:

$$\begin{cases} \frac{\partial z_C(\omega,t)}{\partial t} = -\omega z_C(\omega,t) + \frac{dx(t)}{dt} \quad \omega[0,\infty) \\ ^{C}D_t^n(x(t)) = \int_0^\infty \mu_{1-n}(\omega)z_C(\omega,t)d\omega \\ \mu_{1-n}(\omega) = \frac{\sin((1-n)\pi)}{\pi}\omega^{-(1-n)} \text{ and } 0 < n < 1 \end{cases} \tag{40}$$

with the initial condition $z_C(\omega, 0)$ and $\omega \in [0, \infty)$.

By using the Laplace transform, we can write

$$\begin{cases} Z_C(\omega,s) = \frac{z_C(\omega,0)}{s+\omega} + \frac{L\left\{\frac{dx(t)}{dt}\right\}}{s+\omega} \quad \omega \in [0,\infty) \\ \text{with } L\left\{\frac{dx(t)}{dt}\right\} = sX(s) - x(0) \end{cases}$$

Thus, we obtain

$$L\left\{^{C}D_t^n(x(t))\right\} = \int_0^\infty \mu_{1-n}(\omega)Z_C(\omega,s)d\omega = s^n X(s) - \frac{x(0)}{s^{1-n}} + \int_0^\infty \frac{\mu_{1-n}(\omega)}{s+\omega}z_C(\omega,0)d\omega \tag{41}$$

We can conclude that the usual initial condition of the Caputo derivative is wrong because it does not take into account the transients of its associated integrator $\frac{1}{s^{1-n}}$, i.e., its distributed initial conditions $z_C(\omega,0)$.

Notice that the same conclusions apply to the Riemann–Liouville derivative [34].

Consequently, the Caputo derivative approach to the integration of FDE/FDS must be rejected because it provides wrong solutions to fractional initial-value problems. This approach is wrong for two main reasons:

- The exact initial conditions of the Caputo derivative are $x(0)$ and the distributed state variable initial condition $z_C(\omega,0)$.
- The technique based on the Caputo derivative is not natural because the true and physical initial conditions are those of the fractional integrator $\frac{1}{s^n}$ of Equation (30), i.e., $z(\omega,0)$, such as in the integer order case.

Numerical simulations of the Caputo and Riemann–Liouville derivatives exhibiting the role of their initial conditions are available in [38] and in Volume 1 of [34].

6. Fractional Differential Systems Transients

6.1. Integration of a FDE

We demonstrated in the previous section that it is necessary to use the frequency-distributed model of the fractional integrator to take into account the transients of the free response. Thus, we have to apply the same approach for the integration of any FDE/FDS initial value problem (6,7). As we have shown, the solution is provided by the fractional integral equation, which is equivalent to the integer order case (2):

$$x(t) = I_t^n(f(x(t), u(t))) + x_0(t), \tag{42}$$

where $x_0(t)$ is the initialization function of the Riemann–Liouville integral

$$x_0(t) = \int_0^\infty \mu_n(\omega)z(\omega,0)e^{-\omega t}d\omega. \tag{43}$$

Notice that (42) is a Volterra integral equation.

Fundamentally, Figure 3 displays the graphical representation of the integral Equation (42), which underlines the closed-loop behavior of the FDE/FDS based on the fractional integrator $\frac{1}{s^n}$ with the initialization function $x_0(t)$.

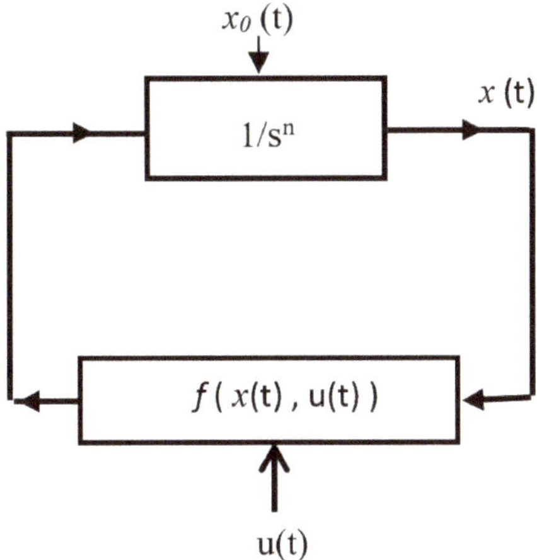

Figure 3. Closed-loop model of the FDE/FDS.

However, we can also use the frequency-distributed model of the fractional integrator, where its input is $v(t) = D^n(x(t)) = f(x(t), u(t))$, which leads to the distributed representation of the FDE:

$$\begin{cases} \frac{\partial z(\omega,t)}{\partial t} = -\omega z(\omega,t) + f(x(t), u(t)) \\ x(t) = \int_0^\infty \mu_n(\omega) z(\omega,t) d\omega \\ \mu(\omega) = \frac{\sin(n\pi)}{\pi} \omega^{-n} \quad 0 < n < 1 \end{cases} \tag{44}$$

In this case, the solution $z(\omega, t)$ is provided by the distributed integer-order integral equation:

$$\begin{aligned} z(\omega, t) &= \int_0^t [-\omega z(\omega,t) + f(x(\tau), u(\tau))]d\tau + z(\omega, 0) \quad \forall \omega \in [0, \infty) \\ &= I_t^1(-\omega z(\omega,t) + f(x(t), u(t))) + z(\omega, 0) \end{aligned} \tag{45}$$

where $z(\omega, 0)$ is the initial condition of the integer order integral.

We can represent (see Figure 4) this frequency distributed system graphically, where frequency varies from $\omega = 0$ to $\omega = +\infty$, according to its Laplace transform. This graph corresponds to Figure 3, where the fractional integrator is replaced by its distributed graph of Figure 1.

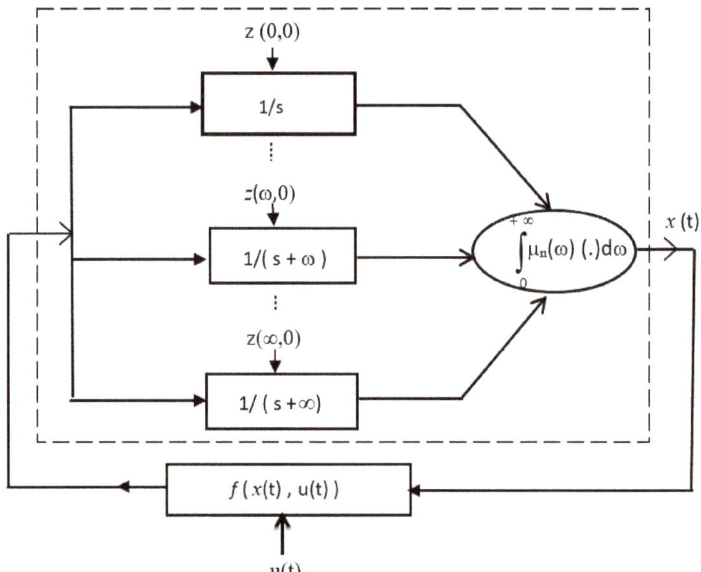

Figure 4. Closed-loop model of the FDE/FDS based on the fractional integrator distributed-frequency model.

Equations (42) and (43) and the graph of Figure 3 focus on the pseudo-state variable $x(t)$, whereas Equation (44) and the graph of Figure 4 focus on the internal state variables $z(\omega, t)$ of the integrator, which are in fact those of the FDE/FDS.

Notice that, for the isolated integrator (Figure 1), with input $v(t)$ and output $x(t)$, the state variables are decoupled and evolve independently. On the other hand, in system (44), i.e., in the graph of Figure 4, the state variables are coupled by the relation $v(t) = f(x(t), u(t))$. This means that the evolution of the state variable $z(\omega, t)$ (for the particular value) depends on all the other state variables $z(\xi, t) \; \xi \in [0, \infty)$. Namely, the original FDE/FDS (6) has been transformed into an infinite-dimension system of first-order differential Equations (44).

These are the two "faces" of the same problem:

- Equation (42) and Figure 3 correspond to the pseudo-state variable $x(t)$, directly taking into account the fractional order n.
- Equation (44) and Figure 4 correspond to the system of distributed state variables $z(\omega, t)$, whose solution is obtained through an integer-order approach. The pseudo-state variable $x(t)$ is provided by a weighted integral, where $\mu_n(\omega)$ is the link between the integer order and fractional order domains.

In Figures 3 and 4, there is no hypothesis about the nature of $f(x(t), u(t))$ which can be either linear or nonlinear. In the nonlinear case, the integral formulation of the FDE/FDS initial-value problem leads to (42) or (45). The solution of these integral equations can be obtained with Picard's method, which is currently used to solve nonlinear fractional order problems (see for instance [3]).

Of course, the nonlinear case is a very wide topic, and our objective is not to treat it in this paper. In fact, the linear case is also of fundamental interest, and we propose to revisit it, essentially to formulate free responses.

6.2. FDS Transients and the Mittag–Leffler Function

First, we consider the elementary FDE initial-value problem:

$$D^n(x(t)) = ax(t) \quad 0 < n < 1 \tag{46}$$

with the distributed initial condition $z(\omega, 0) \quad \forall \omega \in [0, \infty)$.

System (46) can be transformed into the distributed frequency one:

$$\begin{cases} \dfrac{\partial z(\omega, t)}{\partial t} = -\omega z(\omega, t) + ax(t) \\ x(t) = \displaystyle\int_0^\infty \mu_n(\omega) z(\omega, t) d\omega \end{cases} \tag{47}$$

with the initial condition $z(\omega, 0)$.

Using the Laplace transform, we can write

$$sZ(\omega, s) - z(\omega, 0) = -\omega Z(\omega, s) + aX(s),$$

i.e., $X(s) = \dfrac{s^n}{s^n - a} \displaystyle\int_0^\infty \mu_n(\omega) \dfrac{z(\omega, 0)}{s + \omega} d\omega$

$$\text{or } X(s) = s \frac{s^{n-1}}{s^n - a} X_0(s), \tag{48}$$

with $X_0(s) = \displaystyle\int_0^\infty \mu_n(\omega) \dfrac{z(\omega, 0)}{s + \omega} d\omega$

So,

$$x_0(t) = L^{-1} \left\{ \int_0^\infty \mu_n(\omega) \frac{z(\omega, 0)}{s + \omega} d\omega \right\} = \int_0^\infty \mu_n(\omega) z(\omega, 0) e^{-\omega t} d\omega \tag{49}$$

Let us remember that

$$L\{E_{n,1}(at^n)\} = \frac{s^{n-1}}{s^n - a} \tag{50}$$

where $E_{n,1}(at^n)$ is the Mittag–Leffler function:

$$E_{n,1}(at^n) = \sum_{k=0}^\infty \frac{(at^n)^k}{\Gamma(nk + 1)}, \tag{51}$$

which is the generalization of the exponential function, since for $n = 1$ we obtain $E_{1,1}(at^1) = \sum\limits_{k=0}^{\infty} \frac{(at^1)^k}{k!} = e^{at}$.

Consequently, we obtain

$$x(t) = L^{-1}\{X(s)\} = \frac{d}{dt}\{E_{n,1}(at^n) * x_0(t)\}. \tag{52}$$

We can easily generalize this result to any linear FDS initial value problem:

$$D^n(\underline{x}(t)) = A\underline{x}(t) + \underline{B}u(t)\ 0 < n < 1, \tag{53}$$

with the distributed initial condition $\underline{z}(\omega,0)\ \ \forall \omega \in [0,\infty)$.

Defining the matrix Mittag–Leffler function:

$$E_{n,1}(At^n) = \sum\limits_{k=0}^{\infty} \frac{(At^n)^k}{\Gamma(nk+1)}, \tag{54}$$

we can write

$$\underline{x}(t) = \frac{d}{dt}\{E_{n,1}(At^n) * \underline{x}_0(t)\} + \int\limits_0^t E_{n,1}(A(t-\tau)^n)\underline{B}\tilde{u}(\tau)d\tau, \tag{55}$$

where $\tilde{u}(\tau) = D^{1-n}(u(\tau))$ and $\underline{x}_0(t) = \int\limits_0^{\infty} \mu_n(\omega)\underline{z}(\omega,0)e^{-\omega t}d\omega$.

Obviously, the free response $\frac{d}{dt}\{E_{n,1}(At^n) * \underline{x}_0(t)\}$ of $\underline{x}(t)$ is more complex than the wrong usual one derived from (8), i.e., $\underline{x}(t) = E_{n,1}(At^n)\underline{x}(0)$.

Moreover, there is a major difficulty, i.e., the convolution between the Mittag–Leffler function and $x_0(t)$, which is not a straightforward operation. Thus, the interest of this expression is essentially theoretical.

Notice also that the matrix Mittag–Leffler function does not verify the semi-group properties of the matrix exponential function [53,54].

Remark 4: The use of the Caputo derivative is based on the assumption $x_0(t) = x(0) = cte\ \ \forall t$. Is this requirement always wrong? In addition, if it is correct, what does it mean?

Consider the following system

$$D^n(x(t)) = ax(t) + bu(t)\ 0 < n < 1$$

and suppose that it is at rest at $t = 0$ and that $u(t) = UH(t)$.

Since, in this case,

$Z(\omega,s) = \frac{aX(s)+b\frac{U}{s}}{s+\omega}$ and $X(s) = \int\limits_0^{\infty} \mu_n(\omega)Z(\omega,s)d\omega$, we obtain $Z(\omega,s) = b\frac{s^n}{s^n}\frac{1}{as+\omega}\frac{U}{s}$.

Thus

$$z(\omega,\infty) = \lim\limits_{s\to 0} sZ(\omega,s) = 0\ \ \forall \omega \neq 0 \text{ and } z(0,\infty) = \lim\limits_{s\to 0}\frac{bU}{s^{1-n}} = \infty \tag{56}$$

On the other hand, we can also write $x(\infty) = -\frac{b}{a}U$ for a stable system, i.e., $a < 0$ in our case.

Assume that we have applied the same step input $UH(t)$, but, at $t = -\infty$, this means that at $t = 0$, we obtained $x(0) = -\frac{b}{a}U$ for a long time in the past ($t < 0$).

Consequently, the condition $x_0(t) = x(0) = cte\ \ \forall t$ requires that the system has been at rest for a very long time.

Then

$$x_0(t) = \int_0^\infty \mu_n(\omega)z(\omega,0)e^{-\omega t}d\omega = \int_0^\infty \mu_n(\omega)z(0,0)e^{-0t}d\omega = x(0) = cte. \tag{57}$$

Since $z(\omega,\infty) = 0 \forall \omega \neq 0$.

Theoretically, the condition $x_0(t) = x(0) = cte \;\; \forall t$ can be achieved, but it is completely unrealistic. Moreover, it would require infinite energy [34].

Remark 5: Consider again the revisited definition of the Caputo derivative (see Section 5.6):

$$^C D_t^n(x(t)) = I_t^{1-n}\left(\frac{dx(t)}{dt}\right) \Rightarrow
\begin{cases}
\frac{\partial z_C(\omega,t)}{\partial t} = -\omega z_C(\omega,t) + \frac{dx(t)}{dt} & \omega \in [0,\infty) \\
^C D_t^n(x(t)) = \int_0^\infty \mu_{1-n}(\omega)z_C(\omega,t)d\omega \\
\mu_{1-n}(\omega) = \frac{\sin((1-n)\pi)}{\pi}\omega^{-(1-n)} \text{ and } 0 < n < 1
\end{cases}$$

with $z_C(\omega,t) = z_C(\omega,0)$ at $t = 0$.

We have demonstrated (41) that:

$$L\left\{^C D_t^n(x(t))\right\} = s^n X(s) - \frac{x(0)}{s^{1-n}} + \int_0^\infty \frac{\mu_{1-n}(\omega)}{s+\omega}z_C(\omega,0)d\omega.$$

If $x(t) = cte$, for a long time in the past ($t < 0$), then $\frac{dx(t)}{dt} = 0$. This means that $z_C(\omega,t) = 0 \;\; \forall\, \omega$ for a long time in the past ($t < 0$), and consequently $z_C(\omega,0) = 0 \;\; \forall\, \omega$.

Then, we can write $L\left\{^C D_t^n(x(t))\right\} = s^n X(s) - \frac{x(0)}{s^{1-n}}$.

Obviously, the previous conditions are very restrictive, and the usual initial condition is wrong as soon as there is a variation of $x(t)$.

Remark 6: We have demonstrated (52) that:

$$x(t) = \frac{d}{dt}\{E_{n,1}(at^n) * x_0(t)\} = E_{n,1}(at^n) * \frac{d}{dt}x_0(t)$$

Since $x_0(t) = x(0) = cte \;\; \forall t$, we can write $x_0(t) = x(0)H(t)$, so $\frac{d}{dt}x_0(t) = x(0)\delta(t)$.

Then,
$$x(t) = E_{n,1}(at^n) * x(0)\delta(t) = E_{n,1}(at^n)x(0) \tag{58}$$

is the usual result corresponding to the Caputo derivative assumption.

The conclusion is that many results obtained with the Caputo derivative approach are not necessarily wrong, but they require the previous very restrictive assumptions related to $x(0)$.

6.3. FDS Transients Expressed with the Distributed Exponential Function

Previously, the problem of fractional transients has been focused on the pseudo-state variable dynamics, with no insight in the distributed state variable. So, we propose now to express the dynamics of $z(\omega,t)$. Let us start again with the elementary system (46):

$$D^n(x(t)) = ax(t) \quad 0 < n < 1$$

with the distributed initial condition $z(\omega,0) \;\; \forall \omega \in [0,\infty)$.

The corresponding distributed frequency model (47) leads to:

$$\frac{\partial z(\omega,t)}{\partial t} = -\omega z(\omega,t) + a \int_0^\infty \mu_n(\xi)z(\xi,t)d\xi \quad \forall \omega \in [0,\infty). \tag{59}$$

Notice that we have to separate the current frequency ω from all the other ones $\xi \in [0,\infty)$. This means that the behavior of $z(\omega,t)$ depends on all the behaviors of the other state variables $z(\xi,t)$.

Let us define $\delta(\xi)$, which is the frequency Dirac impulse verifying $\int_0^\infty \delta(\xi)d\xi = 1$.

Then,

$$\int_0^\infty \omega\delta(\xi-\omega)z(\xi,t)d\xi = \omega\int_0^\infty \delta(\xi-\omega)z(\xi,t)d\xi = \omega z(\omega,t). \tag{60}$$

Let us define

$$\psi(\omega,\xi) = -\omega\delta(\xi-\omega) + a\mu_n(\xi) \tag{61}$$

Then, (59) can be expressed as:

$$\frac{\partial z(\omega,t)}{\partial t} = -\omega\int_0^\infty \delta(\xi-\omega)z(\xi,t)d\xi + a\int_0^\infty \mu_n(\xi)z(\xi,t)d\xi = \int_0^\infty \psi(\omega,\xi)z(\xi,t)d\xi \tag{62}$$

The solution $z(\omega,t)$ requires the integration of the integer-order distributed system (59) with the initial condition $z(\omega,0) \quad \forall \omega \in [0,\infty)$. Basically, the solution verifies the integral relation:

$$\begin{aligned}
z(\omega,t) &= \int_0^t \left[\int_0^\infty \Psi(\omega,\zeta)z(\zeta,\tau)d\zeta\right]d\tau + z(\omega,0) \\
&= {}_0I_t^1\left[\int_0^\infty \Psi(\omega,\zeta)z(\zeta,\tau)d\zeta\right] + z(\omega,0) \quad \omega \in [0,\infty)
\end{aligned} \tag{63}$$

This integration is performed with Picard's method, which is an iterative technique (3). At the first iteration, $z(\omega,t)$ is approximated by $z(\omega,0)$.

Since $z(\omega,0)$ and $\int_0^\infty \psi(\omega,\xi)d\xi$ are constants for ${}_0I_t^1$, we obtain

$$z_1(\omega,t) = z(\omega,0) + {}_0I_t^1\left[\int_0^\infty \psi(\omega,\zeta)z(\omega,0)d\zeta\right] = z(\omega,0) + z(\omega,0)t\int_0^\infty \psi(\omega,\zeta)d\zeta. \tag{64}$$

At the second iteration, $z(\omega,t)$ is approximated by $z_1(\omega,t)$. So, we obtain

$$z_2(\omega,t) = z(\omega,0) + z(\omega,0)t\int_0^\infty \psi(\omega,\xi)d\xi + z(\omega,0)\frac{t^2}{2}\left[\int_0^\infty \psi(\omega,\xi)d\xi\right]^2 \tag{65}$$

Additionally, at iteration k, we obtain

$$z_k(\omega,t) = \sum_{j=0}^k \frac{t^j}{j!}\left[\int_0^\infty \psi(\omega,\xi)d\xi\right]^j z(\omega,0) \tag{66}$$

Thus,
$$z(\omega, t) = \lim_{k \to \infty} \sum_{j=0}^{k} \frac{t^j}{j!} \left[\int_0^\infty \psi(\omega, \xi) d\xi \right]^j z(\omega, 0), \tag{67}$$

i.e.,
$$z(\omega, t) = \left[\sum_{k=0}^\infty \frac{t^k}{k!} \left[\int_0^\infty \psi(\omega, \xi) d\xi \right]^k \right] z(\omega, 0) \tag{68}$$

Notice that, in the integer order case, $\frac{dx(t)}{dt} = ax(t)$, so $\psi(\omega, \xi) \equiv a$.

Then,
$$x(t) = \left[\sum_{k=0}^\infty a^k \frac{t^k}{k!} \right] x(0) = e^{at} x(0). \tag{69}$$

Thus, in the distributed case, we can define the distributed exponential function:

$$\phi(t) = \exp\left(t \int_0^\infty \psi(\omega, \xi) d\xi \right) = \sum_{k=0}^\infty \frac{t^k}{k!} \left[\int_0^\infty \psi(\omega, \xi) d\xi \right]^k \tag{70}$$

and
$$z(\omega, t) = \exp\left(t \int_0^\infty \psi(\omega, \xi) d\xi \right) z(\omega, 0) = \phi(t) z(\omega, 0) \quad \omega \in [0, \infty), \tag{71}$$

which is the distributed generalization of Equation (69).

Then, we can generalize the distributed exponential to the linear multidimensional non-commensurate order case:

$$D^{\underline{n}}(\underline{x}(t)) = A\underline{x}(t) \quad \dim(\underline{x}(t)) = N, \tag{72}$$

where $\underline{n}^T = [n_1 \dots n_i \dots n_N] \ 0 < n_i \leq 1$.

Moreover, $0 < n_i \leq 1$ means that the FDS system may include integer-order derivatives, since, with real systems, the dynamics are caused either by integer-order or fractional-order derivatives.

So, (72) corresponds to the integer-order frequency-distributed differential system:

$$\begin{cases} \frac{\partial \underline{z}(\omega, t)}{\partial t} = -\omega \underline{z}(\omega, t) + A \int_0^\infty [\mu_{\underline{n}}(\zeta)] \underline{z}(\zeta, t) d\zeta \quad \dim(\underline{z}(t)) = N \\ \underline{x}(t) = \int_0^\infty [\mu_{\underline{n}}(\zeta)] \underline{z}(\zeta, t) d\zeta \end{cases} \tag{73}$$

with the initial condition $\underline{z}(\omega, 0) \quad \forall \omega \in [0, \infty)$,

$$\text{and } [\mu_{\underline{n}}(\zeta)] = \begin{bmatrix} \mu_{n_1}(\zeta) & & 0 \\ & \mu_{n_i}(\zeta) & \\ 0 & & \mu_{n_N}(\zeta) \end{bmatrix} \tag{74}$$

Let us define the matrix

$$\Psi(\omega, \xi) = -\omega \delta(\xi - \omega) I + A[\mu_{\underline{n}}(\xi)] \tag{75}$$

where I is the Identity matrix with appropriate dimension.

Then, the differential system (73) can be expressed as

$$\frac{\partial \underline{z}(\omega, t)}{\partial t} = \int_0^\infty \Psi(\omega, \xi) \underline{z}(\xi, t). \tag{76}$$

Therefore, the distributed exponential $\exp\left(t\int_0^\infty \psi(\omega,\xi)d\xi \right)$ function is replaced by the distributed exponential matrix:

$$\Phi(t) = \exp\left(t\int_0^\infty \Psi(\omega,\xi)d\xi \right) = \sum_{k=0}^\infty \frac{t^k}{k!}\left[\int_0^\infty \Psi(\omega,\xi)d\xi \right]^k, \tag{77}$$

and the solution of (76) is:

$$\underline{z}(\omega,t) = \exp\left(t\int_0^\infty \Psi(\omega,\xi)d\xi \right)\underline{z}(\omega,0) = \Phi(t)\underline{z}(\omega,0) \; \omega \in [0,\infty). \tag{78}$$

Notice that, contrary to the Mittag–Leffler matrix, the distributed exponential matrix shares the same semi-group properties (see chapter 2 of [54]) as its integer order analog, i.e.,

$$\Phi(t,t_0) = \Phi(t,\tau)\Phi(\tau,t_0) \quad t_0 < \tau < t \tag{79}$$

Finally, the solution of the linear multidimensional non-commensurate order FDS (72) with the initial condition $\underline{z}(\omega,0) \;\; \forall \omega \in [0,\infty)$ is similar to the integer-order case (5), i.e.,

$$\underline{z}(\omega,t) = \Phi(t)\underline{z}(\omega,0) + \int_0^t \Phi(t-\tau)\underline{B}u(\tau)d\tau \; \omega \in [0,\infty), \tag{80}$$

where $\Phi(t)$ is the distributed exponential transition matrix.

6.4. Computation of the Distributed Exponential Function

The interest of the distributed exponential function, either in scalar or matrix form, is essentially theoretical. Practically, in order to compute it, the distributed exponential function requires frequency discretization, i.e., a discretization of the continuous distributed integrator model:

$$\begin{cases} \frac{\partial z(\omega,t)}{\partial t} = -\omega z(\omega,t) + v(t) \; \omega \in [0,\infty) \\ x(t) = \int_0^\infty \mu(\omega)z(\omega,t)d\omega \end{cases} \tag{81}$$

Direct discretization is possible, but indirect discretization is more efficient [34,35]. It is based on the approximation of the frequency response of $\frac{1}{s^n}$ $0 < n < 1$, on a frequency interval $[\omega_{min}, \omega_{max}]$ with J first-order cells, associated with an integer-order integrator at =0. So, the global model is composed of $J + 1$ cells (or modes) such as:

$$\begin{cases} \frac{dz_j(t)}{dt} = -\omega_j z_j(t) + v(t) \quad j = 0 \text{ to } J \\ x(t) = \sum_{j=0}^{J} c_j z_j(t) \end{cases} \tag{82}$$

and the elementary system $D^n(x(t)) = ax(t)$ becomes:

$$\begin{cases} \frac{dz_j(t)}{dt} = -\omega_j z_j(t) + a\sum_{j=0}^{J} c_j z_j(t) \; j = 0 \text{ to } J \\ x(t) = \sum_{j=0}^{J} c_j z_j(t) \end{cases} \tag{83}$$

where ω_j varies from $\omega_0 = 0$ to ω_J.

Let us define:

$$\underline{z}_I^T(t) = \left[z_0(t)\ldots z_j(t)\ldots z_J(t)\right] \quad \dim(\underline{z}_I) = J+1$$

$$A_I = \begin{bmatrix} -\omega_0 & & \\ & -\omega_j & \\ & & -\omega_J \end{bmatrix} \quad \underline{B}_I^T = [1\ldots 1\ldots 1] \quad \underline{C}_I = [c_0\ldots c_j\ldots c_J] \tag{84}$$

Then, the previous system (83) is transformed into the following one

$$\begin{cases} \frac{d\underline{z}_I(t)}{dt} = A_I\underline{z}_I(t) + a\underline{B}_I\underline{C}_I\underline{z}_I(t) = A_{syst}\underline{z}_I(t) \\ x(t) = \underline{C}_I\underline{z}_I(t) \end{cases} \text{ with } A_{syst} = A_I + a\underline{B}_I\underline{C}_I \tag{85}$$

Thus, the solution of system (85) with the initial condition $\underline{z}_I^T(0) = \left[z_0(0)\ldots z_j(0)\ldots z_J(0)\right]$ is expressed as

$$\underline{Z}(t) = e^{A_{syst}t}\underline{Z}(0) = \phi_d(t)\underline{Z}(0), \tag{86}$$

where $\phi_d(t) = e^{A_{syst}t}$ is the matrix exponential corresponding to the frequency discretization of the distributed exponential function $\phi(t) = \exp\left(t\int_0^\infty \psi(\omega, \xi)d\xi\right)$.

Numerical examples of the distributed exponential function are available in chapter 9 volume 1 of [34].

7. Conclusions

In this paper, it has been proved, thanks to an elementary counter example, that the solution of the FDE initial value problem cannot be provided by the well-known Caputo derivative approach. The origin of this error relies on ignoring the dynamics of the fractional integrator, which has caused confusion between pseudo initial conditions of the Caputo derivative and the initialization function of the Riemann–Liouville integral.

Furthermore, it has been demonstrated that it is necessary to take into account two complementary models of the fractional integrator: the classical model used to express the pseudo state variable $x(t)$ is an input/output model, whereas the distributed model is an infinite-dimension state variable model, which is adapted to the formulation of free responses thanks to the initial values of its distributed state variable $z(\omega, t)$, as proved by the counter-example.

These two models of the fractional integrator generate two complementary approaches to the modeling of FDE and FDS one being focused on the pseudo-state variable, whereas the other permits to express internal transients linked to the distributed state variable. Two expressions of their free responses have been formulated in the linear case, one with the help of a convolution of the Mittag–Leffler function with the initialization function, and the other thanks to the definition of a distributed exponential function, which is a straightforward generalization of the integer-order case.

Beyond the misuse of fractional derivatives to solve the initial-value problem, the main conclusion of this paper is that any fractional-order system or one characterized by a long memory phenomenon is an infinite dimension system, whatever its input/output representation, linear or nonlinear.

Consequently, the use of an internal distributed representation is not an option but is necessary to correctly express initialization problems and dynamical transients.

Author Contributions: N.M. and J.-C.T. contributed equally to the paper. All authors have read and agreed to the published version of the manuscript.

Funding: This research received no external funding.

Conflicts of Interest: The authors declare no conflict of interest.

References

1. Coddington, E.A.; Levinson, N. *Theory of Ordinary Differential Equations*; Mc Graw Hill: New York, NY, USA, 1955.
2. Tenenbaum, M.; Pollard, H. *Ordinary Differential Equations*; Harper and Row: New York, NY, USA, 1963.
3. Podlubny, I. *Fractional Differential Equations*; Academic Press: San Diego, CA, USA, 1999.
4. Diethelm, K. *The Analysis of Fractional Differential Equations*; Lecture Notes in Mathematics; Springer: Berlin/Heidelberg, Germany, 2010.
5. Kilbas, A.A.; Srivastava, H.M.; Trujillo, J.J. *Theory and Applications of Fractional Differential Equations*; Elsevier: Amsterdam, The Netherlands, 2006.
6. Caputo, M. *Elasticita e Dissipazione*; Zanichelli: Bologna, Italy, 1969.
7. Fukunaga, M.; Shimizu, N. Role of pre-histories in the initial value problems of fractional viscoelastic equations. *Nonlinear Dyn.* **2004**, *38*, 207–220. [CrossRef]
8. Du, M.; Wang, Z. Initialized fractional differential equations with Riemann-Liouville fractional order derivative. In Proceedings of the ENOC 2011 Conference, Rome, Italy, 24–29 July 2011.
9. Du, M.; Wang, Z. Correcting the initialization of models with fractional derivatives via history dependent conditions. *Acta Mech. Sin.* **2016**, *32*, 320–325. [CrossRef]
10. Hartley, T.T.; Lorenzo, C.F. The error incurred in using the Caputo derivative Laplace transform. In Proceedings of the ASME IDET-CIE Conferences, San Diego, CA, USA, 30 August–2 September 2009.
11. Sabatier, J.; Merveillaut, M.; Malti, R.; Oustaloup, A. How to impose physically coherent initial conditions to a fractional system? *Commun. Non Linear Sci. Numer. Simul.* **2010**, *15*, 1318–1326. [CrossRef]
12. Ortigueira, M.D. *Fractional Calculus for Scientists and Engineers*; Springer Science: New York, NY, USA, 2011.
13. Trigeassou, J.C.; Maamri, N.; Oustaloup, A. The Caputo derivative and the infinite state approach. In Proceedings of the 6th Workshop on Fractional Differentiation and its Applications, Grenoble, France, 4–6 February 2013.
14. Sabatier, J.; Farges, C. Comments on the description and initialization of fractional partial differential equations using Riemann-Liouville's and Caputo's definitions. *J. Comput. Appl. Math.* **2018**, *339*, 30–39. [CrossRef]
15. Sabatier, J.; Farges, C. Initial value problems should not be associated to fractional model descriptions whatever the derivative definition used. *AIMS Math.* **2021**, *6*, 11318–11329. [CrossRef]
16. Lorenzo, C.F.; Hartley, T.T. Initialization in fractional order systems. In Proceedings of the European Control Conference (ECC'01), Porto, Portugal, 4–7 September 2001; pp. 1471–1476.
17. Hartley, T.T.; Lorenzo, C.F. The initialization response of linear fractional order system with constant History function. In Proceedings of the 2009 ASME/IDETC, San Diego, CA, USA, 30 August–2 September 2009.
18. Hartley, T.T.; Lorenzo, C.F. The initialization response of linear fractional order system with ramp History function. In Proceedings of the 2009 ASME/IDETC, San Diego, CA, USA, 30 August–2 September 2009.
19. Lorenzo, C.F.; Hartley, T.T. Initialization of fractional differential equation. *ASME J. Comput. Nonlinear Dyn.* **2007**, *4806*, 1341–1347.
20. Lorenzo, C.; Hartley, T. Time-varying initialization and Laplace transform of the Caputo derivative: With order between zero and one. In Proceedings of the IDETC/CIE FDTA'2011 Conference, Washington DC, USA, 14–17 August 2011.
21. Hartley, T.T.; Lorenzo, C.F.; Trigeassou, J.C.; Maamri, N. Equivalence of history function based and infinite dimensional state initializations for fractional order operators. *ASME J. Comput. Nonlinear Dyn.* **2013**, *8*, 041014. [CrossRef]
22. Trigeassou, J.C.; Maamri, N.; Oustaloup, A. The infinite state approach: Origin and necessity. *Comput. Math. Appl.* **2013**, *66*, 892–907. [CrossRef]
23. Tari, M.; Maamri, N.; Trigeassou, J.C. Initial conditions and initialization of fractional systems. *ASME J. Comput. Nonlinear Dyn.* **2016**, *11*, 041014. [CrossRef]
24. Maamri, N.; Tari, M.; Trigeassou, J.C. Improved initialization of fractional order systems. In Proceedings of the 20th World IFAC Congress, Toulouse, France, 14 July 2017; pp. 8567–8573.
25. Khalil, R.; Yousef, A.; Sababdeh, M. A new definition of fractional derivative. *J. Comput. Appl. Math.* **2014**, *264*, 65–70. [CrossRef]
26. Caputo, M.; Fabrizio, M. A new definition of fractional derivative without singular kernel. *Prog. Fract. Differ. Appl.* **2015**, *1*, 73–85.
27. Atanackovic, T.M.; Pilipovic, S.; Zorica, D. Properties of the Caputo-Fabrizio fractional derivative and its distributional setting. *Fract. Calc. Appl. Anal.* **2018**, *21*, 29–44. [CrossRef]
28. Martinez, F.; Othman Mohammed, P.; Napoles Valdes, J.E. Non-conformable fractional Laplace transform. *Kragujev. J. Math.* **2022**, *46*, 341–354. [CrossRef]
29. Ortigueira, M.D.; Machado, J.T. A critical analysis of the Caputo-Fabrizio operator. *Commun. Nonlinear Sci Numer. Simulat.* **2018**, *59*, 608–611. [CrossRef]
30. Tarasov, V.E. No nonlocality, no fractional derivative. *Commun. Nonlinear Sci Numer. Simulat.* **2018**, *62*, 157–163. [CrossRef]
31. Montseny, G. Diffusive Representation of Pseudo Differential Time Operators. *ESAIM Proc.* **1998**, *5*, 159–175. [CrossRef]
32. Heleschewitz, D.; Matignon, D. Diffusive realizations of fractional integro-differential operators: Structural analysis under approximation. In Proceedings of the Conference IFAC, System, Structure and Control, Nantes, France, 8–10 July 1998; Volume 2, pp. 243–248.
33. Trigeassou, J.C.; Maamri, N.; Sabatier, J.; Oustaloup, A. Transients of fractional order integrator and derivatives. In *Special Issue: "Fractional Systems and Signals" of Signal, Image and Video Processing*; Springer: Berlin/Heidelberg, Germany, 2012; Volume 6, pp. 359–372.

34. Trigeassou, J.C.; Maamri, N. *Analysis, Modeling and Stability of Fractional Order Differential Systems: The Infinite State Approach*; John Wiley and Sons: Hoboken, NJ, USA, 2019; Volumes 1 and 2.
35. Trigeassou, J.C.; Poinot, T.; Lin, J.; Oustaloup, A.; Levron, F. Modeling and identification of a non integer order system. In Proceedings of the ECC'99 European Control Conference, Karlsruhe, Germany, 31 August–3 September 1999.
36. Lin, J.; Poinot, T.; Trigeassou, J.C. Parameter estimation of fractional systems: Application to the modeling of a lead-acid battery. In Proceedings of the 12th IFAC Symposium on System Identification, Santa Barbara, CA, USA, 21–23 June 2000.
37. Benchellal, A.; Poinot, T.; Trigeassou, J.C. Approximation and identification of diffusive interface by fractional models. *Signal Process.* **2006**, *86*, 2712–2727. [CrossRef]
38. Trigeassou, J.C.; Maamri, N.; Sabatier, J.; Oustaloup, A. State variables and transients of fractional order differential systems. *Comput. Math. Appl.* **2012**, *64*, 3117–3140.
39. Trigeassou, J.C.; Maamri, N.; Sabatier, J.; Oustaloup, A. A Lyapunov approach to the stability of fractional differentiel equations. *Signal Process.* **2011**, *91*, 437–445. [CrossRef]
40. Du, B.; Wei, Y.; Liang, S.; Wang, Y. Estimation of exact initial states of fractional order systems. *Nonlinear Dyn.* **2016**, *86*, 2061–2070. [CrossRef]
41. Yuan, J.; Zhang, Y.; Liu, J.; Shi, B. Equivalence of initialized fractional integrals and the diffusive model. *ASME J. Comput. Nonlinear Dyn.* **2018**, *13*, 034501. [CrossRef]
42. Zhao, Y.; Wei, Y.; Chen, Y.; Wang, Y. A new look at the fractional initial value problem: The aberration phenomenon. *ASME J. Comput. Nonlinear Dyn.* **2018**, *13*, 121004. [CrossRef]
43. Yuan, J.; Shi, B.; Ji, W. Adaptive sliding mode control of a novel class of fractional chaotic systems. *Adv. Math. Phys.* **2013**, *13*. [CrossRef]
44. Wang, B.; Ding, J.; Wu, F.; Zhu, D. Robust finite time control of fractional order nonlinear systems via frequency distributed model. *Nonlinear Dyn.* **2016**, *85*, 2133–2142. [CrossRef]
45. Chen, Y.; Wei, Y.; Zhou, X.; Wang, Y. Stability for nonlinear fractional order systems: An indirect approach. *Nonlinear Dyn.* **2017**, *89*, 1011–1018. [CrossRef]
46. Wei, Y.; Sheng, D.; Chen, Y.; Wang, Y. Fractional order chattering free robust adaptive backstepping control technique. *Nonlinear Dyn.* **2019**, *95*, 2383–2394. [CrossRef]
47. Hinze, M.; Schmidt, A.; Leine, R.L. Numerical solution of fractional order ordinary differential equations using the reformulated infinite state representation. *Fract. Calc. Appl. Anal.* **2019**, *22*, 1321–1350. [CrossRef]
48. Hinze, M.; Schmidt, A.; Leine, R.L. The direct method of Lyapunov for nonlinear dynamical systems with fractional damping. *Nonlinear Dyn.* **2020**, *102*, 2017–2037. [CrossRef]
49. Maamri, N.; Trigeassou, J.C. Integration of fractional differential equations without fractional derivatives. In Proceedings of the ICSC20 Conference, Caen France, 24–26 November 2021.
50. Lindelöf, E. Sur l'application de la méthode des approximations successives aux équations différentielles ordinaires du premier ordre. *C.R.A.C.* **1894**, *116*, 454–457.
51. Picard, E. Mémoire sur la théorie des équations aux dérivées partielles et la méthodes des approximations successives. *J. Math. Pures Appl.* **1890**, *4*, 145–210.
52. Kailath, T. *Linear Systems*; Prentice Hall Inc.: Englewood Cliffs, NJ, USA, 1980.
53. Monje, C.A.; Chen, Y.Q.; Vinagre, B.M.; Xue, D.; Feliu, V. *Fractional Order Systems and Control*; Springer: London, UK, 2010.
54. Curtain, R.F.; Zwart, H.J. *An Introduction to Infinite Dimensional Linear Systems Theory*; Springer: New York, NY, USA, 1995.

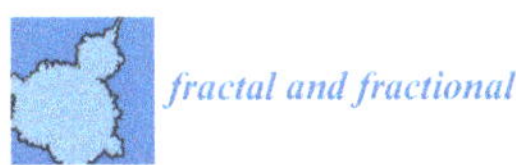

fractal and fractional

Article

Synchronization of Fractional-Order Uncertain Delayed Neural Networks with an Event-Triggered Communication Scheme

M. Hymavathi [1], M. Syed Ali [1], Tarek F. Ibrahim [2,3], B. A. Younis [4], Khalid I. Osman [5] and Kanit Mukdasai [6,*

[1] Department of Mathematics, Thiruvalluvar University, Vellore 632115, Tamilnadu, India
[2] Department of Mathematics, Faculty of Sciences and Arts (Mahayel), King Khalid University, Abha 62529, Saudi Arabia
[3] Mathematics Department, Faculty of Science, Mansoura University, Mansoura 35516, Egypt
[4] Department of Mathematics, Faculty of Science and Arts in Zahran Alganoob, King Khalid University, Abha 62529, Saudi Arabia
[5] Department of Mathematics, Faculty of Sciences and Arts in Sarat Abeda, King Khalid University, Abha 62529, Saudi Arabia
[6] Department of Mathematics, Faculty of Sciences, Khon Kaen University, Khon Kaen 40002, Thailand
* Correspondence: kanit@kku.ac.th; Tel.: +66-43-009700 (ext. 42310)

Abstract: In this paper, the synchronization of fractional-order uncertain delayed neural networks with an event-triggered communication scheme is investigated. By establishing a suitable Lyapunov–Krasovskii functional (LKF) and inequality techniques, sufficient conditions are obtained under which the delayed neural networks are stable. The criteria are given in terms of linear matrix inequalities (LMIs). Based on the drive–response concept, the LMI approach, and the Lyapunov stability theorem, a controller is derived to achieve the synchronization. Finally, numerical examples are presented to confirm the effectiveness of the main results.

Keywords: fractional order; synchronization; event triggered; uncertain

Citation: Hymavathi, M.; Syed Ali, M.; Ibrahim, T.F.; Younis, B.A.; Osman, K.I.; Mukdasai, K. Synchronization of Fractional-Order Uncertain Delayed Neural Networks with an Event-Triggered Communication Scheme. *Fractal Fract.* **2022**, *6*, 641. https://doi.org/10.3390/fractalfract6110641

Academic Editors: Thach Ngoc Dinh, Shyam Kamal, Rajesh Kumar Pandey and Riccardo Caponetto

Received: 12 August 2022
Accepted: 27 October 2022
Published: 2 November 2022

Publisher's Note: MDPI stays neutral with regard to jurisdictional claims in published maps and institutional affiliations.

1. Introduction

Fractional calculus is a mathematical theory that has been studied and applied in different fields for the past 300 years. Compared with traditional integer-order systems, fractional-order (FO) derivatives provide an excellent tool for the description of memory and inherent properties of various materials and processes, with applications in many areas, such as heat conduction, electronics, and abnormal diffusion [1,2]. As a result, fractional calculus has attracted increasing attention from physicists and engineers [3–7]. Moreover, fractional calculus has been applied to numerous neural network models [8,9]. Hence, the research on fractional neural networks (NNs) is important for practical applications, and many important results on chaotic dynamics, stability analysis, stabilization, synchronization, dissipativity, and passivity have been reported [10–16]. This popularity is due to the fact that fractional calculus has the ability to include memory when describing complex systems and gives a more precise characterization than the standard integer-order approach. A key characteristic is that the FO derivatives require an infinite number of terms, whereas the integer-order derivatives only indicate a finite series. Consequently, the integer derivatives are local operators, whereas the FO derivative has the memory of all past events.

In the real world, there are different types of uncertainty that can attenuate the performance of the system and affect its stability. These uncertainties may result from parameter variations and external disturbances. If a structural process is observed experimentally, it is not possible to assign precise values to the observed events. This means data uncertainty occurs, which may result from scale-dependent impacts that are not considered, which create inaccuracies in the estimations and incomplete sets of observations. In this manner, the estimated results are more or less described by the data uncertainty that begins with imprecision. In addition, the parameter uncertainties are unavoidable while displaying a neural network, which creates unstable results. It is known that a precise physical model of an engineering plant is difficult to build because of the uncertainties and noises. In actual operation, due to the existence of some external or internal uncertain disturbances, system states sometimes are not always fully accessible [17–26].

Generally speaking, an event-triggered control strategy is more appealing than the traditional time-triggered one from an economic perspective, since the control input is updated only when the predetermined triggering condition is reached. Since the event-triggered control approach can reduce information exchange in systems, event-triggered synchronization or consensus for fractional-order systems has received increasing attention in recent years. Recently, there has been significant research on the event-triggered control (ETC) strategy [27–29]. Compared to the time-driven consensus, the event-triggered consensus is more realistic. The event-triggered controller introduced in the field of networked control systems has the advantage of using limited communication network resources efficiently. Recently, an event-triggered scheme (ETS) provided an effective way of determining when the sampling action should be carried out and when the packet should be transmitted. A number of researchers have recommended event-triggered control. To deal with network congestion, the ETS has been proposed to improve data transmission efficiency. In the past few years, event-triggered control has proved to be an efficient way to reduce the transmitted data in the networks, which can relieve the burden of network bandwidth. Thus event-triggered control strategies have been employed to study networked systems [30–32].

In addition, in many practical applications, the system is expected to reach synchronization as quickly as possible. Synchronization is an important phenomenon in the real world, which exists widely in practical systems, as well as in nature. The problem of achieving synchronization in a neural network is another research hotspot. Different kinds of synchronization, such as pinning synchronization [33], local synchronization [34,35], lag synchronization [36], and impulsive synchronization [37] have been considered in the literature. Recently synchronization has also attracted attention in the field of complex networks systems [38,39]. Synchronization techniques require communication among nodes, which creates network congestion and wastes network resources. Moreover, the treatment of the synchronization problem of fractional-order systems with input quantization is quite limited in the literature. Numerous consequence have been described for the synchronization-based event-triggered problem [40–42]. As collective behaviors, consensus and synchronization are important in nature.

There is no doubt that the Lyapunov functional method provides an effective approach to analyze the stability of integer-order nonlinear systems. The synchronization and stabilization of fractional Caputo neural network (FCNNs) were proved by constructing a simple quadratic Lyapunov function and calculating its fractional derivative. The contributions of this article are listed below:

1. The synchronization of fractional-order uncertain delayed neural networks with an event-triggered communication scheme is investigated.

2. A fractional integral, which is suitable for the considered fractional-order error system, is proposed.

3. A Lyapunov–Krasovskii (L–K) functional is established, and the conditions corresponding to asymptotic stability are derived for the design of an event-triggered controller based on linear matrix inequalities (LMIs).

4. The derived conditions are expressed in terms of linear matrix inequalities (LMIs), which can be checked numerically via the LMI toolbox very efficiently.

5. Numerical examples are provided to demonstrate the effectiveness and applicability of the proposed stability results.

The following notations are used in this paper. $\mathscr{R}$ and $\mathscr{R}^n$ denote the set of real numbers and the n-dimensional real spaces, respectively; $\mathscr{R}^{n \times n}$ denotes the set of $n \times n$ matrices. $\mathscr{I}$ denotes the identity matrix of appropriate dimension. The super script "$\mathscr{T}$"

denotes the matrix transposition. "(-1)" represents the matrix inverse. $\mathscr{X} > 0$ ($\mathscr{X} < 0$) means that $\mathscr{X}$ is positive definite (negative definite). I represents the identity matrix and zero matrix with compatible dimensions. In symmetric block matrices or a long matrix expression, we use an asterisk (*) to represent a term that is induced by symmetry. $\mathscr{L}_2[0, \infty)$ denotes the space of square-integrable vector functions over $[0, \infty)$.

2. Preliminaries

In this section, we recall the basic definition and some properties concerning fractional-order calculus. In addition, definition, remark, assumption and some lemmas are presented.

Definition 1 ([43]). *The Caputo fractional derivative of order β for a function $\mathfrak{f}(t)$ is defined as*

$$D^\beta f(t) = \frac{1}{\Gamma(m - \beta)} \int_0^t \frac{f^m(\gamma)}{(t - \gamma)^{\beta - m + 1}} d\gamma,$$

where $t \geq 0$, and $m - 1 < \beta < m \in Z^+$. In particular, when $\beta \in (0, 1)$,

$$D^\beta f(t) = \frac{1}{\Gamma(1 - \beta)} \int_0^t \frac{f'(\gamma)}{(t - \gamma)^\beta} d\gamma.$$

Lemma 1 ([44]). *Let a vector-valued function $\varrho(t) \in \mathscr{R}^n$ be differentiable. Then, for any $t > 0$, one has*

$$\mathscr{D}^\alpha \left(\varrho^T(t) \mathscr{S} \varrho(t) \right) \leq 2 \varrho^T(t) \mathscr{S} \mathscr{D}^\alpha \varrho(t), 0 < \alpha < 1.$$

Lemma 2 ([45]). *For the given positive scalar $\lambda > 0$, $\mathfrak{l}, \mathfrak{r} \in \mathscr{R}^m$ and matrix $\mathscr{D}$,*

$$\mathfrak{l}^T \mathscr{D} \mathfrak{r} \leq \frac{\lambda^{-1}}{2} \mathfrak{l}^T \mathscr{D} \mathscr{D}^T \mathfrak{l} + \frac{\lambda}{2} \mathfrak{r}^T \mathfrak{r}.$$

Lemma 3 ([46]). *If $\mathscr{N} > 0$, and the given matrices are $\mathscr{S}, \mathscr{Q}, \mathscr{N}$, then*

$$\begin{bmatrix} \mathscr{Q} & \mathscr{S}^T \\ \mathscr{S} & -\mathscr{N} \end{bmatrix} < 0,$$

if and only if

$$\mathscr{Q} + \mathscr{S}^T \mathscr{N}^{-1} \mathscr{S} < 0.$$

Lemma 4 ([47]). *For a vector function $\Xi : [t_1, t_2] \to \mathscr{R}^n$ and any positive definite matrix $\mathscr{P}$, we have*

$$\left(\int_{t_1}^{t_2} \Xi(s) \partial s \right)^T \mathscr{P} \left(\int_{t_1}^{t_2} \Xi(s) \partial s \right) \leq (t_2 - t_1) \int_{t_1}^{t_2} \Xi^{\mathscr{T}}(s) \mathscr{P} \Xi(s) \partial s.$$

Assumption 1. *Let $\mathfrak{g}_i(\cdot)$ be continuous and bounded; $\mathscr{X}_s^-$ and $\mathscr{X}_s^+$ are constants,*

$$\mathscr{X}_s^- \leq \frac{\mathfrak{g}_s(\mathfrak{r}_1) - \mathfrak{g}_s(\mathfrak{r}_2)}{\mathfrak{r}_1 - \mathfrak{r}_2} \leq \mathscr{X}_s^+, s = 1, 2, \ldots, n,$$

where $\mathfrak{r}_1, \mathfrak{r}_2 \in \mathbb{R}$ and $\mathfrak{r}_1 \neq \mathfrak{r}_2$.

Remark 1. *From the literature survey, it is clear that most of the results on fractional order neural networks (FONNs) are derived with fractional-order Lyapunov stability criteria having quadratic terms. However, in this paper, we introduce the integral term $\mathscr{D}^{(-\alpha+1)} \int_{t-\eta}^t \mathfrak{e}^{\mathscr{T}}(s) \mathscr{R}_2 \mathfrak{e}(s) \partial s$ in the Lyapunov functional candidate, which is solved by utilizing the properties of Caputo fractional-order derivatives and integrals. The Lyapunov functional is novel, as it contains the quadratic term. By* *applying fractional-order derivatives in the error system of the FCNNs under suitable adaptive update laws, a new sufficient condition can be derived in terms of solvable LMIs.*

3. Main Results

Consider the following uncertain delayed neural network described by

$$
\mathscr{D}^{\alpha}\mathfrak{w}_i(t) = -(\mathfrak{r}_i + \Delta\mathfrak{r}_i(t))\mathfrak{w}_i(t) + \sum_{j=1}^{n}(\mathfrak{c}_{ij} + \Delta\mathfrak{c}_{ij}(t))\mathfrak{h}_j(\mathfrak{w}_j(t))
$$
$$
+ \sum_{j=1}^{n}(\mathfrak{b}_{ij} + \Delta\mathfrak{b}_{ij}(t))\mathfrak{h}_j(\mathfrak{w}_j(t - \sigma_j(t)))
$$
$$
+ \sum_{j=1}^{n}(\mathfrak{a}_{ij} + \Delta\mathfrak{a}_{ij}(t))\int_{t-\eta}^{t}\mathfrak{w}_j(\mathfrak{s})\partial\mathfrak{s} + \mathfrak{p}_i(t). \tag{1}
$$

Conveniently, we write the master system as

$$
\mathscr{D}^{\alpha}\mathfrak{w}(t) = -(\mathscr{R} + \Delta\mathscr{R}(t))\mathfrak{w}(t) + (\mathscr{C} + \Delta\mathscr{C}(t))\mathfrak{h}(\mathfrak{w}(t)) + (\mathscr{B} + \Delta\mathscr{B}(t))\mathfrak{h}(\mathfrak{w}(t - \sigma(t)))
$$
$$
+ (\mathscr{A} + \Delta\mathscr{A}(t))\int_{t-\eta}^{t}(\mathfrak{w}(\mathfrak{s}))\partial\mathfrak{s} + \mathscr{P}(t), \tag{2}
$$

in which $\mathfrak{w}(t) = (\mathfrak{w}_1(t), \mathfrak{w}_2(t), \ldots, \mathfrak{w}_n(t))^{\mathscr{T}} \in \mathscr{R}^n$, is the state vector associated with n neurons, the diagonal matrix $\mathfrak{r}_i(t) = diag\{\mathfrak{r}_1(t), \mathfrak{r}_2(t), \ldots, \mathfrak{r}_n(t)\}$, and $\mathscr{C}(t)$, $\mathscr{B}(t)$, and $\mathscr{A}(t)$ are the known constant matrices of appropriate dimensions; the symbol Δ denotes the uncertain term, and $\Delta\mathscr{C}(t)$, $\Delta\mathscr{B}(t)$, and $\Delta\mathscr{A}(t)$ are known matrices that represent the time-varying parameter uncertainties. $\mathfrak{h}(\mathfrak{w}(t))$ is the neuron activation function.

Next, we consider the corresponding slave system as follows:

$$
\mathscr{D}^{\alpha}\mathfrak{v}_i(t) = -(\mathfrak{r}_i + \Delta\mathfrak{r}_i(t))\mathfrak{v}_i(t) + \sum_{j=1}^{n}(\mathfrak{c}_{ij} + \Delta\mathfrak{c}_{ij}(t))\mathfrak{h}_j(\mathfrak{v}_j(t))
$$
$$
+ \sum_{j=1}^{n}(\mathfrak{b}_{ij} + \Delta\mathfrak{b}_{ij}(t))\mathfrak{h}_j(\mathfrak{v}_j(t - \sigma_j(t)))
$$
$$
+ \sum_{j=1}^{n}(\mathfrak{a}_{ij} + \Delta\mathfrak{a}_{ij}(t))\int_{t-\eta}^{t}\mathfrak{v}_j(\mathfrak{s})\partial\mathfrak{s} + \mathfrak{p}_i(t) + \mathfrak{h}\mathfrak{q}_i(t). \tag{3}
$$

The compact form of (3) is

$$
\mathscr{D}^{\alpha}\mathfrak{v}(t) = -(\mathscr{R} + \Delta\mathscr{R}(t))\mathfrak{v}(t) + (\mathscr{C} + \Delta\mathscr{C}(t))\mathfrak{h}(\mathfrak{v}(t)) + (\mathscr{B} + \Delta\mathscr{B}(t))\mathfrak{h}(\mathfrak{v}(t - \sigma(t)))
$$
$$
+ (\mathscr{A} + \Delta\mathscr{A}(t))\int_{t-\eta}^{t}\mathfrak{v}(\mathfrak{s})\partial\mathfrak{s} + \mathscr{P}(t) + \mathscr{H}\mathscr{Q}(t). \tag{4}
$$

Now, we introduce the $\mathfrak{e}(t) = \mathfrak{v}(t) - \mathfrak{w}(t)$:

$$
\mathscr{D}^{\alpha}\mathfrak{e}(t) = -(\mathscr{R} + \Delta\mathscr{R}(t))\mathfrak{e}(t) + (\mathscr{C} + \Delta\mathscr{C}(t))\mathfrak{h}(\mathfrak{e}(t)) + (\mathscr{B} + \Delta\mathscr{B}(t))\mathfrak{h}(\mathfrak{e}(t - \sigma(t)))
$$
$$
+ (\mathscr{A} + \Delta\mathscr{A}(t))\int_{t-\eta}^{t}\mathfrak{e}(\mathfrak{s})\partial\mathfrak{s} + \mathscr{H}\mathscr{Q}(t). \tag{5}
$$

The purpose of this paper is to design a controller $\mathscr{Q}(t) = \mathscr{K}\mathfrak{e}(t)$, such that the slave system (3) synchronizes with the master system (1), and $\mathscr{K}$ is the controller gain to be determined.

Without distributed delays in the system (1), it is easy to obtain the error system

$$
\mathscr{D}^{\alpha}\mathfrak{e}(t) = -(\mathscr{R} + \Delta\mathscr{R}(t))\mathfrak{e}(t) + (\mathscr{C} + \Delta\mathscr{C}(t))\mathfrak{h}(\mathfrak{e}(t)) + (\mathscr{B} + \Delta\mathscr{B}(t))\mathfrak{h}(\mathfrak{e}(t - \sigma(t)))
$$
$$
+ \mathscr{H}\mathscr{K}\mathfrak{e}(t). \tag{6}
$$

Theorem 1. *The FNNs (1) and (3) are globally asymptotically synchronized under the event-triggered control scheme, for the given scalars $\delta_1, \delta_2, \delta_3, \delta_4, \delta_5$, and μ_1, and if there exist symmetric positive definite matrices $\mathscr{R}_1 > 0, \mathscr{R}_2 > 0$, such that a feasible solution exists for the following LMIs,*

$$
\Omega = \begin{bmatrix}
\Omega_{11} & \mathscr{R}_1 \mathscr{J}_{\mathfrak{r}} & \mathscr{R}_1 \mathscr{J}_{\mathfrak{c}} & \mathscr{R}_1 \mathscr{J}_{\mathfrak{b}} & \mathscr{R}_1 \mathscr{C} & \mathscr{R}_1 \mathscr{B} & 0 \\
* & -\delta_1 \mathscr{I} & 0 & 0 & 0 & 0 & 0 \\
* & * & -\delta_2 \mathscr{I} & 0 & 0 & 0 & 0 \\
* & * & * & -\delta_3 \mathscr{I} & 0 & 0 & 0 \\
* & * & * & * & -\delta_4 \mathscr{I} & 0 & 0 \\
* & * & * & * & * & -\delta_5 \mathscr{I} & 0 \\
* & * & * & * & * & * & \Omega_{66}.
\end{bmatrix} < 0, \tag{7}
$$

where

$$
\Omega_{11} = -2\mathscr{R}_1 \mathscr{R} + \delta_1 \mathscr{L}_r^{\mathscr{T}} L_{\mathfrak{r}} + \delta_2 \phi^{\mathscr{T}} \mathscr{L}_c^{\mathscr{T}} \mathscr{L}_c \phi + \delta_4 \phi^{\mathscr{T}} \phi + \mathscr{R}_2 + \mathscr{R}_1 \mathscr{H} \mathscr{K},
$$
$$
\Omega_{66} = \delta_3 \phi^T \mathscr{L}_{\mathfrak{b}}^T \mathscr{L}_{\mathfrak{b}} \phi + \delta_5 \phi^T \phi - \mathscr{R}_2 (1 - \mu).
$$

Proof. Now, let us define the Lyapunov–Krasovskii functional as follows:

$$
\mathscr{V}(t) = \mathscr{V}_1(t) + \mathscr{V}_2(t), \tag{8}
$$

where

$$
\mathscr{V}_1(t) = \mathfrak{e}^{\mathscr{T}}(t) \mathscr{R}_1 \mathfrak{e}(t),
$$
$$
\mathscr{V}_2(t) = \mathscr{D}^{(-\alpha+1)} \int_{t-\sigma(t)}^{t} \mathfrak{e}^{\mathscr{T}}(\mathfrak{s}) \mathscr{R}_2 \mathfrak{e}(\mathfrak{s}) \partial \mathfrak{s}.
$$

By using Lemma 2, we have,

$$
2\mathfrak{e}^{\mathscr{T}}(t) \mathscr{R}_1 \Delta \mathscr{R}(t) \mathfrak{e}(t) \leq 2\mathfrak{e}^{\mathscr{T}}(t) \mathscr{R}_1 \mathscr{J}_{\mathfrak{d}} \mathscr{K}(t) \mathscr{L}_{\mathfrak{d}} \mathfrak{e}(t),
$$
$$
\leq \delta_1^{-1} \mathfrak{e}^{\mathscr{T}}(t) \mathscr{R}_1 \mathscr{J}_{\mathfrak{d}} \mathscr{J}_{\mathfrak{d}}^{\mathscr{T}} \mathscr{R}_1^{\mathscr{T}} \mathfrak{e}(t)
$$
$$
+ \delta_1 \mathfrak{e}^{\mathscr{T}}(t) \mathscr{L}_{\mathfrak{r}}^{\mathscr{T}} \mathscr{L}_{\mathfrak{r}} \mathfrak{e}(t), \tag{9}
$$
$$
2\mathfrak{e}^{\mathscr{T}}(t) \mathscr{R}_1 \Delta \mathscr{C}(t) \mathfrak{h}(\mathfrak{e}(t)) \leq 2\mathfrak{e}^{\mathscr{T}}(t) \mathscr{R}_1 \mathscr{J}_{\mathfrak{c}} \mathscr{K}(t) \mathscr{L}_{\mathfrak{c}} \mathfrak{h}(\mathfrak{e}(t)),
$$
$$
\leq \delta_2^{-1} \mathfrak{e}^{\mathscr{T}}(t) \mathscr{R}_1 \mathscr{J}_{\mathfrak{c}} \mathscr{J}_{\mathfrak{c}}^{T} \mathscr{R}_1^{\mathscr{T}} \mathfrak{e}(t)
$$
$$
+ \delta_2 \mathfrak{e}^{\mathscr{T}}(t) \phi^{\mathscr{T}} \mathscr{L}_{\mathfrak{c}}^{\mathscr{T}} \mathscr{L}_{\mathfrak{c}} \phi \mathfrak{e}(t), \tag{10}
$$
$$
2\mathfrak{e}^{\mathscr{T}}(t) \mathscr{R}_1 \Delta \mathscr{B}(t) \mathfrak{h}(\mathfrak{e}(t - \sigma(t))) \leq 2\mathfrak{e}^{\mathscr{T}}(t) \mathscr{R}_1 \mathscr{J}_{\mathfrak{b}} \mathscr{K}(t) \mathscr{L}_{\mathfrak{b}} \mathfrak{h}(\mathfrak{e}(t - \sigma(t))),
$$
$$
\leq \delta_3^{-1} \mathfrak{e}^{\mathscr{T}}(t) \mathscr{R}_1 \mathscr{J}_{\mathfrak{b}} \mathscr{J}_{\mathfrak{b}}^{\mathscr{T}} \mathscr{R}_1^{\mathscr{T}} \mathfrak{e}(t)
$$
$$
+ \delta_3 \mathfrak{e}^{\mathscr{T}}(t - \sigma(t)) \phi^{\mathscr{T}} \mathscr{L}_{\mathfrak{b}}^{\mathscr{T}} \mathscr{L}_{\mathfrak{b}} \phi \mathfrak{e}(t - \sigma(t)), \tag{11}
$$
$$
2\mathfrak{e}^{\mathscr{T}}(t) \mathscr{R}_1 \mathscr{C} \mathfrak{h}(\mathfrak{e}(t))) \leq \delta_4^{-1} \mathfrak{e}^{\mathscr{T}}(t) \mathscr{R}_1 \mathscr{C} \mathscr{C}^{T} \mathscr{R}_1^{T} \mathfrak{e}(t)
$$
$$
+ \delta_4 \mathfrak{e}^{\mathscr{T}}(t) \phi^{\mathscr{T}} \phi \mathfrak{e}(t),
$$
$$
2\mathfrak{e}^{\mathscr{T}}(t) \mathscr{R}_1 \mathscr{B} \mathfrak{h}(\mathfrak{e}(t - \sigma(t)))) \leq \delta_5^{-1} \mathfrak{e}^{\mathscr{T}}(t) \mathscr{R}_1 \mathscr{B} \mathscr{B}^{T} \mathscr{R}_1^{T} \mathfrak{e}(t)
$$
$$
+ \delta_5 \mathfrak{e}^{\mathscr{T}}(t - \sigma(t)) \phi^{\mathscr{T}} \phi \mathfrak{e}(t - \sigma(t)). \tag{12}
$$

Then, with the support of Lemma 1 and the linearity nature of the Caputo fractional-order derivative, the fractional derivative along the trajectories of the system state is acquired as follows

$$
\begin{aligned}
D^{\alpha}\mathcal{V}(t) &\leq 2\mathfrak{e}^{\mathcal{T}}(t)\mathscr{R}_1\mathscr{D}^{\alpha}\mathfrak{e}(t), \\
&\leq 2\mathfrak{e}^{\mathcal{T}}(t)\mathscr{R}_1\big[-(\mathscr{R}+\Delta\mathscr{R}(t))\mathfrak{e}(t)+(\mathscr{C}+\Delta\mathscr{C}(t))\mathfrak{h}(\mathfrak{e}(t)) \\
&\quad +(\mathscr{B}+\Delta\mathscr{B}(t))\mathfrak{h}(\mathfrak{e}(t-\sigma(t))+\mathscr{H}\mathscr{K}\mathfrak{e}(t)\big], \\
&\leq -2\mathfrak{e}^{\mathcal{T}}(t)\mathscr{R}_1\mathscr{R}\mathfrak{e}(t)+\delta_1^{-1}\mathfrak{e}^{\mathcal{T}}(t)\mathscr{R}_1\mathscr{J}_{\mathfrak{d}}\mathscr{J}_{\mathfrak{d}}^{\mathcal{T}}\mathscr{R}_1^{\mathcal{T}}\mathfrak{e}(t) \\
&\quad +\delta_1\mathfrak{e}^{\mathcal{T}}(t)\mathscr{L}_{\mathfrak{d}}^{\mathcal{T}}\mathscr{L}_{\mathfrak{d}}\mathfrak{e}(t)+2\mathfrak{e}^{\mathcal{T}}(t)\mathscr{R}_1\mathscr{C}\mathfrak{h}(\mathfrak{e}(t)) \\
&\quad +\delta_2^{-1}\mathfrak{e}^{\mathcal{T}}(t)\mathscr{R}_1\mathscr{J}_{\mathfrak{c}}\mathscr{J}_{\mathfrak{c}}^{\mathcal{T}}\mathscr{R}_1^{\mathcal{T}}\mathfrak{e}(t)+\delta_2\mathfrak{e}^{\mathcal{T}}(t)\phi^{\mathcal{T}}\mathscr{L}_{\mathfrak{c}}^{\mathcal{T}}\mathscr{L}_{\mathfrak{c}}\phi\mathfrak{e}(t) \\
&\quad +2\mathfrak{e}^{\mathcal{T}}(t)\mathscr{R}_1\mathscr{B}\mathfrak{h}(\mathfrak{e}(t-\sigma(t)))+\delta_3^{-1}\mathfrak{e}^{\mathcal{T}}(t)\mathscr{R}_1\mathscr{J}_{\mathfrak{b}}\mathscr{J}_{\mathfrak{b}}^{\mathcal{T}}\mathscr{R}_1^{\mathcal{T}}\mathfrak{e}(t) \\
&\quad +\delta_3\mathfrak{e}^{\mathcal{T}}(t-\sigma(t))\phi^{\mathcal{T}}\mathscr{L}_{\mathfrak{b}}^{\mathcal{T}}\mathscr{L}_{\mathfrak{b}}\phi\mathfrak{e}(t-\sigma(t)) \\
&\quad +\delta_4^{-1}\mathfrak{e}^{\mathcal{T}}(t)\mathscr{R}_1\mathscr{C}^{T}\mathscr{C}\mathscr{R}_1^{T}\mathfrak{e}(t)+\delta_4\mathfrak{e}^{\mathcal{T}}(t)\phi^{T}\phi\mathfrak{e}(t) \\
&\quad +\delta_5^{-1}\mathfrak{e}^{\mathcal{T}}(t)\mathscr{R}_1\mathscr{B}^{T}\mathscr{B}\mathscr{R}_1^{T}\mathfrak{e}(t)+\delta_5\mathfrak{e}^{\mathcal{T}}(t-\sigma(t))\phi^{T}\phi\mathfrak{e}(t-\sigma(t)) \\
&\quad +\mathfrak{e}^{\mathcal{T}}(t)\mathscr{R}_2\mathfrak{e}(t)-\mathfrak{e}^{\mathcal{T}}(t-\sigma(t))\mathscr{R}_2\mathfrak{e}(t-\sigma(t))(1-\mu).
\end{aligned} \tag{13}
$$

From (9)–(13), the following can be obtained.

$$
D^{\alpha}\mathcal{V}(t)\leq \zeta^{T}(t)\Omega\zeta(t), \tag{14}
$$

where

$$
\zeta(t)=col[\mathfrak{e}(t),\ \mathfrak{e}(t-\sigma(t)))].
$$

From the aforementioned part, we know that matrix inequality (7) guarantees $\Omega<0$.

Thereby, the master system (1) is synchronized with the slave system (3). The proof of Theorem 1 is complete. $\square$

Theorem 2. *The FNNs (1) and (3) are globally asymptotically synchronized, for given scalars* $\delta_1,\delta_2,\delta_3,\delta_4,\delta_5,$ *and* $\sigma,$ *if there exist symmetric positive definite matrices* $\mathscr{R}_1>0,\mathscr{R}_2>0,$ *such that the following LMIs hold:*

$$
\pi=\begin{bmatrix}
\pi_{11} & \mathscr{J}_{\mathfrak{r}} & \mathscr{J}_{\mathfrak{c}} & \mathscr{J}_{\mathfrak{b}} & \mathscr{C} & \mathscr{B} & 0 \\
* & -\delta_1\mathscr{I} & 0 & 0 & 0 & 0 & 0 \\
* & * & -\delta_2\mathscr{I} & 0 & 0 & 0 & 0 \\
* & * & * & -\delta_3\mathscr{I} & 0 & 0 & 0 \\
* & * & * & * & -\delta_4\mathscr{I} & 0 & 0 \\
* & * & * & * & * & -\delta_5\mathscr{I} & 0 \\
* & * & * & * & * & * & \pi_{66}
\end{bmatrix}<0, \tag{15}
$$

where

$$
\begin{aligned}
\pi_{11} &= -2\mathscr{R}_1\mathscr{X}_1+\mathscr{X}_1\delta_1\mathscr{L}_r^{\mathcal{T}}\mathscr{L}_{\mathfrak{r}}\mathscr{X}_1+\mathscr{X}_1\delta_2\phi^{\mathcal{T}}\mathscr{L}_{\mathfrak{c}}^{\mathcal{T}}\mathscr{L}_{\mathfrak{c}}\phi\mathscr{X}_1 \\
&\quad +\mathscr{X}_1\delta_4\phi^{\mathcal{T}}\phi\mathscr{X}_1+\mathscr{X}_1\mathscr{R}_2\mathscr{X}_1+\mathscr{H}\mathscr{Y}_1, \\
\pi_{66} &= \delta_3\phi^{T}\mathscr{L}_{\mathfrak{b}}^{T}\mathscr{L}_{\mathfrak{b}}\phi+\delta_5\phi^{T}\phi-\mathscr{R}_2(1-\mu),
\end{aligned} \tag{16}
$$

and the other parameters are the same as in Theorem 1; among them, the gain matrix is defined with $\mathscr{R}_1^{-1}=\mathscr{X}_1.$

Proof. We pre- and post-multiply Ω by $\{\mathscr{R}_1^{-1},\mathscr{I},\mathscr{I},\mathscr{I},\mathscr{I},\mathscr{I},\mathscr{I}\}$ and $\mathscr{R}_1^{-1}=\mathscr{X}_1$

$$\Phi = \begin{bmatrix} \Phi_{11} & \mathscr{J}_{\mathfrak{r}} & \mathscr{J}_{\mathfrak{c}} & \mathscr{J}_{\mathfrak{b}} & \mathscr{C} & \mathscr{B} & 0 \\ * & -\delta_1 \mathscr{I} & 0 & 0 & 0 & 0 & 0 \\ * & * & -\delta_2 \mathscr{I} & 0 & 0 & 0 & 0 \\ * & * & * & -\delta_3 \mathscr{I} & 0 & 0 & 0 \\ * & * & * & * & -\delta_4 \mathscr{I} & 0 & 0 \\ * & * & * & * & * & -\delta_5 \mathscr{I} & 0 \\ * & * & * & * & * & * & \Phi_{66} \end{bmatrix} < 0, \tag{17}$$

where

$$\Phi_{11} = -2\mathscr{R}_1 \mathscr{X}_1 + \mathscr{X}_1 \delta_1 \mathscr{L}_r^{\mathscr{T}} L_{\mathfrak{r}} \mathscr{X}_1 + \mathscr{X}_1 \delta_2 \phi^{\mathscr{T}} \mathscr{L}_c^{\mathscr{T}} \mathscr{L}_c \phi \mathscr{X}_1$$
$$+ \mathscr{X}_1 \delta_4 \phi^{\mathscr{T}} \phi \mathscr{X}_1 + \mathscr{X}_1 \mathscr{R}_2 \mathscr{X}_1 + \mathscr{H} \mathscr{K} \mathscr{X}_1,$$
$$\Phi_{66} = \delta_3 \phi^T \mathscr{L}_{\mathfrak{b}}^T \mathscr{L}_{\mathfrak{b}} \phi + \delta_5 \phi^T \phi - \mathscr{R}_2 (1 - \mu).$$

At the same time, the controller gain matrix $\mathscr{K}$ can be obtained as $\mathscr{Y}_1 = \mathscr{K} \mathscr{X}_1$,

$$\pi = \begin{bmatrix} \pi_{11} & \mathscr{J}_{\mathfrak{r}} & \mathscr{J}_{\mathfrak{c}} & \mathscr{J}_{\mathfrak{b}} & \mathscr{C} & \mathscr{B} & 0 \\ * & -\delta_1 \mathscr{I} & 0 & 0 & 0 & 0 & 0 \\ * & * & -\delta_2 \mathscr{I} & 0 & 0 & 0 & 0 \\ * & * & * & -\delta_3 \mathscr{I} & 0 & 0 & 0 \\ * & * & * & * & -\delta_4 \mathscr{I} & 0 & 0 \\ * & * & * & * & * & -\delta_5 \mathscr{I} & 0 \\ * & * & * & * & * & * & \pi_{66} \end{bmatrix} < 0. \tag{18}$$

Hence, (15) guarantees that

$$\pi < 0. \tag{19}$$

Thereby, the master system (1) is synchronized with the slave system (3). The proof of Theorem 2 is complete. $\square$

Remark 2. *Specifically, when there are no uncertainties in the given system, the neural network (6) reduces to*

$$\mathscr{D}^{\alpha} \mathfrak{e}(t) = -\mathscr{R} \mathfrak{e}(t) + \mathscr{C} \mathfrak{h}(\mathfrak{e}(t)) + \mathscr{B} \mathfrak{h}(\mathfrak{e}(t - \sigma(t)))$$
$$+ \mathscr{A} \int_{t-\eta}^{t} \mathfrak{e}(\mathfrak{s}) \partial \mathfrak{s} + \mathscr{H} \mathscr{K} \mathfrak{e}(t). \tag{20}$$

Corollary 1. *The scalars are $\delta_4, \delta_5, \eta, \epsilon$, and σ, and if there exist symmetric positive definite matrices $\mathscr{R}_1 > 0, \mathscr{R}_2 > 0$, a feasible solution exists for the following LMIs:*

$$\beta < 0. \tag{21}$$

Proof. Now, let us define the Lyapunov–Krasovskii functional as follows:

$$\mathscr{V}(t) = \mathscr{V}_1(t) + \mathscr{V}_2(t), \tag{22}$$

where

$$\mathscr{V}_1(t) = \mathfrak{e}^{\mathscr{T}}(t) \mathscr{R}_1 \mathfrak{e}(t),$$
$$\mathscr{V}_2(t) = \mathscr{D}^{(-\alpha+1)} \int_{t-\sigma(t)}^{t} \mathfrak{e}^{\mathscr{T}}(\mathfrak{s}) \mathscr{R}_2 \mathfrak{e}(\mathfrak{s}) ds.$$

By using Lemma 2, we have

$$2\mathfrak{e}^{\mathscr{T}}(t)\mathscr{R}_1\mathscr{C}\mathfrak{h}(\mathfrak{e}(t))) \le \delta_4^{-1}\mathfrak{e}^{\mathscr{T}}(t)\mathscr{R}_1\mathscr{C}\mathscr{C}^T\mathscr{R}_1^T\mathfrak{e}(t)$$
$$+ \delta_4\mathfrak{e}^{\mathscr{T}}(t)\phi^{\mathscr{T}}\phi\mathfrak{e}(t), \tag{23}$$

$$2\mathfrak{e}^{\mathscr{T}}(t)\mathscr{R}_1\mathscr{B}\mathfrak{h}(e(t-\sigma(t)))) \le \delta_5^{-1}\mathfrak{e}^{\mathscr{T}}(t)\mathscr{R}_1\mathscr{B}\mathscr{B}^T\mathscr{R}_1^T\mathfrak{e}(t)$$
$$+ \delta_5\mathfrak{e}^{\mathscr{T}}(t-\sigma(t))\phi^{\mathscr{T}}\phi\mathfrak{e}(t-\sigma(t)). \tag{24}$$

Further, the above term is computed in view of the procedure in [47], and by employing Lemma 2.1 in [47] and the Cauchy matrix inequality, we have

$$2\mathfrak{e}^{\mathscr{T}}(t)\mathscr{R}_1\mathscr{A}(t)\int_{t-\eta}^{t}\mathfrak{e}(\mathfrak{s}))\eth\mathfrak{s} \le \eta\mathfrak{e}^{\mathscr{T}}(t)\mathscr{R}_1\mathscr{A}\mathscr{R}_1^{-1}\mathscr{A}^T\mathscr{R}_1\mathfrak{e}(t)$$

$$+ \frac{1}{\eta}\left(\int_{t-\eta}^{t}\mathfrak{e}(\mathfrak{s}))\eth\mathfrak{s}\right)^T\mathscr{R}_1\left(\int_{t-\eta}^{t}\mathfrak{e}(\mathfrak{s}))\eth\mathfrak{s}\right),$$

$$\le \eta\mathfrak{e}^{\mathscr{T}}(t)\mathscr{R}_1\mathscr{A}\mathscr{R}_1^{-1}\mathscr{A}^T\mathscr{R}_1\mathfrak{e}(t)$$

$$+ \frac{1}{\eta}\left(\int_{t-\eta}^{t}\mathfrak{e}^T(\mathfrak{s}))\mathscr{R}_1\mathfrak{e}(\mathfrak{s}))\eth\mathfrak{s}\right),$$

$$\le \eta\mathfrak{e}^{\mathscr{T}}(t)\mathscr{R}_1\mathscr{A}\mathscr{R}_1^{-1}\mathscr{A}^T\mathscr{R}_1\mathfrak{e}(t)$$

$$+ \frac{1}{\eta}\left(\int_{-\eta}^{o}\mathfrak{e}^T(t+\mathfrak{s}))\mathscr{R}_1\mathfrak{e}(t+\mathfrak{s}))\eth\mathfrak{s}\right), \tag{25}$$

since $\mathscr{V}(t+\mathfrak{s},\mathfrak{x}(t+\mathfrak{s})) \le \epsilon\mathscr{V}(t,\mathfrak{x}(t))$

$$2\mathfrak{e}^{\mathscr{T}}(t)\mathscr{R}_1\mathscr{A}(t)\int_{t-\eta}^{t}\mathfrak{e}(\mathfrak{s}))\eth\mathfrak{s} \le \eta\mathfrak{e}^{\mathscr{T}}(t)\mathscr{R}_1\mathscr{A}\mathscr{R}_1^{-1}\mathscr{A}^T\mathscr{R}_1\mathfrak{e}(t) + \eta\epsilon\mathfrak{e}^{\mathscr{T}}(t)\mathscr{R}_1\mathfrak{e}(t). \tag{26}$$

Then, with the support of Lemma 1 and the linearity nature of the Caputo fractional-order derivative, the fractional derivative along the trajectories of the system state is acquired as follows

$$D^{\alpha}\mathscr{V}(t) \le 2\mathfrak{e}^{\mathscr{T}}(t)\mathscr{R}_1\mathscr{D}^{\alpha}\mathfrak{e}(t),$$

$$\le 2\mathfrak{e}^{\mathscr{T}}(t)\mathscr{R}_1\big[-\mathscr{R}\mathfrak{e}(t) + \mathscr{C}\mathfrak{h}(\mathfrak{e}(t)) + \mathscr{B}\mathfrak{h}(\mathfrak{e}(t-\sigma(t)))$$

$$+ 2\mathfrak{e}^{\mathscr{T}}(t)\mathscr{R}_1\mathscr{A}(t)\int_{t-\eta}^{t}\mathfrak{e}(\mathfrak{s}))\eth\mathfrak{s} + \mathscr{K}\mathfrak{e}(t)\big],$$

$$\le -2\mathfrak{e}^{\mathscr{T}}(t)\mathscr{R}_1\mathscr{R}\mathfrak{e}(t) + \delta_4^{-1}\mathfrak{e}^{\mathscr{T}}(t)\mathscr{R}_1\mathscr{C}^T\mathscr{C}\mathscr{R}_1^T\mathfrak{e}(t) + \delta_4\mathfrak{e}^{\mathscr{T}}(t)\phi^T\phi\mathfrak{e}(t)$$

$$+ \delta_5^{-1}\mathfrak{e}^{\mathscr{T}}(t)\mathscr{R}_1\mathscr{B}^T\mathscr{B}\mathscr{R}_1^T\mathfrak{e}(t) + \delta_5\mathfrak{e}^{\mathscr{T}}(t-\sigma(t))\phi^T\phi\mathfrak{e}(t-\sigma(t))$$

$$+ \eta\mathfrak{e}^{\mathscr{T}}(t)\mathscr{R}_1\mathscr{A}\mathscr{R}_1^{-1}\mathscr{A}^T\mathscr{R}_1\mathfrak{e}(t) + \eta\epsilon\mathfrak{e}^{\mathscr{T}}(t)\mathscr{R}_1\mathfrak{e}(t)$$

$$+ \mathfrak{e}^T(t)\mathscr{R}_2\mathfrak{e}(t) - \mathfrak{e}^T(t-\sigma(t))\mathscr{R}_2\mathfrak{e}(t-\sigma(t))(1-\mu). \tag{27}$$

From (23)–(27) and applying Lemma 4, we obtain

$$\Theta = \begin{bmatrix} \Theta_{11} & \mathscr{R}_1\mathscr{C} & \mathscr{R}_1\mathscr{B} & \eta\mathscr{R}_1\mathscr{A} & 0 \\ * & -\delta_4\mathscr{I} & 0 & 0 & 0 \\ * & * & -\delta_5\mathscr{I} & 0 & 0 \\ * & * & * & \eta\mathscr{R}_1 & 0 \\ * & * & * & * & \delta_5\phi^T\phi - \mathscr{R}_2 \end{bmatrix} < 0, \tag{28}$$

$$\Theta_{11} = -2\mathscr{R}_1\mathscr{R} + \eta\epsilon\mathscr{R}_1 + \mathscr{R}_2 + \mathscr{R}_1\mathscr{H}\mathscr{K}.$$

We pre- and post-multiply Θ by $\{\mathscr{R}_1^{-1},\mathscr{I},\mathscr{I},\mathscr{R}_1^{-1},\mathscr{I}\}$

$$\Xi = \begin{bmatrix} \Xi_{11} & \mathscr{C} & \mathscr{B} & \eta\mathscr{A}\mathscr{X}_1 & 0 \\ * & -\delta_4\mathscr{I} & 0 & 0 & 0 \\ * & * & -\delta_5\mathscr{I} & 0 & 0 \\ * & * & * & -\eta\mathscr{X}_1 & 0 \\ * & * & * & * & \delta_5\phi^T\phi - \mathscr{R}_2 \end{bmatrix}, \tag{29}$$

where $\Xi_{11} = -2\mathscr{R}\mathscr{X}_1 + \mathscr{X}_1\delta_4\phi^T\phi\mathscr{X}_1 + \mathscr{X}_1\eta\epsilon + \mathscr{X}_1\mathscr{R}_2\mathscr{X}_1 + \mathscr{H}\mathscr{K}\mathscr{X}_1$

$$\varsigma = \begin{bmatrix} \varsigma_{11} & \mathscr{C} & \mathscr{B} & \eta\mathscr{A}\mathscr{X}_1 & 0 \\ * & -\delta_4\mathscr{I} & 0 & 0 & 0 \\ * & * & -\delta_5\mathscr{I} & 0 & 0 \\ * & * & * & -\eta\mathscr{X}_1 & 0 \\ * & * & * & * & \delta_5\phi^T\phi - \mathscr{R}_2 \end{bmatrix}, \tag{30}$$

where $\varsigma_{11} = -2\mathscr{R}\mathscr{X}_1 + \mathscr{X}_1\delta_4\phi^T\phi\mathscr{X}_1 + \mathscr{X}_1\eta\epsilon + \mathscr{X}_1\mathscr{R}_2\mathscr{X}_1 + \mathscr{H}\mathscr{Y}$.

Thereby, the master system (1) is synchronized with the slave system (3). □

4. Event-Triggered Control Scheme

In this section, we introduce an event generator in the controller node by using the following judgment algorithm

$$[\mathfrak{e}((\mathfrak{k}+\mathfrak{j})\mathfrak{h}) - \mathfrak{e}(\mathfrak{k}\mathfrak{h})]^T\Phi[\mathfrak{e}((\mathfrak{k}+\mathfrak{j})\mathfrak{h}) - \mathfrak{e}(\mathfrak{k}\mathfrak{h})] \leq \Sigma\mathfrak{e}^T((\mathfrak{k}+\mathfrak{j})\mathfrak{h})\Phi\mathfrak{e}((\mathfrak{k}+\mathfrak{j})\mathfrak{h}), \tag{31}$$

where Φ is a positive definite matrix to be determined, $\mathfrak{k}, \mathfrak{j} \in \mathscr{L}_+$ and $\mathfrak{k}\mathfrak{h}$ denotes the release instant, $\mathfrak{e}((\mathfrak{k}+\mathfrak{j})\mathfrak{h}) = \mathfrak{v}((\mathfrak{k}+\mathfrak{j})) - \mathfrak{w}((\mathfrak{k}+\mathfrak{j})\mathfrak{h})$ is the error information at the instant $(\mathfrak{k}+\mathfrak{j})\mathfrak{h}$, and $\sigma \in [0,1)$ is a given constant. Cases A and B relate to the following delayed differential equation

$$\mathscr{D}^\alpha\mathfrak{e}(t) = -(\mathscr{R} + \Delta\mathscr{R}(t))\mathfrak{e}(t) + (\mathscr{C} + \Delta\mathscr{C}(t))\mathfrak{h}(\mathfrak{e}(t)) + (\mathscr{B} + \Delta\mathscr{B}(t))\mathfrak{h}(\mathfrak{e}(t - \sigma(t)))$$

$$+ (\mathscr{A} + \Delta\mathscr{A}(t))\int_{t-\eta}^{t}\mathfrak{h}(\mathfrak{e}(\mathfrak{s}))\partial\mathfrak{s} + \mathscr{H}\mathscr{K}\mathfrak{e}(t_\mathfrak{k}\mathfrak{h}), t \in [t_\mathfrak{k}\mathfrak{h} + \tau_\mathfrak{k}, t_{\mathfrak{k}+1}\mathfrak{h} + \tau_{\mathfrak{k}+1}). \tag{32}$$

Case A: if $t_\mathfrak{k}\mathfrak{h} + \mathfrak{h} + \bar{\tau} \geq t_{\mathfrak{k}+1}\mathfrak{h} + \tau_{\mathfrak{k}+1}$, we can define $\tau(t)$ as

$$\tau(t) = t - t_\mathfrak{k}\mathfrak{h}, t \in [t_\mathfrak{k}\mathfrak{h} + \tau_\mathfrak{k}, t_{\mathfrak{k}+1}\mathfrak{h} + \tau_{\mathfrak{k}+1}).$$

It can be seen that

$$\tau_\mathfrak{k} \leq \tau(t) \leq (t_{\mathfrak{k}+1} - t_\mathfrak{k})\mathfrak{h} + t_{\mathfrak{k}+1} \leq \mathfrak{h} + \bar{\tau}.$$

Case B: if $t_\mathfrak{k}\mathfrak{h} + \mathfrak{h} + \bar{\tau} < t_{\mathfrak{k}+1}\mathfrak{h} + \tau_{\mathfrak{k}+1}$, since $t_\mathfrak{k} \leq \bar{\tau}$, we can easily demonstrate that a positive constant m exists such that $t_\mathfrak{k}\mathfrak{h} + m\mathfrak{h} + \bar{\tau} < t_{\mathfrak{k}+1}\mathfrak{h} + \tau_{\mathfrak{k}+1} \leq t_\mathfrak{k}\mathfrak{h} + (m+1)\mathfrak{h} + \bar{\tau}$. For the time intervals $[t_\mathfrak{k}\mathfrak{h} + \tau_\mathfrak{k}, t_{\mathfrak{k}+1}\mathfrak{h} + \tau_{\mathfrak{k}+1})$, we divide them as $\mathscr{F}_0 = [t_\mathfrak{k}\mathfrak{h} + \tau_\mathfrak{k}, t_\mathfrak{k}\mathfrak{h} + \mathfrak{h} + \bar{\tau})$, $\mathscr{F}_i = [t_\mathfrak{k}\mathfrak{h} + i\mathfrak{h} + \bar{\tau}, t_\mathfrak{k}\mathfrak{h} + i\mathfrak{h} + \mathfrak{h} + \bar{\tau})$, and $\mathscr{F}_m = [t_\mathfrak{k}\mathfrak{h} + m\mathfrak{h} + \bar{\tau}, t_{\mathfrak{k}+1}\mathfrak{h} + \tau_{\mathfrak{k}+1})$, and we define $\tau(t)$ as

$$\tau(t) = t - t_\mathfrak{k}(t) - i\mathfrak{h}, it \in \mathscr{F}_i, i = 0, 1, \ldots, m.$$

It is easy to prove that $0 \leq \tau_\mathfrak{k} \leq \tau((t)) \leq \mathfrak{h} + \bar{\tau} = \tau_M, t \in [t_\mathfrak{k}\mathfrak{h} + \tau_\mathfrak{k}, t_{\mathfrak{k}+1}\mathfrak{h} + \tau_{\mathfrak{k}+1})$. Finally, we define

$$\mathfrak{e}_\mathfrak{k}(t) = \mathfrak{e}(t_\mathfrak{k}\mathfrak{h}) - \mathfrak{e}(t_\mathfrak{k}\mathfrak{h} + i\mathfrak{h}), t \in \mathscr{F}_i, i = 0, 1, \ldots, m. \tag{33}$$

For case A, $m = 0$, we have $\mathfrak{e}_\mathfrak{k}(t) = 0$ from (33). Based on the analysis above, the event generator (31) can be rewritten as

$$\mathfrak{e}_\mathfrak{k}^T(t)\Phi\mathfrak{e}_\mathfrak{k}((t)) \leq \Sigma\mathfrak{e}^T(t - \tau(t))\Phi\mathfrak{e}^T(t - \tau(t), t \in [t_\mathfrak{k}\mathfrak{h} + \tau_\mathfrak{k}, t_{\mathfrak{k}+1}\mathfrak{h} + \tau_{\mathfrak{k}+1}).$$

Then, the system is reduced to

$$\mathscr{D}^\alpha \mathfrak{e}(t) = -\mathscr{R}\mathfrak{e}(t) + \mathscr{C}\mathfrak{h}(\mathfrak{e}(t)) + \mathscr{B}\mathfrak{h}(\mathfrak{e}(t-\sigma(t)))$$
$$+ \mathscr{A}\int_{t-\eta}^{t} \mathfrak{e}(\mathfrak{s})\partial\mathfrak{s} + \mathscr{H}\mathscr{K}\mathfrak{e}(t) + \mathscr{H}\mathscr{K}\mathfrak{e}(t-\tau(t)). \tag{34}$$

Theorem 3. *For the given scalars* $\delta_1, \delta_2, \delta_3, \delta_4, \delta_5, \mu_1,$ *and* σ *and the diagonal matrices* $\mathscr{L}_1, \mathscr{L}_2,$ *and* $\mathscr{L}_3,$ *if there exist symmetric positive definite matrices* $\mathscr{R}_1 > 0, \mathscr{R}_2 > 0,$ *then a feasible solution exists for the following LMIs:*

$$\xi < 0. \tag{35}$$

Proof. Now, let us define the Lyapunov–Krasovskii functional as follows:

$$\mathscr{V}(t) = \mathscr{V}_1(t) + \mathscr{V}_2(t), \tag{36}$$

where

$$\mathscr{V}_1(t) = \mathfrak{e}^{\mathscr{T}}(t)\mathscr{R}_1\mathfrak{e}(t),$$
$$\mathscr{V}_2(t) = \mathscr{D}^{(-\alpha+1)}\int_{t-\sigma(t)}^{t} \mathfrak{e}^{\mathscr{T}}(\mathfrak{s})\mathscr{R}_2\mathfrak{e}(\mathfrak{s})ds.$$

Using Lemma 2, we have

$$2\mathfrak{e}^{\mathscr{T}}(t)\mathscr{R}_1\Delta\mathscr{R}(t)\mathfrak{e}(t) \le 2\mathfrak{e}^{\mathscr{T}}(t)\mathscr{R}_1\mathscr{J}_\mathfrak{d}\mathscr{K}(t)\mathscr{L}_\mathfrak{d}\mathfrak{e}(t),$$
$$\le \delta_1^{-1}\mathfrak{e}^{\mathscr{T}}(t)\mathscr{R}_1\mathscr{J}_\mathfrak{d}\mathscr{J}_\mathfrak{d}^{\mathscr{T}}\mathscr{R}_1^{\mathscr{T}}\mathfrak{e}(t)$$
$$+ \delta_1\mathfrak{e}^{\mathscr{T}}(t)\mathscr{L}_\mathfrak{r}^{\mathscr{T}}\mathscr{L}_\mathfrak{r}\mathfrak{e}(t), \tag{37}$$
$$2\mathfrak{e}^{\mathscr{T}}(t)\mathscr{R}_1\Delta\mathscr{C}(t)\mathfrak{h}(\mathfrak{e}(t)) \le 2\mathfrak{e}^{\mathscr{T}}(t)\mathscr{R}_1\mathscr{J}_\mathfrak{c}\mathscr{K}(t)\mathscr{L}_\mathfrak{c}\mathfrak{h}(\mathfrak{e}(t)),$$
$$\le \delta_2^{-1}\mathfrak{e}^{\mathscr{T}}(t)\mathscr{R}_1\mathscr{J}_\mathfrak{c}\mathscr{J}_\mathfrak{c}^{\mathsf{T}}\mathscr{R}_1^{\mathscr{T}}\mathfrak{e}(t)$$
$$+ \delta_2\mathfrak{e}^{\mathscr{T}}(t)\phi^{\mathscr{T}}\mathscr{L}_\mathfrak{c}^{\mathscr{T}}\mathscr{L}_\mathfrak{c}\phi\mathfrak{e}(t), \tag{38}$$
$$2\mathfrak{e}^{\mathscr{T}}(t)\mathscr{R}_1\Delta\mathscr{B}(t)\mathfrak{h}(\mathfrak{e}(t-\sigma(t))) \le 2\mathfrak{e}^{\mathscr{T}}(t)\mathscr{R}_1\mathscr{J}_\mathfrak{b}\mathscr{K}(t)\mathscr{L}_\mathfrak{b}\mathfrak{h}(\mathfrak{e}(t-\sigma(t))),$$
$$\le \delta_3^{-1}\mathfrak{e}^{\mathscr{T}}(t)\mathscr{R}_1\mathscr{J}_\mathfrak{b}\mathscr{J}_\mathfrak{b}^{\mathscr{T}}\mathscr{R}_1^{\mathscr{T}}\mathfrak{e}(t)$$
$$+ \delta_3\mathfrak{e}^{\mathscr{T}}(t-\sigma(t))\phi^{\mathscr{T}}\mathscr{L}_\mathfrak{b}^{\mathscr{T}}\mathscr{L}_\mathfrak{b}\phi\mathfrak{e}(t-\sigma(t)). \tag{39}$$

Then, with the support of Lemma 1 and the linearity nature of the Caputo fractional-order derivative, the fractional derivative along the trajectories of the system state is acquired as follows

$$D^\alpha\mathscr{V}(t) \le 2\mathfrak{e}^{\mathscr{T}}(t)\mathscr{R}_1\mathscr{D}^\alpha\mathfrak{e}(t),$$
$$\le 2\mathfrak{e}^{\mathscr{T}}(t)\mathscr{R}_1\big[-(\mathscr{R}+\Delta\mathscr{R}(t))\mathfrak{e}(t) + (\mathscr{C}+\Delta\mathscr{C}(t))\mathfrak{h}(\mathfrak{e}(t))$$
$$+ (\mathscr{B}+\Delta\mathscr{B}(t))\mathfrak{h}(\mathfrak{e}(t-\sigma(t))) + \mathscr{H}\mathscr{K}\mathfrak{e}(t) + \mathscr{H}\mathscr{K}\mathfrak{e}(t-\tau(t))\big],$$
$$\le -2\mathfrak{e}^{\mathscr{T}}(t)\mathscr{R}_1\mathscr{R}\mathfrak{e}(t) + \delta_1^{-1}\mathfrak{e}^{\mathscr{T}}(t)\mathscr{R}_1\mathscr{J}_\mathfrak{d}\mathscr{J}_\mathfrak{d}^{\mathscr{T}}\mathscr{R}_1^{\mathscr{T}}\mathfrak{e}(t) + \delta_1\mathfrak{e}^{\mathscr{T}}(t)\mathscr{L}_\mathfrak{d}^{\mathscr{T}}\mathscr{L}_\mathfrak{d}\mathfrak{e}(t)$$
$$+ 2\mathfrak{e}^{\mathscr{T}}(t)\mathscr{R}_1\mathscr{C}\mathfrak{h}(\mathfrak{e}(t)) + \delta_2^{-1}\mathfrak{e}^{\mathscr{T}}(t)\mathscr{R}_1\mathscr{J}_\mathfrak{c}\mathscr{J}_\mathfrak{c}^{\mathscr{T}}\mathscr{R}_1^{\mathscr{T}}\mathfrak{e}(t) + \delta_2\mathfrak{e}^{\mathscr{T}}(t)\phi^{\mathscr{T}}\mathscr{L}_\mathfrak{c}^{\mathscr{T}}\mathscr{L}_\mathfrak{c}\phi\mathfrak{e}(t)$$
$$+ 2\mathfrak{e}^{\mathscr{T}}(t)\mathscr{R}_1\mathscr{B}\mathfrak{h}(\mathfrak{e}(t-\sigma(t))) + \delta_3^{-1}\mathfrak{e}^{\mathscr{T}}(t)\mathscr{R}_1\mathscr{J}_\mathfrak{b}\mathscr{J}_\mathfrak{b}^{\mathscr{T}}\mathscr{R}_1^{\mathscr{T}}\mathfrak{e}(t)$$
$$+ \delta_3\mathfrak{e}^{\mathscr{T}}(t-\sigma(t))\phi^{\mathscr{T}}\mathscr{L}_\mathfrak{b}^{\mathscr{T}}\mathscr{L}_\mathfrak{b}\phi\mathfrak{e}(t-\sigma(t))$$
$$+ \mathfrak{e}^{\mathscr{T}}(t)\mathscr{R}_2\mathfrak{e}(t) - \mathfrak{e}^{\mathscr{T}}(t-\sigma(t))\mathscr{R}_2\mathfrak{e}(t-\sigma(t))(1-\mu). \tag{40}$$

From Assumption 1, we have

$$\begin{bmatrix} \mathfrak{e}(t) \\ \mathfrak{h}(\mathfrak{e}(t)) \end{bmatrix}^T \begin{bmatrix} -\mathscr{L}_1\Gamma_2 & \mathscr{L}_1\Gamma_1 \\ * & -\mathscr{L}_1 \end{bmatrix} \begin{bmatrix} \mathfrak{e}(t) \\ \mathfrak{h}(\mathfrak{e}(t)) \end{bmatrix} \leq 0 \tag{41}$$

$$\begin{bmatrix} \mathfrak{e}(t-\sigma(t)) \\ \mathfrak{h}(\mathfrak{e}(t-\sigma(t))) \end{bmatrix}^T \begin{bmatrix} -\mathscr{L}_2\Gamma_2 & \mathscr{L}_2\Gamma_1 \\ * & -\mathscr{L}_2 \end{bmatrix} \begin{bmatrix} \mathfrak{e}(t-\sigma(t)) \\ \mathfrak{h}(\mathfrak{e}(t-\sigma(t))) \end{bmatrix} \leq 0 \tag{42}$$

$$\begin{bmatrix} \mathfrak{e}(t-\tau(t)) \\ \mathfrak{h}(\mathfrak{e}(t-\tau(t))) \end{bmatrix}^T \begin{bmatrix} -\mathscr{L}_3\Gamma_2 & \mathscr{L}_3\Gamma_1 \\ * & -\mathscr{L}_3 \end{bmatrix} \begin{bmatrix} \mathfrak{e}(t-\tau(t)) \\ \mathfrak{h}(\mathfrak{e}(t-\tau(t))) \end{bmatrix} \leq 0. \tag{43}$$

From (37)–(43), we obtain

$$\begin{aligned}
D^\alpha \mathscr{V}(t) &\leq 2\mathfrak{e}^{\mathscr{T}}(t)\mathscr{R}_1\mathscr{D}^\alpha \mathfrak{e}(t), \\
&\leq 2\mathfrak{e}^{\mathscr{T}}(t)\mathscr{R}_1\big[-(\mathscr{R}+\Delta\mathscr{R}(t))\mathfrak{e}(t) + (\mathscr{C}+\Delta\mathscr{C}(t))\mathfrak{h}(\mathfrak{e}(t)) \\
&\quad + (\mathscr{B}+\Delta\mathscr{B}(t))\mathfrak{h}(\mathfrak{e}(t-\sigma(t)) + \mathscr{H}\mathscr{K}\mathfrak{e}(t) + \mathscr{H}\mathscr{K}\mathfrak{e}(t-\tau(t))\big], \\
&\leq -2\mathfrak{e}^{\mathscr{T}}(t)\mathscr{R}_1\mathscr{R}\mathfrak{e}(t) + \delta_1^{-1}\mathfrak{e}^{\mathscr{T}}(t)\mathscr{R}_1\mathscr{J}_{\mathfrak{d}}\mathscr{J}_{\mathfrak{d}}^{\mathscr{T}}\mathscr{R}_1^{\mathscr{T}}\mathfrak{e}(t) + \delta_1\mathfrak{e}^{\mathscr{T}}(t)\mathscr{L}_{\mathfrak{d}}^{\mathscr{T}}\mathscr{L}_{\mathfrak{d}}\mathfrak{e}(t) \\
&\quad + 2\mathfrak{e}^{\mathscr{T}}(t)\mathscr{R}_1\mathscr{C}\mathfrak{h}(\mathfrak{e}(t)) + \delta_2^{-1}\mathfrak{e}^{\mathscr{T}}(t)\mathscr{R}_1\mathscr{J}_{\mathfrak{c}}\mathscr{J}_{\mathfrak{c}}^{\mathscr{T}}\mathscr{R}_1^{\mathscr{T}}\mathfrak{e}(t) + \delta_2\mathfrak{e}^{\mathscr{T}}(t)\phi^{\mathscr{T}}\mathscr{L}_{\mathfrak{c}}^{\mathscr{T}}\mathscr{L}_{\mathfrak{c}}\phi\mathfrak{e}(t) \\
&\quad + 2\mathfrak{e}^{\mathscr{T}}(t)\mathscr{R}_1\mathscr{B}\mathfrak{h}(\mathfrak{e}(t-\sigma(t))) + \delta_3^{-1}\mathfrak{e}^{\mathscr{T}}(t)\mathscr{R}_1\mathscr{J}_{\mathfrak{b}}\mathscr{J}_{\mathfrak{b}}^{\mathscr{T}}\mathscr{R}_1^{\mathscr{T}}\mathfrak{e}(t) \\
&\quad + \delta_3\mathfrak{e}^{\mathscr{T}}(t-\sigma(t))\phi^{\mathscr{T}}\mathscr{L}_{\mathfrak{b}}^{\mathscr{T}}\mathscr{L}_{\mathfrak{b}}\phi\mathfrak{e}(t-\sigma(t)) \\
&\quad + \delta_4^{-1}\mathfrak{e}^{\mathscr{T}}(t)\mathscr{R}_1\mathscr{C}^T\mathscr{C}\mathscr{R}_1^T\mathfrak{e}(t) + \delta_4\mathfrak{e}^{\mathscr{T}}(t)\phi^T\phi\mathfrak{e}(t) \\
&\quad + \delta_5^{-1}\mathfrak{e}^{\mathscr{T}}(t)\mathscr{R}_1\mathscr{B}^T\mathscr{B}\mathscr{R}_1^T\mathfrak{e}(t) + \delta_5\mathfrak{e}^{\mathscr{T}}(t-\sigma(t))\phi^T\phi\mathfrak{e}(t-\sigma(t)) \\
&\quad + \mathfrak{e}^{\mathscr{T}}(t)\mathscr{R}_2\mathfrak{e}(t) - \mathfrak{e}^{\mathscr{T}}(t-\sigma(t))\mathscr{R}_2\mathfrak{e}(t-\sigma(t)) \\
&\quad + \begin{bmatrix} \mathfrak{e}(t) \\ \mathfrak{h}(\mathfrak{e}(t)) \end{bmatrix}^T \begin{bmatrix} -\mathscr{L}_1\Gamma_2 & \mathscr{L}_1\Gamma_1 \\ * & -\mathscr{L}_1 \end{bmatrix} \begin{bmatrix} \mathfrak{e}(t) \\ \mathfrak{h}(\mathfrak{e}(t)) \end{bmatrix} \\
&\quad + \begin{bmatrix} \mathfrak{e}(t-\sigma(t)) \\ \mathfrak{h}(\mathfrak{e}(t-\sigma(t))) \end{bmatrix}^T \begin{bmatrix} -\mathscr{L}_2\Gamma_2 & \mathscr{L}_2\Gamma_1 \\ * & -\mathscr{L}_2 \end{bmatrix} \begin{bmatrix} \mathfrak{e}(t-\sigma(t)) \\ \mathfrak{h}(\mathfrak{e}(t-\sigma(t))) \end{bmatrix} \\
&\quad + \begin{bmatrix} \mathfrak{e}(t-\tau(t)) \\ \mathfrak{h}(\mathfrak{e}(t-\tau(t))) \end{bmatrix}^T \begin{bmatrix} -\mathscr{L}_3\Gamma_2 & \mathscr{L}_3\Gamma_1 \\ * & -\mathscr{L}_3 \end{bmatrix} \begin{bmatrix} \mathfrak{e}(t-\tau(t)) \\ \mathfrak{h}(\mathfrak{e}(t-\tau(t))) \end{bmatrix}.
\end{aligned}$$

Then,

$$\Lambda = \begin{bmatrix}
\Lambda_{11} & \Lambda_{12} & 0 & \Lambda_{14} & \Lambda_{15} & 0 & \Lambda_{17} & \Lambda_{18} & \Lambda_{19} \\
* & \Lambda_{22} & 0 & 0 & 0 & 0 & 0 & 0 & 0 \\
* & * & \Lambda_{33} & \Lambda_{34} & 0 & 0 & 0 & 0 & 0 \\
* & * & * & \Lambda_{44} & 0 & 0 & 0 & 0 & 0 \\
* & * & * & * & \Lambda_{55} & 0 & 0 & 0 & 0 \\
* & * & * & * & * & \Lambda_{66} & 0 & 0 & 0 \\
* & * & * & * & * & * & \Lambda_{77} & 0 & 0 \\
* & * & * & * & * & * & * & \Lambda_{88} & 0 \\
* & * & * & * & * & * & * & * & \Lambda_{99}
\end{bmatrix} < 0, \tag{44}$$

where

$$\begin{aligned}
\Lambda_{11} &= -2\mathscr{R}_1\mathscr{R} + \delta_1\mathscr{L}_r^{\mathscr{T}}\mathrm{L}_{\mathfrak{r}} + \delta_2\phi^{\mathscr{T}}\mathscr{L}_{\mathfrak{c}}^{\mathscr{T}}\mathscr{L}_{\mathfrak{c}}\phi + \mathscr{R}_2 + 2\mathscr{R}_1\mathscr{H}\mathscr{K} \\
&\quad - \mathscr{L}_1\Gamma_2 - \Phi, \quad \Lambda_{12} = \mathscr{R}_1\mathscr{C} + \mathscr{L}_1\Gamma_1, \quad \Lambda_{14} = \mathscr{R}_1\mathscr{B}, \quad \Lambda_{15} = \mathscr{R}_1\mathscr{H}\mathscr{K}, \\
\Lambda_{17} &= \mathscr{R}_1\mathscr{J}_{\mathfrak{r}}, \quad \Lambda_{18} = \mathscr{R}_1\mathscr{J}_{\mathfrak{c}}, \quad \Lambda_{19} = \mathscr{R}_1\mathscr{J}_{\mathfrak{b}}, \quad \Lambda_{22} = -\mathscr{L}_1, \quad \Lambda_{33} = \delta_3\phi^T\mathscr{L}_{\mathfrak{b}}^T\mathscr{L}_{\mathfrak{b}}\phi \\
&\quad - \mathscr{R}_2(1-\mu) - \mathscr{L}_2\Gamma_2, \quad \Lambda_{34} = \mathscr{L}_2\Gamma_1, \quad \Lambda_{44} = -\mathscr{L}_2, \quad \Lambda_{55} = \Sigma\Phi - \mathscr{L}_3\Gamma_2, \\
\Lambda_{56} &= \mathscr{L}_3\Gamma_1, \quad \Lambda_{66} = -\mathscr{L}_3, \quad \Lambda_{77} = -\delta_1\mathscr{I}, \quad \Lambda_{88} = -\delta_2\mathscr{I}, \quad \Lambda_{99} = -\delta_3\mathscr{I}.
\end{aligned}$$

We pre- and post-multiply Λ with $\{\mathscr{R}_1^{-1}, \mathscr{I}, \mathscr{I}, \mathscr{I}, \mathscr{R}_1^{-1}, \mathscr{I}, \mathscr{I}, \mathscr{I}, \mathscr{I}\}$

$$
Y = \begin{bmatrix}
Y_{11} & Y_{12} & 0 & Y_{14} & Y_{15} & 0 & Y_{17} & Y_{18} & Y_{19} \\
* & Y_{22} & 0 & 0 & 0 & 0 & 0 & 0 & 0 \\
* & * & Y_{33} & Y_{34} & 0 & 0 & 0 & 0 & 0 \\
* & * & * & Y_{44} & 0 & 0 & 0 & 0 & 0 \\
* & * & * & * & Y_{55} & 0 & 0 & 0 & 0 \\
* & * & * & * & * & Y_{66} & 0 & 0 & 0 \\
* & * & * & * & * & * & Y_{77} & 0 & 0 \\
* & * & * & * & * & * & * & Y_{88} & 0 \\
* & * & * & * & * & * & * & * & Y_{99}
\end{bmatrix} < 0, \tag{45}
$$

where

$$
\begin{aligned}
Y_{11} &= -2\mathscr{R}\mathscr{X}_1 + \mathscr{X}_1\delta_1\mathscr{L}_r^{\mathscr{T}}L_{\mathfrak{r}}\mathscr{X}_1 + \mathscr{X}_1\delta_2\phi^{\mathscr{T}}\mathscr{L}_c^{\mathscr{T}}\mathscr{L}_{\mathfrak{c}}\phi\mathscr{X}_1 + \mathscr{X}_1\mathscr{R}_2\mathscr{X}_1 \\
&\quad + 2\mathscr{H}\mathscr{K}\mathscr{X}_1 - \mathscr{X}_1\mathscr{L}_1\Gamma_2\mathscr{X}_1 - \mathscr{X}_1\phi\mathscr{X}_1, \; Y_{12} = \mathscr{C} + \mathscr{X}_1\mathscr{L}_1\Gamma_1, \; Y_{14} = \mathscr{B}, \\
Y_{15} &= \mathscr{H}\mathscr{K}\mathscr{X}_1, \; Y_{17} = \mathscr{J}_{\mathfrak{r}}, \; Y_{18} = \mathscr{J}_{\mathfrak{c}}, \; Y_{19} = \mathscr{J}_{\mathfrak{b}}, \; Y_{22} = -\mathscr{L}_1, \; Y_{33} = \delta_3\phi^T\mathscr{L}_{\mathfrak{b}}^T\mathscr{L}_{\mathfrak{b}}\phi \\
&\quad - \mathscr{R}_2(1-\mu) - \mathscr{L}_2\Gamma_2, \; Y_{34} = \mathscr{L}_2\Gamma_1, \; Y_{44} = -\mathscr{L}_2, \; Y_{55} = \mathscr{X}_1\Sigma\Phi\mathscr{X}_1 - \mathscr{X}_1\mathscr{L}_3\Gamma_2\mathscr{X}_1, \\
Y_{56} &= \mathscr{L}_3\Gamma_1, \; Y_{66} = -\mathscr{L}_3, \; Y_{77} = -\delta_1\mathscr{I}, \; Y_{88} = -\delta_2\mathscr{I}, \; Y_{99} = -\delta_3\mathscr{I}.
\end{aligned}
$$

At the same time, the controller gain matrix $\mathscr{K}$ can be obtained as $\mathscr{Y}_1 = \mathscr{K}\mathscr{X}_1$

$$
\zeta = \begin{bmatrix}
\zeta_{11} & \zeta_{12} & 0 & \zeta_{14} & \zeta_{15} & 0 & \zeta_{17} & \zeta_{18} & \zeta_{19} \\
* & \zeta_{22} & 0 & 0 & 0 & 0 & 0 & 0 & 0 \\
* & * & \zeta_{33} & \zeta_{34} & 0 & 0 & 0 & 0 & 0 \\
* & * & * & \zeta_{44} & 0 & 0 & 0 & 0 & 0 \\
* & * & * & * & \zeta_{55} & 0 & 0 & 0 & 0 \\
* & * & * & * & * & \zeta_{66} & 0 & 0 & 0 \\
* & * & * & * & * & * & \zeta_{77} & 0 & 0 \\
* & * & * & * & * & * & * & \zeta_{88} & 0 \\
* & * & * & * & * & * & * & * & \zeta_{99}
\end{bmatrix} < 0, \tag{46}
$$

where

$$
\begin{aligned}
\zeta_{11} &= -2\mathscr{R}\mathscr{X}_1 + \mathscr{X}_1\delta_1\mathscr{L}_r^{\mathscr{T}}L_{\mathfrak{r}}\mathscr{X}_1 + \mathscr{X}_1\delta_2\phi^{\mathscr{T}}\mathscr{L}_c^{\mathscr{T}}\mathscr{L}_{\mathfrak{c}}\phi\mathscr{X}_1 + \mathscr{X}_1\mathscr{R}_2\mathscr{X}_1 + 2\mathscr{H}\mathscr{Y} \\
&\quad - \mathscr{X}_1\mathscr{L}_1\Gamma_2\mathscr{X}_1 - \mathscr{X}_1\phi\mathscr{X}_1, \; \zeta_{12} = \mathscr{C} + \mathscr{X}_1\mathscr{L}_1\Gamma_1, \; \zeta_{14} = \mathscr{B}, \; \zeta_{15} = \mathscr{H}\mathscr{Y}, \; \zeta_{17} = \mathscr{J}_{\mathfrak{r}}, \\
\zeta_{18} &= \mathscr{J}_{\mathfrak{c}}, \; \zeta_{19} = \mathscr{J}_{\mathfrak{b}}, \; \zeta_{22} = -\mathscr{L}_1, \; \zeta_{33} = \delta_3\phi^T\mathscr{L}_{\mathfrak{b}}^T\mathscr{L}_{\mathfrak{b}}\phi - \mathscr{R}_2(1-\mu) - \mathscr{L}_2\Gamma_2, \\
\zeta_{34} &= \mathscr{L}_2\Gamma_1, \; \zeta_{44} = -\mathscr{L}_2, \; \zeta_{55} = \mathscr{X}_1\Sigma\Phi\mathscr{X}_1 - \mathscr{X}_1\mathscr{L}_3\Gamma_2\mathscr{X}_1, \; \zeta_{56} = \mathscr{L}_3\Gamma_1, \; \zeta_{66} = -\mathscr{L}_3, \\
\zeta_{77} &= -\delta_1\mathscr{I}, \; \zeta_{88} = -\delta_2\mathscr{I}, \; \zeta_{99} = -\delta_3\mathscr{I}.
\end{aligned}
$$

$$
D^{\alpha}\mathscr{V}(\mathfrak{t}) \leq \varphi^T(\mathfrak{t})\zeta\varphi(\mathfrak{t}), \tag{47}
$$

where

$$
\begin{aligned}
\varphi(\mathfrak{t}) = col[&\mathfrak{e}(\mathfrak{t}), \; \mathfrak{h}(\mathfrak{e}(\mathfrak{t})), \; \mathfrak{e}(t-\sigma(t)), \; \mathfrak{h}(\mathfrak{e}(t-\sigma(t))), \; \mathfrak{e}(t-\tau(t)), \\
&\mathfrak{h}(\mathfrak{e}(t-\tau(t)))].
\end{aligned}
$$

By the Lypunov stability theory analysis, the event-triggered synchronization of the fractional-order uncertain neural networks' error system (34) is globally asymptotic stable if LMI (35) holds. This completes the proof. $\square$

5. Numerical Example

Example 1. *Consider the following uncertain neural networks (5) with time-varying delays described by*

$$\mathscr{D}^{\alpha}\mathfrak{e}(\mathfrak{t}) = -(\mathscr{R} + \Delta\mathscr{R}(t))\mathfrak{e}(\mathfrak{t}) + (\mathscr{C} + \Delta\mathscr{C}(t))\mathfrak{h}(\mathfrak{e}(\mathfrak{t})) + (\mathscr{B} + \Delta\mathscr{B}(t))\mathfrak{h}(\mathfrak{e}(\mathfrak{t} - \sigma(\mathfrak{t}))$$
$$+ \mathscr{H}\mathscr{Q}(\mathfrak{t}), \tag{48}$$

with the following parameters

$$\mathscr{C} = \begin{bmatrix} -1.5241 & 1.2489 & 1.6844 & 1.2946 & 1.8722 \\ -1.2567 & 1.1247 & 1.4211 & 1.6522 & 1.2807 \\ 1.5427 & 1.1227 & -1.4567 & 1.0425 & 1.1727 \\ 1.2514 & -1.1077 & 1.2404 & 1.6507 & 1.2701 \\ 1.9472 & -1.1174 & -1.2567 & 1.9989 & 1.2486 \end{bmatrix},$$

$$\mathscr{B} = \begin{bmatrix} -1.4932 & 1.5968 & 1.2567 & 1.0567 & 1.2674 \\ 1.2942 & 1.9942 & -1.6911 & 1.2849 & 1.5677 \\ 1.0977 & 1.4217 & -1.2415 & 1.5661 & 1.5717 \\ 1.2567 & -1.0741 & 1.2961 & 1.2247 & 1.2702 \\ 1.0047 & 1.2742 & 1.4274 & 1.6611 & 1.4428 \end{bmatrix},$$

$$\mathscr{H} = \begin{bmatrix} -1.5432 & 1.0968 & 1.2987 & 1.0097 & 1.9974 \\ 1.6542 & 1.5642 & -1.3411 & 1.7649 & 1.5767 \\ 1.2377 & 1.3417 & -1.9815 & 1.3461 & 1.5887 \\ 1.8767 & -1.8741 & 1.6561 & 1.9847 & 1.2092 \\ 1.3247 & 1.2652 & 1.4094 & 1.6871 & 1.4488 \end{bmatrix},$$

$$\mathscr{R} = \begin{bmatrix} 0.7289 & 0 & 0 & 0 & 0 \\ 0 & 0.7289 & 0 & 0 & 0 \\ 0 & 0 & 0.7289 & 0 & 0 \\ 0 & 0 & 0 & 0.7289 & 0 \\ 0 & 0 & 0 & 0 & 0.7289 \end{bmatrix}, \mathscr{I} = \begin{bmatrix} 1 & 0 & 0 & 0 & 0 \\ 0 & 1 & 0 & 0 & 0 \\ 0 & 0 & 1 & 0 & 0 \\ 0 & 0 & 0 & 1 & 0 \\ 0 & 0 & 0 & 0 & 1 \end{bmatrix}.$$

$$\mathscr{J}_r = \begin{bmatrix} 0.4428 & 0 & 0 & 0 & 0 \\ 0 & 0.4428 & 0 & 0 & 0 \\ 0 & 0 & 0.4428 & 0 & 0 \\ 0 & 0 & 0 & 0.4428 & 0 \\ 0 & 0 & 0 & 0 & 0.4428 \end{bmatrix},$$

$$\mathscr{L}_d = \begin{bmatrix} 1.7782 & 0 & 0 & 0 & 0 \\ 0 & 1.7782 & 0 & 0 & 0 \\ 0 & 0 & 1.7782 & 0 & 0 \\ 0 & 0 & 0 & 1.7782 & 0 \\ 0 & 0 & 0 & 0 & 1.7782 \end{bmatrix},$$

$$\mathscr{J}_c = \begin{bmatrix} 0.5242 & 0 & 0 & 0 & 0 \\ 0 & 0.5242 & 0 & 0 & 0 \\ 0 & 0 & 0.5242 & 0 & 0 \\ 0 & 0 & 0 & 0.5242 & 0 \\ 0 & 0 & 0 & 0 & 0.5242 \end{bmatrix}.$$

$$\mathscr{L}_c = \begin{bmatrix} 2.8976 & 0 & 0 & 0 & 0 \\ 0 & 2.8976 & 0 & 0 & 0 \\ 0 & 0 & 2.8976 & 0 & 0 \\ 0 & 0 & 0 & 2.8976 & 0 \\ 0 & 0 & 0 & 0 & 2.8976 \end{bmatrix},$$

$$\mathscr{L}_b = \begin{bmatrix} 1.8974 & 0 & 0 & 0 & 0 \\ 0 & 1.8974 & 0 & 0 & 0 \\ 0 & 0 & 1.8974 & 0 & 0 \\ 0 & 0 & 0 & 1.8974 & 0 \\ 0 & 0 & 0 & 0 & 1.8974 \end{bmatrix},$$

$$\mathscr{J}_b = \begin{bmatrix} 0.2995 & 0 & 0 & 0 & 0 \\ 0 & 0.2995 & 0 & 0 & 0 \\ 0 & 0 & 0.2995 & 0 & 0 \\ 0 & 0 & 0 & 0.2995 & 0 \\ 0 & 0 & 0 & 0 & 0.2995 \end{bmatrix},$$

$$\phi = \begin{bmatrix} 0.2494 & 0 & 0 & 0 & 0 \\ 0 & 0.2494 & 0 & 0 & 0 \\ 0 & 0 & 0.2494 & 0 & 0 \\ 0 & 0 & 0 & 0.2494 & 0 \\ 0 & 0 & 0 & 0 & 0.2494 \end{bmatrix}.$$

Moreover, the activation functions are $\mathfrak{f}(\mathfrak{e}(\mathfrak{t})) = tanh(\mathfrak{e}(\mathfrak{t}))$ *and* $\mathfrak{f}(\mathfrak{e}(\mathfrak{t} - \sigma(\mathfrak{t}))) = sinh(\mathfrak{e}(\mathfrak{t}))$. *Solving the LMI conditions provided in (7) based on the MATLAB toolbox returns the following feasible solutions:*

$$\mathscr{R}_1 = \begin{bmatrix} 0.0284 & 0.0154 & 0.0180 & -0.0127 & -0.0074 \\ 0.0154 & 0.0244 & 0.0120 & 0.0070 & -0.0260 \\ 0.0180 & 0.0120 & 0.0209 & -0.0054 & -0.0102 \\ -0.0127 & 0.0070 & -0.0054 & 0.0904 & -0.0873 \\ -0.0074 & -0.0260 & -0.0102 & -0.0873 & 0.1118 \end{bmatrix},$$

$$\mathscr{R}_2 = \begin{bmatrix} 36.6572 & 0.0000 & 0.0000 & -0.0000 & -0.0000 \\ 0.0000 & 36.6572 & 0.0000 & 0.0000 & -0.0000 \\ 0.0000 & 0.0000 & 36.6572 & -0.0000 & -0.0000 \\ -0.0000 & 0.0000 & -0.0000 & 36.6572 & -0.0000 \\ -0.0000 & -0.0000 & -0.0000 & -0.0000 & 36.6572 \end{bmatrix}.$$

The gain matrix of the designed controller can be obtained as:

$$\mathscr{K} = \begin{bmatrix} -9.2914 & 0.0000 & 0.0000 & -0.0000 & -0.0000 \\ 0.0000 & -9.2914 & 0.0000 & 0.0000 & -0.0000 \\ 0.0000 & 0.0000 & -9.2914 & -0.0000 & -0.0000 \\ -0.0000 & 0.0000 & -0.0000 & -9.2914 & -0.0000 \\ -0.0000 & -0.0000 & -0.0000 & -0.0000 & -9.2914 \end{bmatrix}.$$

$\delta_1 = 20.2099, \delta_2 = 20.2097, \delta_3 = 20.2099, \delta_4 = 20.2099,$ *and* $\delta_5 = 20.2099,$ *which preserves system (48) as synchronous.*

Example 2. *Consider the following uncertain neural networks with time-varying delays described by*

$$\mathscr{D}^\alpha \mathfrak{e}(\mathfrak{t}) = -(\mathscr{R} + \Delta\mathscr{R}(\mathfrak{t}))\mathfrak{e}(\mathfrak{t}) + (\mathscr{C} + \Delta\mathscr{C}(\mathfrak{t}))\mathfrak{h}(\mathfrak{e}(\mathfrak{t})) + (\mathscr{B} + \Delta\mathscr{B}(\mathfrak{t}))\mathfrak{h}(\mathfrak{e}(\mathfrak{t} - \sigma(\mathfrak{t})))$$
$$+ \mathscr{H}\mathscr{Q}\mathscr{K}(\mathfrak{t}) \tag{49}$$

$$\mathscr{C} = \begin{bmatrix} -1.7841 & 1.2499 & 1.6876 & 1.9046 & 1.8092 \\ -1.3367 & 1.3447 & 1.4541 & 1.6982 & 1.7807 \\ 1.2327 & 1.1447 & -1.4897 & 1.0895 & 1.5627 \\ 1.8714 & -1.7677 & 1.2094 & 1.9807 & 1.7801 \\ 1.3472 & -1.8974 & -1.6667 & 1.5689 & 1.2986 \end{bmatrix},$$

$$\mathscr{B} = \begin{bmatrix} -1.9832 & 1.5878 & 1.6767 & 1.0567 & 1.2674 \\ 1.3442 & 1.9482 & -1.9811 & 1.2899 & 1.5097 \\ 1.9877 & 1.4977 & -1.6615 & 1.5687 & 1.5787 \\ 1.6767 & -1.6741 & 1.2977 & 1.2277 & 1.2982 \\ 1.9847 & 1.2892 & 1.8774 & 1.6666 & 1.4499 \end{bmatrix},$$

$$\mathscr{H} = \begin{bmatrix} -1.7632 & 1.0878 & 1.2897 & 1.7897 & 1.9674 \\ 1.9942 & 1.3342 & -1.8711 & 1.7999 & 1.6767 \\ 1.9877 & 1.3817 & -1.5615 & 1.7861 & 1.4587 \\ 1.6567 & -1.6741 & 1.9561 & 1.8747 & 1.2702 \\ 1.6647 & 1.2652 & 1.4564 & 1.6771 & 1.6788 \end{bmatrix},$$

$$\mathscr{R} = \begin{bmatrix} 0.2389 & 0 & 0 & 0 & 0 \\ 0 & 0.2389 & 0 & 0 & 0 \\ 0 & 0 & 0.2389 & 0 & 0 \\ 0 & 0 & 0 & 0.2389 & 0 \\ 0 & 0 & 0 & 0 & 0.2389 \end{bmatrix},$$

$$\mathscr{J}_r = \begin{bmatrix} 0.7628 & 0 & 0 & 0 & 0 \\ 0 & 0.7628 & 0 & 0 & 0 \\ 0 & 0 & 0.7628 & 0 & 0 \\ 0 & 0 & 0 & 0.7628 & 0 \\ 0 & 0 & 0 & 0 & 0.7628 \end{bmatrix},$$

$$\mathscr{L}_d = \begin{bmatrix} 1.9882 & 0 & 0 & 0 & 0 \\ 0 & 1.9882 & 0 & 0 & 0 \\ 0 & 0 & 1.9882 & 0 & 0 \\ 0 & 0 & 0 & 1.9882 & 0 \\ 0 & 0 & 0 & 0 & 1.9882 \end{bmatrix},$$

$$\mathscr{J}_c = \begin{bmatrix} 0.9087 & 0 & 0 & 0 & 0 \\ 0 & 0.9087 & 0 & 0 & 0 \\ 0 & 0 & 0.9087 & 0 & 0 \\ 0 & 0 & 0 & 0.9087 & 0 \\ 0 & 0 & 0 & 0 & 0.9087 \end{bmatrix},$$

$$\mathscr{L}_c = \begin{bmatrix} 2.5676 & 0 & 0 & 0 & 0 \\ 0 & 2.5676 & 0 & 0 & 0 \\ 0 & 0 & 2.5676 & 0 & 0 \\ 0 & 0 & 0 & 2.5676 & 0 \\ 0 & 0 & 0 & 0 & 2.5676 \end{bmatrix},$$

$$\mathscr{L}_b = \begin{bmatrix} 1.0987 & 0 & 0 & 0 & 0 \\ 0 & 1.0987 & 0 & 0 & 0 \\ 0 & 0 & 1.0987 & 0 & 0 \\ 0 & 0 & 0 & 1.0987 & 0 \\ 0 & 0 & 0 & 0 & 1.0987 \end{bmatrix},$$

$$\mathscr{J}_b = \begin{bmatrix} 0.8765 & 0 & 0 & 0 & 0 \\ 0 & 0.8765 & 0 & 0 & 0 \\ 0 & 0 & 0.8765 & 0 & 0 \\ 0 & 0 & 0 & 0.8765 & 0 \\ 0 & 0 & 0 & 0 & 0.8765 \end{bmatrix},$$

$$\phi = \begin{bmatrix} 0.2476 & 0 & 0 & 0 & 0 \\ 0 & 0.2476 & 0 & 0 & 0 \\ 0 & 0 & 0.2476 & 0 & 0 \\ 0 & 0 & 0 & 0.2476 & 0 \\ 0 & 0 & 0 & 0 & 0.2476 \end{bmatrix},$$

$$\mathscr{I} = \begin{bmatrix} 1 & 0 & 0 & 0 & 0 \\ 0 & 1 & 0 & 0 & 0 \\ 0 & 0 & 1 & 0 & 0 \\ 0 & 0 & 0 & 1 & 0 \\ 0 & 0 & 0 & 0 & 1 \end{bmatrix}.$$

Moreover, the activation functions are $\mathfrak{f}(\mathfrak{e}(t)) = tanh(\mathfrak{e}(t))$ *and* $\mathfrak{f}(\mathfrak{e}(t - \sigma(t))) = sinh(\mathfrak{e}(t))$. *Solving the LMI conditions provided in (15) based on the MATLAB toolbox returns the following feasible solutions:*

$$\mathscr{X}_1 = \begin{bmatrix} 0.0346 & 0.0132 & 0.0158 & -0.0124 & -0.0074 \\ 0.0132 & 0.0310 & 0.0070 & 0.0006 & -0.0156 \\ 0.0158 & 0.0070 & 0.0301 & -0.0123 & -0.0014 \\ -0.0124 & 0.0006 & -0.0123 & 0.1428 & -0.1226 \\ -0.0074 & -0.0156 & -0.0014 & -0.1226 & 0.1419 \end{bmatrix},$$

$$\mathscr{R}_2 = \begin{bmatrix} 34.3231 & 0.0000 & 0.0000 & -0.0000 & -0.0000 \\ 0.0000 & 34.3231 & 0.0000 & 0.0000 & -0.0000 \\ 0.0000 & 0.0000 & 34.3231 & -0.0000 & -0.0000 \\ -0.0000 & 0.0000 & -0.0000 & 34.3231 & -0.0000 \\ -0.0000 & -0.0000 & -0.0000 & -0.0000 & 34.3231 \end{bmatrix},$$

$$\mathscr{Y} = \begin{bmatrix} -10.4053 & 0.0000 & 0.0000 & -0.0000 & -0.0000 \\ 0.0000 & -10.4053 & 0.0000 & 0.0000 & -0.0000 \\ 0.0000 & 0.0000 & -10.4053 & -0.0000 & -0.0000 \\ -0.0000 & 0.0000 & -0.0000 & -10.4053 & -0.0000 \\ -0.0000 & -0.0000 & -0.0000 & -0.0000 & -10.4053 \end{bmatrix}.$$

The gain matrix of the designed controller can be obtained as:

$$\mathscr{K} = \begin{bmatrix} -5.0940 & 1.0518 & 1.7014 & -1.6104 & -1.5249 \\ 1.0518 & -4.4548 & 0.0090 & -1.0158 & -1.3140 \\ 1.7014 & 0.0090 & -4.7390 & -0.8662 & -0.7067 \\ -1.6104 & -1.0158 & -0.8662 & -4.3173 & -3.9362 \\ -1.5249 & -1.3140 & -0.7067 & -3.9362 & -4.3671 \end{bmatrix}.$$

$\delta_1 = 21.1589, \delta_2 = 21.1589, \delta_3 = 21.1567, \delta_4 = 21.1590,$ *and* $\delta_5 = 21.1583$, *which preserves (49) as synchronous.*

Example 3. *Consider the following neural networks (20), with the following parameters*

$$\mathscr{C} = \begin{bmatrix} -1.5041 & 1.0489 & 1.0844 & 1.9946 & 1.8762 \\ -1.7567 & 1.5247 & 1.7211 & 1.4522 & 1.2877 \\ 1.0427 & 1.8227 & -1.5567 & 1.9425 & 1.1877 \\ 1.6514 & -1.0077 & 1.2904 & 1.6507 & 1.7601 \\ 1.9872 & -1.6174 & -1.6567 & 1.9989 & 1.0986 \end{bmatrix},$$

$$\mathscr{A} = \begin{bmatrix} -1.0941 & 1.9889 & 1.6544 & 1.2096 & 1.1722 \\ -1.1567 & 1.6547 & 1.4871 & 1.6672 & 1.7807 \\ 1.5727 & 1.1347 & -1.4987 & 1.0765 & 1.6727 \\ 1.2514 & -1.8777 & 1.2094 & 1.6597 & 1.9701 \\ 1.9272 & -1.8874 & -1.6767 & 1.8089 & 1.9486 \end{bmatrix},$$

$$\mathscr{B} = \begin{bmatrix} -1.4872 & 1.5878 & 1.8767 & 1.6667 & 1.9074 \\ 1.8742 & 1.9452 & -1.9911 & 1.9049 & 1.8877 \\ 1.0877 & 1.4987 & -1.2315 & 1.7761 & 1.0917 \\ 1.0567 & -1.3441 & 1.9861 & 1.0947 & 1.8902 \\ 1.6047 & 1.2872 & 1.4874 & 1.0911 & 1.0928 \end{bmatrix},$$

$$\mathscr{R} = \begin{bmatrix} 0.1459 & 0 & 0 & 0 & 0 \\ 0 & 0.1459 & 0 & 0 & 0 \\ 0 & 0 & 0.1459 & 0 & 0 \\ 0 & 0 & 0 & 0.1459 & 0 \\ 0 & 0 & 0 & 0 & 0.1459 \end{bmatrix},$$

$$\mathscr{H} = \begin{bmatrix} -1.5782 & 1.0068 & 1.7687 & 1.0097 & 1.0974 \\ 1.6942 & 1.8742 & -1.0911 & 1.6749 & 1.8767 \\ 1.6377 & 1.9817 & -1.4515 & 1.9861 & 1.7687 \\ 1.8097 & -1.8651 & 1.0961 & 1.0947 & 1.0992 \\ 1.3677 & 1.2698 & 1.4874 & 1.8671 & 1.9888 \end{bmatrix},$$

$$\mathscr{I} = \begin{bmatrix} 1 & 0 & 0 & 0 & 0 \\ 0 & 1 & 0 & 0 & 0 \\ 0 & 0 & 1 & 0 & 0 \\ 0 & 0 & 0 & 1 & 0 \\ 0 & 0 & 0 & 0 & 1 \end{bmatrix}, \phi = \begin{bmatrix} 0.0987 & 0 & 0 & 0 & 0 \\ 0 & 0.0987 & 0 & 0 & 0 \\ 0 & 0 & 0.0987 & 0 & 0 \\ 0 & 0 & 0 & 0.0987 & 0 \\ 0 & 0 & 0 & 0 & 0.0987 \end{bmatrix}.$$

Moreover, the activation functions are $\mathfrak{f}(\mathfrak{e}(\mathfrak{t})) = tanh(\mathfrak{e}(\mathfrak{t}))$ and $\mathfrak{f}(\mathfrak{e}(t - \sigma(t))) = sinh(\mathfrak{e}(\mathfrak{t}))$. Solving the LMI conditions provided in (21) based on the MATLAB toolbox returns the following feasible solutions:

$$\mathscr{R}_1 = \begin{bmatrix} 0.6253 & 0.2376 & 0.3124 & -0.1046 & -0.2094 \\ 0.2376 & 0.4979 & 0.1106 & -0.1841 & -0.0731 \\ 0.3124 & 0.1106 & 0.4894 & 0.1397 & -0.3460 \\ -0.1046 & -0.1841 & 0.1397 & 1.4658 & -1.2618 \\ -0.2094 & -0.0731 & -0.3460 & -1.2618 & 1.5839 \end{bmatrix},$$

$$\mathscr{R}_2 = \begin{bmatrix} 29.0877 & 0.0002 & 0.0001 & 0.0005 & -0.0009 \\ 0.0002 & 29.0933 & -0.0006 & 0.0017 & -0.0018 \\ 0.0001 & -0.0006 & 29.0873 & 0.0002 & -0.0008 \\ 0.0005 & 0.0017 & 0.0002 & 29.0905 & -0.0026 \\ -0.0009 & -0.0018 & -0.0008 & -0.0026 & 29.0847 \end{bmatrix}.$$

The gain matrix of the designed controller and trigger parameters can be obtained as follows:

$$\mathscr{K} = \begin{bmatrix} -6.6909 & 0.0173 & 0.0082 & 0.0374 & -0.0673 \\ 0.0173 & -6.4671 & -0.0464 & 0.1301 & -0.1326 \\ 0.0082 & -0.0464 & -6.7093 & 0.0127 & -0.0577 \\ 0.0374 & 0.1301 & 0.0127 & -6.5897 & -0.1818 \\ -0.0673 & -0.1326 & -0.0577 & -0.1818 & -6.8241 \end{bmatrix}.$$

$\delta_4 = 4.3607$ *and* $\delta_5 = 4.5189$. *Therefore, preserves system (20) is synchronous.*

6. Conclusions

In this paper, the synchronization problem was investigated for neural networks. It is well known that the Lyapunov direct method is the most effective method to analyze the stability of neural networks; the authors gave an important inequality on the Caputo derivative of quadratic functions, which plays an important role in analyzing the stability of fractional-order systems. By using Lyapunov functionals and analytical techniques, we obtained some sufficient conditions, and we derived event triggering to guarantee the synchronization of the delayed neural networks. We appled the Lyapunov functional method and the LMI approach to establish the synchronization criteria for the fractional-order nerual network matrix. A linear matrix inequality approach was developed to solve the problem. Numerical examples were given to demonstrate the effectiveness of the proposed schemes. Future work will focus on event-triggered control for fractional-order systems with time-delay and measurement noises. In addition, more effective event-triggered schemes such as an adaptive one, a dynamic one, and a hybrid one will also be considered for the stability analysis of fractional-order systems.

Author Contributions: Conceptualization, M.H., M.S.A., T.F.I. and B.A.Y.; methodology, K.I.O., M.H., M.S.A., T.F.I. and B.A.Y.; software, M.H., M.S.A. and K.I.O.; validation, M.H., M.S.A., T.F.I. and K.M.; formal analysis, K.I.O., M.H., M.S.A., T.F.I. and K.M.; investigation, M.H., T.F.I., B.A.Y. and K.I.O. resources, M.H., M.S.A., T.F.I. and K.M.; writing—review and editing, M.H., M.S.A., T.F.I., B.A.Y. and K.M.; visualization, M.H., M.S.A., T.F.I. and B.A.Y.; supervision, K.I.O., M.H., M.S.A., T.F.I. and B.A.Y.; project administration, M.H., M.S.A., T.F.I. and K.M.; funding acquisition, M.S.A. and K.M. All authors have read and agreed to the published version of the manuscript.

Funding: The authors extend their appreciation to the Deanship of Scientific Research at King Khalid University for funding this work through large groups (project under grant number RGP.2/47/43/1443). Moreover, this research received funding support from the NSRF via the Program Management Unit for Human Resources & Institutional Development, Research and Innovation [grant number B05F650018].

Data Availability Statement: Not applicable.

Conflicts of Interest: The authors declare no conflict of interest.

References

1. Podlubny, I. Fractional differential equations. In *Mathematics in Science and Engineering*; Academic Press: San Diego, CA, USA, 1999; Volume 198.
2. Monje, C.A.; Chen, Y.; Vinagre, B.M.; Xue, D.; Feliu, V. Fractional-Order systems and controls. In *Advances in Industrial Control*; Springer: London, UK, 2010.
3. Wang, Y.; Gu, L.; Xu, Y.; Cao, X. Practical tracking control of robot manipulators with continuous fractional-order nonsingular terminal sliding mode. *IEEE Trans. Ind. Electron.* **2016**, *63*, 6194–6204. [CrossRef]
4. Li, C.; Chen, G. Chaos and hyperchaos in the fractional-order rossler equations. *Phys. A Stat. Mech. Its Appl.* **2004**, *341*, 55–61. [CrossRef]
5. Debnath, L.; Feyman, R.P. Recent applications of fractional calculus to science and engineering. *Int. J. Math. Math. Sci.* **2003**, *54*, 3413–3442. [CrossRef]
6. Ding, Z.; Zeng, Z.; Zhang, H.; Wang, L.; Wang, L. New results on passivity of fractional-order uncertain neural networks. *Neurocomputing* **2019**, *351*, 51–59. [CrossRef]
7. Rajivganthi, C.; Rihan, F.; Laxshmanan, S.; Rakkiappan, R.; Muthuumar, P. Synchronization of memristor-based delayed BAM neural networks with fractional-order derivatives. *Complexity* **2016**, *21*, 412–426. [CrossRef]

8. Bao, H.; Park, J.H.; Cao, J. Non-fragile state estimation for fractional-order delayed memristive BAM neural networks. *Neural Netw.* **2019**, *119*, 190–199. [CrossRef] [PubMed]

9. Song, C.; Fei, S.; Cao, J.; Huang, C. Robust synchronization of fractional-order uncertain chaotic systems based on output feedback sliding mode control. *Mathematics* **2019**, *7*, 599. [CrossRef]

10. Henderson, J.; Ouahab, A. Fractional functional differential inclusions with finite delay. *Nonlinear Anal.* **2009**, *70*, 2091–2105. [CrossRef]

11. Tan, N.; Faruk, O.; Mine, M. Robust stability analysis of fractional order interval polynomials. *ISA Trans.* **2009**, *48*, 166–172. [CrossRef]

12. Ahn, H.S.; Chen, Y. Necessary and sufficient stability condition of fractional-order interval linear systems. *Automatica* **2008**, *44*, 2985–2988. [CrossRef]

13. Dzielinski, A.; Sierociuk, D. Stability of discrete fractional order state—Space system. *J. Vib. Control* **2008**, *14*, 1543–1556. [CrossRef]

14. Liao, Z.; Peng, C.; Li, W.; Wang, Y. Robust stability analysis for a class of fractional order systems with uncertain parameters. *J. Frankl. Inst. Eng. Appl. Math.* **2011**, *348*, 1101–1113. [CrossRef]

15. Chilali, M.; Gahinet, P.; Apkarian, P. Robust pole placement in LMI regions. *Inst. Electr. Electron. Eng. Autom. Control* **1999**, *44*, 2257–2270. [CrossRef]

16. Tang, Y.; Wang, Z.; Fang, J. Pinning control of fractional-order weighted complex networks. *Chaos* **2009**, *19*, 013112 [CrossRef]

17. Luo, J.; Tian, W.; Zhong, S.; Shi, K.; Gu, X.M.; Wang, W. Improved delay-probability-dependent results for stochastic neural networks with randomly occurring uncertainties and multiple delays. *Int. J. Syst. Sci.* **2018**, *49*, 2039–2059. [CrossRef]

18. Shi, K.; Tang, Y.; Liu, X.; Zhong, S. Secondary delay-partition approach on robust performance analysis for uncertain time-varying Lurie nonlinear control system. *Optim. Control Appl. Methods* **2017**, *38*, 1208–1226. [CrossRef]

19. Syed Ali, M.; Balasubramaniam, P. Global exponential stability for uncertain stochastic fuzzy BAM neural networks with time-varying delays. *Chaos Solitons Fractals* **2009**, *42*, 2191–2199. [CrossRef]

20. Zeng, D.; Shi, K.; Zhang, R.; Zhong, S. Novel mean square exponential stability criterion of uncertain stochastic interval type-2 fuzzy neural networks with multiple time-varying delays. In Proceedings of the 2017 36th Chinese Control Conference, Dalian, China, 26–28 July 2017.

21. Syed Ali, M.; Esther Rani, M. Passivity analysis of uncertain stochastic neural networks with time-varying delays and Markovian jumping parameters. *Netw. Comput. Neural Syst.* **2015**, *26*, 73–96. [CrossRef]

22. Shi, K.; Tang, Y.; Zhong, S.; Yin, C.; Huang, X.; Wang, W. Nonfragile asynchronous control for uncertain chaotic Lurie network systems with Bernoulli stochastic process. *Int. J. Robust Nonlinear Control* **2018**, *28*, 1693–1714. [CrossRef]

23. Luo, J.; Tian, W.; Zhong, S.; Shi, K.; Wang, W. Non-fragile asynchronous event-triggered control for uncertain delayed switched neural networks. *Nonlinear Anal. Hybrid Syst.* **2018**, *29*, 54–73. [CrossRef]

24. Saravanakumar, R.; Syed Ali, M.; Huang, H.; Cao, J.; Joo, Y.H. Robust H_∞ state-feedback control for nonlinear uncertain systems with mixed time-varying delay. *Int. J. Control Autom. Syst.* **2018**, *16*, 225–233. [CrossRef]

25. Chen, H.; Shi, K.; Zhong, S.; Liu, X. Error state convergence on master-slave generalized uncertain neural networks using robust nonlinear H_∞ Control theory. *IEEE Trans. Syst. Man Cybern. Syst.* **2020**, *50*, 2042–2055. [CrossRef]

26. Saravanakumar, R.; Syed Ali, M.; Karimi, H.R. Robust H_∞ control for a class of uncertain stochastic Markovian jump systems (SMJSs) with interval and distributed time-varying delays. *Int. J. Syst.* **2017**, *48* 862–872. [CrossRef]

27. Zhang, X.M.; Han, Q.L.; Zhang, B.L. An overview and deep investigation on sampled-data-based event-triggered control and filtering for networked systems. *IEEE Trans. Inf. Theory* **2017**, *13*, 4–16. [CrossRef]

28. Ge, X.H.; Han, Q.L.; Wang, Z.D. A threshold-parameter-dependent approach to designing distributed event-triggered H_∞ consensus filters over sensor networks. *IEEE Trans. Cybern.* **2019**, *49*, 1148–1159. [CrossRef]

29. Syed Ali, M.; Vadivel, R.; Saravanakumar, R. Event-triggered state estimation for Markovian jumping impulsive neural networks with interval time varying delays. *Int. J. Control* **2017**, *92*, 270–290.

30. Dimarogonas, D.V.; Frazzoli, E.; Johansson, K.H. Distributed event triggered control for multi-agent systems. *IEEE Trans. Autom. Control* **2012**, *57*, 1291–1297. [CrossRef]

31. Xie, D.; Xu, S.; Li, Z. Event-triggered consensus control for second-order multi-agent systems. *IET Control Theory Appl.* **2015**, *9*, 667–680. [CrossRef]

32. Zhou, B.; Liao, X.; Huang, T. Leader-following exponential consensus of general linear multi-agent systems via event-triggered control with combinational measurements, *Appl. Math. Lett.* **2015**, *40*, 35–39.

33. Qin, J.; Zheng, W.X.; Gao, H. On pinning synchronisability of complex networks with arbitrary topological structure. *Int. J. Syst. Sci.* **2011**, *42*, 1559–1571. [CrossRef]

34. Chakravartula, S.; Indic, P.; Sundaram, B.; Killingback, T. Emergence of local synchronization in neuronal networks with adaptive couplings. *PLoS ONE* **2017**, *12*, e0178975.

35. Liu, X.; Cao, J. Local synchronization of one-to-one coupled neural networks with discontinuous activations. *Cogn. Neurodyn.* **2011**, *5*, 13–20. [CrossRef] [PubMed]

36. Yang, Y.; Cao, J. Exponential lag synchronization of a class of chaotic delayed neural networks with impulsive effects. *Phys. A Stat. Mech. Its Appl.* **2007**, *386*, 492–502. [CrossRef]

37. He, W.; Qian, F.; Cao, J.; Han, Q.L. Impulsive synchronization of two nonidentical chaotic systems with time-varying delay. *Phys. Lett. A* **2011**, 498–504. [CrossRef]

38. Pan, L.; Cao, J.; Hu, J. Synchronization for complex networks with Markov switching via matrix measure approach. *Appl. Math. Model.* **2015**, *39*, 5636–5649. [CrossRef]
39. Du, C.; Liu L.; Shi, S. Synchronization of fractional-order complex chaotic system using active control method. In Proceedings of the 2019 IEEE 10th Annual Ubiquitous Computing, Electronics and Mobile Communication Conference (UEMCON), New York, NY, USA, 10–12 October 2019; pp. 817–823.
40. Xu, Z.; He, W. Quantized synchronization of master-slave systems under event-triggered control against DoS attacks. In Proceedings of the IECON 2020 the 46th Annual Conference of the IEEE Industrial Electronics Society, Singapore, 18–21 October 2020; pp. 3568–3573.
41. Du, S.; Dong, L.; Ho, D.W.C. Event-triggered control for output synchronization of heterogeneous network with input saturation constraint. In Proceedings of the IECON 2017—43rd Annual Conference of the IEEE Industrial Electronics Society, Beijing, China, 29 October–1 November 2017; pp. 5785–5790.
42. Yang, Y.; Long, Y. Event-triggered sampled-data synchronization of complex networks with time-varying coupling delays. *Adv. Differ. Equ.* **2020**, *312*, 2020. [CrossRef]
43. Lundstrom, B.; Higgs, M.; Spain, W.; Fairhall, A. Fractional differentiation by neocortical pyramidal neurons. *Nat. Neurosci.* **2008**, *11*, 1335–1342. [CrossRef]
44. Aguila-Camacho, N.; Duarte-Mermoud, M.A.; Gallegos, J.A. Lyapunov functions for fractional order systems. *Commun. Nonlinear Sci. Numer. Simul.* **2014**, *19*, 2951–2957. [CrossRef]
45. Zhang, S.; Yu, Y.; Yu, J. LMI conditions for global stability of fractional-order neural networks. *IEEE Trans. Neural Netw.* **2017**, *28*, 2423–2433. [CrossRef]
46. Boyd, B.; Ghoui, L.; Feron, E.; Balakrishnan, V. *Linear Matrix Inequalities in System and Control Theory*; SIAM: Philadephia, PA, USA, 1994.
47. Dinh, C.; Mai, H.V.T.; Duong, T.H. New results on stability and stabilization of delayed caputo fractional order systems with convex polytopic uncertainties. *J. Syst. Sci. Complex* **2020**, *33*, 563–583. [CrossRef]

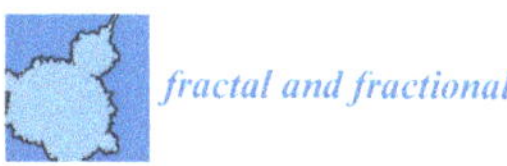

fractal and fractional

MDPI

Article

Finite Difference–Collocation Method for the Generalized Fractional Diffusion Equation

Sandeep Kumar [1], Rajesh K. Pandey [1,*], Kamlesh Kumar [2], Shyam Kamal [3] and Thach Ngoc Dinh [4,*]

[1] Department of Mathematical Sciences, Indian Institute of Technology (BHU) Varanasi, Varanasi 221005, Uttar Pradesh, India; sandeepkumar.rs.mat18@iitbhu.ac.in

[2] Department of Mathematics, Manav Rachana University, Faridabad 121004, Haryana, India; kkp.iitbhu@gmail.com

[3] Department of Electrical Engineering, Indian Institute of Technology (BHU) Varanasi, Varanasi 221005, Uttar Pradesh, India; shyamkamal.eee@iitbhu.ac.in

[4] Conservatoire National des Arts et Métiers (CNAM), Cedric-Laetitia, 292 Rue St-Martin, CEDEX 03, 75141 Paris, France

* Correspondence: rkpandey.mat@iitbhu.ac.in (R.K.P.); ngoc-thach.dinh@lecnam.net (T.N.D.)

Abstract: In this paper, an approximate method combining the finite difference and collocation methods is studied to solve the generalized fractional diffusion equation (GFDE). The convergence and stability analysis of the presented method are also established in detail. To ensure the effectiveness and the accuracy of the proposed method, test examples with different scale and weight functions are considered, and the obtained numerical results are compared with the existing methods in the literature. It is observed that the proposed approach works very well with the generalized fractional derivatives (GFDs), as the presence of scale and weight functions in a generalized fractional derivative (GFD) cause difficulty for its discretization and further analysis.

Keywords: generalized Caputo derivate; fractional diffusion equation; finite difference method; collocation method; error; stability and convergence analysis

Citation: Kumar, S.; Pandey, R.K.; Kumar, K.; Kamal, S.; Dinh, T.N. Finite Difference–Collocation Method for the Generalized Fractional Diffusion Equation. *Fractal Fract.* **2022**, *6*, 387. https://doi.org/10.3390/fractalfract6070387

Academic Editor: Tomasz W. Dłotko

Received: 25 May 2022
Accepted: 6 July 2022
Published: 11 July 2022

Publisher's Note: MDPI stays neutral with regard to jurisdictional claims in published maps and institutional affiliations.

1. Introduction

Fractional calculus (FC) is an important branch of applied mathematics, which deals with the arbitrary order derivative and integration [1–5]. Its applications in different fields are seen in biophysics [6], engineering [7], fluid mechanics and bioengineering [8,9] and other areas including image processing [10–13]. Most of these studies are mainly based on the traditional fractional derivatives, such as the Riemann–Liouville fractional derivative and the Caputo fractional derivative, etc. Recently, in [14] Agrawal discussed a new GFD, which generalized the traditional derivatives using weight and scale functions. The scale functions were used to compress and enlarge the domain for the close observation of physical phenomena, while the weight functions provide flexibility for the researchers to assess physical events at different times. Due to such behaviors of the scale and weight functions, the study of fractional partial differential equations (FPDE) using a GFD has attracted researchers in recent years. Several authors have developed numerical schemes for solving FPDEs involving GFDs. Here, we cite only few of them. In [15,16], Agrawal and coauthors presented the numerical solutions to the Berger's equation and the fractional advection–diffusion equation (FADE) with a generalized time-fractional derivative for the first time. The numerical solutions to these problems were obtained using the finite difference method. Further, the authors of [17,18], studied the time-fractional telegraph equations and fractional advection–diffusion equations in terms of GFDs and developed the higher order schemes to solve such equations. Kumar et al. [19] presented two numerical schemes to approximate the GFDs and obtained the convergence orders as $(2 - \alpha)$ and $(3 - \alpha)$, respectively. Further, these schemes were applied to solve the fractional integro-differential equations defined in terms of GFDs. In [20], Kumari and Pandey proposed an approximation method with a $(4 - \alpha)th$ order of convergence to approximate GFDs and applied it to solve the fractional advection–diffusion equations. Li and Wong [21] discussed a numerical scheme to find the solution of the generalized subdiffusion equation. Here, the authors used the generalized Grunwald–Letnikov approximation method to solve the problem. Some other recently presented methods for solving different types of fractional diffusion equations are summarized as follows: in [22], the authors discussed the matrix method for the reaction–diffusion equation that involved the Mittag–Leffler kernel. Duan et al. [23] solved the convection–diffusion equations using the Shannon–Runge–Kutta–Gill method. In [24], the authors discussed the solution to the fractional subdiffusion and the reaction–subdiffusion equations with a nonlinear source term using the Legendre collocation method. Lin and Xu [25] solved the time-fractional diffusion equation (TFDE) using the spectral method and finite difference method (FDM). In [26], Murio solved the time-fractional advection–diffusion equation (TFADE) by using implicit FDM. Sweilam et al. [27] developed the Crank–Nicolson FDM to solve a linear TFDE defined with a Caputo fractional derivative. In [28], Luchko solved the initial value and boundary value multi-term TFDE using the Fourier series method. Dubey et al. [29] proposed a residual power series method to obtain the solution of homogeneous and nonhomogeneous nonlinear fractional order partial differential equations.

In this paper, we study the generalized fractional diffusion equation (GFDE) obtained from the standard diffusion equation by replacing the first-order time derivative term with a fractional derivative of order γ, $0 < \gamma < 1$, given as,

$$* \mathcal{D}_t^\gamma v(x,t) = \frac{\partial^2 v(x,t)}{\partial x^2} + g(x,t), \ x \in [0,1], \quad t \in [0,\tau], \tag{1}$$

with initial and boundary conditions,

$$\begin{cases} v(x,0) = \eta_1(x), & 0 \leq x \leq 1, \\ v(0,t) = \eta_2(t), \ v(1,t) = \eta_3(t), & 0 \leq t \leq \tau, \end{cases} \tag{2}$$

where $g(x,t)$ is the source/sink function.

The motive of this paper is to construct an efficient method to obtain the numerical solution to the GFDE. The present method is based on the finite difference and collocation methods with Jacobi polynomials as the basis function. The outline of the paper is as follows: In Section 2, we discuss some basic facts about FC and Jacobi polynomials, which are needed throughout this paper. In Section 3, first, the finite difference method is used to discretize the time derivative. Second, on the space variable, we use the collocation method for numerical approximation. Further, we estimate the error and convergence analysis analytically, which ensures the numerical applicability of the proposed method. In Section 5, we present two numerical examples to validate the proposed method. Furthermore, we compare our results with a few other methods from the literature, which are presented in Section 5. Finally, in Section 6, the conclusions are discussed.

2. Preliminaries of Fractional Calculus and Jacobi Polynomials

In this section, we discuss some preliminary facts and basic properties of generalized fractional calculus and Jacobi polynomials [1,30].

2.1. Generalized Fractional Calculus

Definition 1 ([14])**.** *The Caputo derivative of a function $v(\tau)$ of order γ is defined as*

$$(\mathcal{D}_{0+}^\gamma v)(\tau) = (\mathcal{I}^{\delta-\gamma} \mathcal{D}^\delta v)(\tau) = \frac{1}{\Gamma(\delta-\gamma)} \int_0^\tau (\tau-s)^{\delta-\gamma-1} v^{(\delta)}(s)ds, \ \tau > 0, \tag{3}$$

where $\delta - 1 < \gamma \leq \delta$ and $\delta \in \mathbb{Z}^+$.

Definition 2 ([14]). *A generalized fractional derivative of order $\gamma > 0$ of a function $v(\tau)$ with respect to weight function $\omega(\tau)$ and scale function $z(\tau)$ is defined as*

$$(\mathcal{D}^{\gamma}_{[z;\omega;L]}v)(\tau) = [\omega(\tau)]^{-1}\left[\left(\frac{D_t}{z'(\tau)}\right)^{\gamma}(\omega(\tau)v(\tau))\right]. \tag{4}$$

Definition 3. *A generalized fractional integral of a function $v(\tau)$ of fractional order $\gamma > 0$ is defined as*

$$(\mathcal{I}^{\gamma}_{[z;\omega]}v)(\tau) = \frac{[\omega(\tau)]^{-1}}{\Gamma(\gamma)}\int_0^{\tau}\frac{\omega(s)z'(s)v(s)}{[z(\tau)-z(s)]^{1-\gamma}}ds. \tag{5}$$

Definition 4 ([14]). *The left/forward generalized Caputo fractional derivative of order $\gamma > 0$ of function $v(\tau)$ with respect to weight function $\omega(\tau)$ and scale function $z(\tau)$ is defined by*

$$(*\mathcal{D}^{\gamma}_{0+}v)(\tau) = \mathcal{I}^{\delta-\gamma}_{0+;[z;\omega]}\mathcal{D}^{\delta}_{0+;[z;\omega;L]} = \frac{[\omega(\tau)]^{-1}}{\Gamma(\delta-\gamma)}\int_0^{\tau}\frac{(\omega(s)v(s))^{(\delta)}}{(z(\tau)-z(s))^{\gamma+1-\delta}}ds, \tag{6}$$

where $\delta - 1 < \gamma \leq \delta,\ \delta \in \mathbb{Z}^+$.

For a particular choice of the weight and scale functions $(z(\tau) = \tau,\ \omega(\tau) = 1)$, Equation (6) reduces to the Caputo derivative. We have considered the weight and scale sufficiently good such that the integral exists in GFD.

2.2. Jacobi Polynomials

Definition 5. *The Jacobi polynomials [31] $\mathcal{J}_n^{\alpha,\beta}(z)$ for indices $\alpha,\ \beta > -1$ and degree n are the solutions of the Sturm–Liouville problems. These are orthogonal polynomials with respect to the Jacobi weight function $\omega(z) = (1-z)^{\alpha}(1+z)^{\beta}$ in interval $[-1,1]$, defined as follows:*

$$\mathcal{J}_n^{\alpha,\beta}(z) = \frac{\Gamma(\alpha+n+1)}{n!\Gamma(\alpha+\beta+n+1)}\sum_{m=0}^{n}\binom{n}{m}\frac{\Gamma(\alpha+\beta+n+m+1)}{\Gamma(\alpha+m+1)}\left(\frac{z-1}{2}\right)^m. \tag{7}$$

The kth derivative of Jacobi polynomials defined as

$$\frac{d^k}{dz^k}\mathcal{J}_n^{\alpha,\beta}(z) = \frac{\Gamma(\alpha+\beta+n+1+k)}{2^k\Gamma(\alpha+\beta+n+1)}\mathcal{J}_{n-k}^{(\alpha+k,\beta+k)}(z),\ k \in \mathbb{N}, \tag{8}$$

satisfy the recurrence relation

$$\mathcal{J}_{n+1}^{\alpha,\beta}(z) = (\mathcal{A}_n z - \mathcal{B}_n)\mathcal{J}_n^{\alpha,\beta}(z) - \rho_n\mathcal{J}_{n-1}^{\alpha,\beta}(z),\ n \geq 1, \tag{9}$$

where

$$\mathcal{J}_0^{\alpha,\beta}(z) = 1,\ \ \mathcal{J}_1^{\alpha,\beta}(z) = \frac{1}{2}(\alpha+\beta+z) + \frac{1}{2}(\alpha-\beta),$$

$$\mathcal{A}_n = \frac{(2n+\alpha+\beta+1)(2n+\alpha+\beta+2)}{2(n+1)(n+\alpha+\beta+1)},\ \ \mathcal{B}_n = \frac{(2n+\alpha+\beta+1)(\alpha^2-\beta^2)}{2(n+1)(n+\alpha+\beta+1)(2n+\alpha+\beta)},$$

and

$$\rho_n = \frac{(2n+\alpha+\beta+2)(n+\alpha)(n+\alpha)}{(n+1)(n+\alpha+\beta+1)(2n+\alpha+\beta)}.$$

For the transformation of the interval $[-1,1]$ to $[0,1]$, we use the relation, $z = 2x - 1$. The recurrence relation (9) becomes

$$\mathcal{J}_{n+1}^{\alpha,\beta}(x) = (\xi_n x - \mathcal{K}_n)\mathcal{J}_n^{\alpha,\beta}(x) - \rho_n\mathcal{J}_{n-1}^{\alpha,\beta}(x),\ n \geq 1, \tag{10}$$

where

$$\zeta_n = \frac{(2n + \alpha + \beta + 1)(2n + \alpha + \beta + 2)}{(n+1)(n + \alpha + \beta + 1)},$$

$$\mathcal{K}_n = \frac{(2n + \alpha + \beta + 1)(2n^2 + (1 + \beta)(\alpha + \beta) + 2n(\alpha + \beta + 1))}{(n+1)(n + \alpha + \beta)(2n + \alpha + \beta)}.$$

It satisfies the orthogonality relation

$$\int_0^1 \mathcal{J}_n^{\alpha,\beta}(x)\mathcal{J}_m^{\alpha,\beta}(x)w_1^{\alpha,\beta}(x)dx = \delta_{n,m}\mathcal{H}_n^{\alpha,\beta}, \ \alpha, \ \beta > -1, \tag{11}$$

where $\delta_{n,m}$ is the Kronecker delta function, $w_1^{\alpha,\beta}(x) = x^\beta(1-x)^\alpha$ is the weight function, and

$$\mathcal{H}_n^{\alpha,\beta} = \frac{\Gamma(n+1+\beta)\Gamma(n+\alpha+1)}{(2n+1+\alpha+\beta)n!\Gamma(n+1+\alpha+\beta)}. \tag{12}$$

The summation form of the Jacobi polynomials $\mathcal{J}_n^{\alpha,\beta}(x)$ is written as

$$\mathcal{J}_n^{\alpha,\beta}(x) = \sum_{k=0}^{j}(-1)^{k+j}\frac{\Gamma(k+j+1+\alpha+\beta)\Gamma(j+1+\beta)}{k!\Gamma(j+1+\alpha+\beta)\Gamma(k+1+\beta)}, \tag{13}$$

and the pth derivative of Equation (13) in $[0,1]$, can be further rewritten in terms of x as

$$\frac{d^p}{dx^p}\mathcal{J}_n^{\alpha,\beta}(x) = \frac{\Gamma(\alpha+\beta+n+1+p)}{\Gamma(\alpha+\beta+n+1)}\mathcal{J}_{n-p}^{(\alpha+p,\beta+p)}(x), \ \ p \in \mathbb{N}. \tag{14}$$

The value of the shifted Jacobi polynomials at the end points are given as

$$\mathcal{J}_n^{\alpha,\beta}(0) = (-1)^n\frac{\Gamma(1+n+\beta)}{\Gamma(1+\beta)n!}, \ \ \mathcal{J}_n^{\alpha,\beta}(1) = \frac{\Gamma(1+n+\alpha)}{\Gamma(1+\alpha)n!}. \tag{15}$$

Let $v(x)$ be a square integrable function defined on $[0,1]$, then

$$v(x) = \sum_{i=0}^{\infty} c_i\mathcal{J}_i^{\alpha,\beta}(x), \tag{16}$$

where c_i are the unknown coefficients $(c_i, i = 0, 1, 2 \dots)$ determined by the relation

$$c_i = \frac{1}{\mathcal{H}_i^{\alpha,\beta}}\int_0^1 v(x)w^{\alpha,\beta}(x)\mathcal{J}_i^{\alpha,\beta}(x)dx, \ i = 0, 1, 2 \dots \ . \tag{17}$$

Truncating the series in Equation (16) up to $(m+1)$ terms, the approximation of $v(x)$ is given as

$$v_m(x) = \sum_{i-0}^{m} c_i\mathcal{J}_i^{\alpha,\beta}(x). \tag{18}$$

3. Numerical Scheme and Stability Analysis

3.1. Discretization in the Time Direction

For the discretization of the GFD, we follow the discretization process as discussed in Refs. [19,32]. We split the time interval $[0, \tau]$ into $\mathcal{M}$ equal parts having step size $\Delta t = \frac{\mathcal{T}}{\mathcal{M}}$, $\mathcal{M} \in \mathbb{Z}^+$, and choose the node points as $t_r = r(\Delta t)$, $r = 0, 1, 2, \dots, \mathcal{M}$, with starting point $t_0 = 0$. Assuming $w(t) > 0$, $\gamma \subset (0,1)$, and $z(t)$ is a monotonic increasing function on $[0, \mathcal{T}]$, such that $\eta = z(s)$; then, $s = z^{-1}(\eta)$. The discretization of the GFD of $v(t)$ at node t_r is given as

$$
*\mathcal{D}_t^\gamma v(t_r) = \frac{[w(t_r)]^{-1}}{\Gamma(1-\gamma)} \sum_{l=1}^{r} \int_{t_{l-1}}^{t_l} \frac{[w(s)v(s)]'}{[z(t_r)-z(s)]^\gamma} ds,
$$

$$
= \frac{[w(t_r)]^{-1}}{\Gamma(1-\gamma)} \sum_{l=1}^{r} \int_{z(t_{l-1})}^{z(t_l)} \frac{1}{[z(t_r)-\eta]^\gamma} \cdot \frac{d[w(z^{-1}(\eta))v(z^{-1}(\eta))]}{dz^{-1}(\eta)} dz^{-1}(\eta),
$$

$$
= \frac{[w(t_r)]^{-1}}{\Gamma(1-\gamma)} \sum_{l=1}^{r} \frac{w(t_l)v(t_l)-w(t_{l-1})v(t_{l-1})}{z(t_l)-z(t_{l-1})} \int_{z(t_{l-1})}^{z(t_l)} \frac{1}{[z(t_r)-\eta]^\gamma} d\eta + \mathcal{R}_r,
$$

$$
= \frac{[w(t_r)]^{-1}}{\Gamma(2-\gamma)} \sum_{l=1}^{r} q_l[w(t_l)v(t_l)-w(t_{l-1})v(t_{l-1})] + \mathcal{R}_r, \tag{19}
$$

where

$$
q_l = \frac{[z(t_r)-z(t_{l-1})]^{1-\gamma} - [z(t_r)-z(t_l)]^{1-\gamma}}{z(t_l)-z(t_{l-1})}, \ l=1,2,\ldots r, \tag{20}
$$

and $\mathcal{R}_r$ is the truncation error given by

$$
\mathcal{R}_r = \frac{[w(t_r)]^{-1}}{\Gamma(1-\gamma)} \sum_{l=1}^{r} \int_{z(t_{l-1})}^{z(t_l)} \frac{1}{[z(t_r)-\eta]^\gamma} \left[\frac{d[w(z^{-1}(\eta))v(x,z^{-1}(\eta))]}{d\eta} \right.
$$
$$
\left. - \frac{w(t_l)v(t_l)-w(t_{l-1})v(t_{l-1})}{z(t_l)-z(t_{l-1})} \right] d\eta. \tag{21}
$$

Lemma 1. *The coefficient q_l, $l=1,2,\ldots,r$, given by Equation (20), satisfies $q_r > q_{r-1} > \ldots > q_0 > 0$.*

Proof. For the proof of Lemma 1, see Ref [32]. $\square$

3.2. Approximation in Space Direction

We apply the collocation method to approximate the spatial domain of Equation (1) with Jacobi polynomials. We consider the approximate solution $v_\mathcal{N}(x,t)$ of the form,

$$
v_\mathcal{N}(x,t) = \sum_{s_1=0}^{\mathcal{N}} c_{s_1}(t)\mathcal{J}_{s_1}^{\alpha,\beta}(x). \tag{22}
$$

From Equations (22) and (1), we obtain

$$
*\mathcal{D}_t^\gamma v_\mathcal{N}(x,t) = \frac{\partial^2 v_\mathcal{N}(x,t)}{\partial x^2} + g(x,t), \ t \in [0,\tau], \tag{23}
$$

with the initial and boundary conditions of Equation (1),

$$
v_\mathcal{N}(x_0,t) = \sum_{s_1=0}^{\mathcal{N}} c_{s_1}(t)\mathcal{J}_{s_1}^{\alpha,\beta}(x_0), \tag{24}
$$

$$
v_\mathcal{N}(x_\mathcal{N},t) = \sum_{s_1=0}^{\mathcal{N}} c_{s_1}(t)\mathcal{J}_{s_1}^{\alpha,\beta}(x_\mathcal{N}). \tag{25}
$$

From Equations (1), (19), and (22), we have the semi-discretized scheme as follows

$$
\sum_{s_1=0}^{\mathcal{N}} \left[\frac{[w(t_r)]^{-1}}{\Gamma(2-\gamma)} \sum_{l=1}^{r} q_l[w(t_l)c_{s_1}(t_l)-w(t_{l-1})c_{s_1}(t_{l-1})] \right] \mathcal{J}_{s_1}^{\alpha,\beta}(x)
$$
$$
= \sum_{s_1=0}^{\mathcal{N}} c_{s_1}(t_r) \frac{\Gamma(\alpha+\beta+s_1+3)}{\Gamma(\alpha+\beta+s_1+1)} \mathcal{J}_{s_1-2}^{(\alpha+2,\beta+2)}(x)
$$
$$
+ \mathcal{R}_r + \mathcal{R}_s + g(x,t_r), \ r=1,2,\ldots,\mathcal{M}, \tag{26}
$$

where $\mathcal{R}_s$ denotes the error term in the space direction arising due to replacing $v(x,t)$ with $v_{\mathcal{N}}(x,t)$.

Neglecting the error part, we obtain the fully discretized scheme of Equation (1) by the collocation method [32,33]. We choose the collocation points such that the stability is unchanged. So, we choose the collocation point of the form x_i, $i = 1, 2, \ldots \mathcal{N} - 1$, which are the roots of the nth degree Jacobi polynomials, and x_0, $x_{\mathcal{N}}$ are the boundary conditions. Thus, for $(x_i, t_r) \in (0,1) \times [0,\tau]$, $i = 1, \ldots, \mathcal{N}$; $r = 1, \ldots \mathcal{M}$, it holds that

$$\sum_{s_1=0}^{\mathcal{N}} \left[\frac{[w(t_r)]^{-1}}{\Gamma(2-\gamma)} \sum_{l=1}^{r} q_l [w(t_l) c_{s_1}(t_l) \mathcal{J}_{s_1}^{\alpha,\beta}(x_i) - w(t_{l-1}) c_{s_1}(t_{l-1}) \mathcal{J}_{s_1}^{\alpha,\beta}(x_i)] \right]$$
$$= \sum_{s_1=0}^{\mathcal{N}} c_{s_1}(t_r) \frac{\Gamma(\alpha+\beta+s_1+3)}{\Gamma(\alpha+\beta+s_1+1)} \mathcal{J}_{s_1-2}^{(\alpha+2,\beta+2)}(x_i) + g(x_i, t_r), \quad r = 1, 2, \ldots, \mathcal{M}, \tag{27}$$

and the initial and boundary conditions become

$$\begin{cases} v_{\mathcal{N}}(x_i, 0) = \sum_{s=0}^{\mathcal{N}} c_{s_1}(0) \mathcal{J}_{s_1}^{\alpha,\beta}(x_i) = \eta_1(x_i), \\ v_{\mathcal{N}}(x_0, t_r) = \sum_{s_1=0}^{\mathcal{N}} c_{s_1}(t_r) \mathcal{J}_{s_1}^{\alpha,\beta}(x_0) = \eta_2(t_r), \\ v_{\mathcal{N}}(x_{\mathcal{N}}, t_r) = \sum_{s_1=0}^{\mathcal{N}} c_{s_1}(t_r) \mathcal{J}_{s_1}^{\alpha,\beta}(x_{\mathcal{N}}) = \eta_3(t_r). \end{cases} \tag{28}$$

In this way, from Equations (27)–(28) we have a system of $(\mathcal{N}+1)$ linear difference equations in unknown coefficients c_{s_1}, $s_1 = 0, 1, 2 \ldots \mathcal{N}$. We can find the value of the unknown coefficients by solving the system of linear equations using any standard method. Hence, the approximate solution can be found from Equation (22).

4. Error and Convergence Analysis

In this section, we discuss the error and convergence analysis of the proposed numerical method for Equation (1). We prove the error and convergence analysis analytically with the help of the following lemma and theorems.

Lemma 2 ([32]). *The truncation error $\mathcal{R}_r$ defined by Equation (21) satisfies*

$$\mathcal{R}_r \leq \left[\frac{1}{8w_r \Gamma(1-\gamma)} + \frac{\gamma}{2w_r \Gamma(3-\gamma)} \right] \max_{t_0 \leq \eta \leq t_r} |\mathcal{U}''(\eta) \mathcal{L}| \Delta t^{2-\gamma}, \tag{29}$$

where $\mathcal{U}(\eta)$ is the approximating function, w_r is the weight function at node t_r for $r = 1, 2 \ldots \mathcal{M}$, and $\mathcal{L}$ is the Lipschitz constant on the interval $[t_{l-1}, t_l]$.

Proof. For the detailed proof of this lemma, we refer to [32]. $\square$

Theorem 1. *The error in approximation of the function $v(x)$ by the first m terms of the series in Equation (16) is bounded by the sum of the absolute values of all the neglected coefficients in the series, i.e.,*

$$\mathcal{E}_\tau(\mathcal{N}) = |v(x) - v_{\mathcal{N}}(x)| \leq \sum_{i=m+1}^{\infty} |c_i|, \tag{30}$$

$\forall\, v(x), \forall m,$ *and* $x \in [0,1]$.

Proof. The proof is trivial since $|\mathcal{J}_i^{\alpha,\beta}(x)| \leq 1, \forall x \in [0,1]$ and $i \geq 0$. $\square$

Theorem 2. *Let $v(x)$ be the square integrable function defined on $[0,1]$ and $|v(x)| \leq \mathcal{M}_1$, where $\mathcal{M}_1$ is constant. Then, the $v(x)$ can be expanded with an infinite sum of Jacobi polynomials, and the infinite series converges to $v(x)$ uniformly, i.e.,*

$$v(x) = \sum_{i=0}^{\infty} c_i \mathcal{J}_n^{\alpha,\beta}(x), \tag{31}$$

where

$$|c_i| \leq \frac{\mathcal{M}_1 \Gamma(1+\beta)}{\Gamma(2+\alpha+\beta)} \frac{1}{i^3}, \ i > 1,$$

and $\mathcal{E}_m \to 0$.

Proof. From Equations (16) and (18), we have

$$v_m(x) = \sum_{i=0}^{m} c_i \mathcal{J}_i^{\alpha,\beta}(x), \tag{32}$$

where c_i are the unknown coefficients. Furthermore, from Equation (17), we obtain

$$c_i = \frac{1}{\mathcal{H}_i^{\alpha,\beta}} \int_0^1 (x)^\beta (1-x)^\alpha v(x) \mathcal{J}_i^{\alpha,\beta}(x) dx,$$

$$|c_i| = \left| \frac{1}{\mathcal{H}_i^{\alpha,\beta}} \int_0^1 (x)^\beta (1-x)^\alpha v(x) \mathcal{J}_i^{\alpha,\beta}(x) \right| dx,$$

$$\leq \frac{\mathcal{M}_1}{\mathcal{H}_i^{\alpha,\beta}} \int_0^1 \left| (x)^\beta (1-x)^\alpha \mathcal{J}_i^{\alpha,\beta}(x) \right| dx, \tag{33}$$

$$\leq \frac{\mathcal{M}_1}{\mathcal{H}_i^{\alpha,\beta}} \frac{\Gamma(\alpha+i+1)}{i! \Gamma(\alpha+\beta+i+1)} \sum_{m=0}^{i} \binom{i}{m} \frac{\Gamma(\alpha+\beta+i+m+1)}{\Gamma(\alpha+m+1)} \int_0^1 |x^\beta (1-x)^\alpha (x-1)^m| dx,$$

$$\leq \frac{\mathcal{M}_1(2i+1+\alpha+\beta)\Gamma(i+1+\alpha+\beta)\Gamma(1+\beta)}{\Gamma(i+1+\beta)\Gamma(2+\alpha+\beta)} \frac{1}{i^4},$$

$$\leq \frac{\mathcal{M}_1 \Gamma(1+\beta)}{\Gamma(2+\alpha+\beta)} \frac{1}{i^3}.$$

Hence, the series $v_m(x)$ converges to $v(x)$ uniformly. $\square$

Theorem 3. *Let* $h(t)$ *be* $\mathcal{N}$ *times differentiable function defined on interval* $[0, \tau]$. *Let* $v_{\mathcal{N}}(t) = \sum_{j_1}^{\mathcal{N}} c_{j_1} \mathcal{J}_{j_1}^{\alpha,\beta}(t)$ *be the approximation of* $h(t)$, *then*

$$\|h(t) - v_{\mathcal{N}}(t)\| \leq \frac{\mathcal{M} \mathcal{S}^{n+1}}{((\mathcal{N}+1)!)} \sqrt{\mathcal{B}_\tau(1+\alpha, \ 1+\beta)}, \tag{34}$$

where $\mathcal{M} = \max_{t \in [0,\tau]} h^{\mathcal{N}+1}(t)$, $\mathcal{S} = \max\{\tau - t_0, \ t_0\}$ *and* $\mathcal{B}_\tau(1+\alpha, \ 1+\beta)$ *denote the incomplete Beta function. At* $\tau = 1$, *it reduces to the standard Beta function.*

Proof. By the Taylor series expansion, we have

$$h(t) = h(t_0) + h'(t_0)(t-t_0) + \ldots + h^{\mathcal{N}}(t_0)\frac{(t-t_0)^{\mathcal{N}}}{\mathcal{N}!} + h^{\mathcal{N}+1}(t)\frac{(t-t_0)^{\mathcal{N}+1}}{(\mathcal{N}+1)!}, \tag{35}$$

where $t_0 \in [0, \tau]$ and $\zeta \in [t_0, \ t]$. Let

$$\mathcal{P}_{\mathcal{N}}(t) = f(t_0) + f'(t_0)(t-t_0) + \ldots + \frac{f^{\mathcal{N}}(t_0)(t-t_0)^{\mathcal{N}}}{\mathcal{N}!}, \tag{36}$$

then

$$|h(t) - \mathcal{P}_{\mathcal{N}}(t)| = \left| h^{\mathcal{N}+1}(t)\frac{(t-t_0)^{\mathcal{N}+1}}{(\mathcal{N}+1)!} \right|. \tag{37}$$

Since, we assume that $v_{\mathcal{N}}(t)$ is the best square approximation of $h(t)$, we have

$$\|h(t) - v_{\mathcal{N}}(t)\|^2 \leq \|h(t) - \mathcal{P}_{\mathcal{N}}(t)\|^2,$$

$$= \int_0^{\tau} w(t)[h(t) - \mathcal{P}_{\mathcal{N}}(t)]^2 dt,$$

$$= \int_0^{\tau} \left[h^{\mathcal{N}+1}(t) \frac{(t - t_0)^{\mathcal{N}+1}}{(\mathcal{N}+1)!} \right]^2 dt,$$

$$\leq \frac{\mathcal{M}^2}{((\mathcal{N}+1)!)^2} \int_0^{\tau} (t - t_0)^{2n+2} w(t) dt,$$

$$\leq \frac{\mathcal{M}^2 S^{2n+2}}{((\mathcal{N}+1)!)^2} \int_0^{\tau} (t)^{\beta}(1 - t)^{\alpha} dt,$$

$$= \frac{\mathcal{M}^2 S^{2n+2}}{((\mathcal{N}+1)!)^2} \mathcal{B}_{\tau}(1 + \alpha, 1 + \beta). \tag{38}$$

Hence,

$$\|h(t) - v_{\mathcal{N}}(t)\| \leq \frac{\mathcal{M} S^{n+1}}{((\mathcal{N}+1)!)} \sqrt{\mathcal{B}_{\tau}(1 + \alpha,\, 1 + \beta)}. \tag{39}$$

$\square$

Theorem 4. *Let $v(x, t)$ be a continuous function satisfying the conditions (2) for any t, and $g(x, t)$ is continuous. Assuming $v_{\mathcal{N}}(x, t_r) = v_{\mathcal{N}}^r(x) = \sum_{s_1=0}^{\mathcal{N}} c_{s_1}(t_r) \mathcal{J}_{s_1}^{\alpha,\beta}(x)$ is the numerical approximation of the scheme (27), then the scheme (27) is unconditionally stable, and for any $r \geq 0$, it holds that*

$$\|v_{\mathcal{N}}^r(x)\|_{L^2} \leq \frac{w_0}{w_r} \|v_{\mathcal{N}}^0(x)\|_{L^2} + \sum_{l=1}^{r-1} \frac{h_l}{w_r} \|g^l\|_{L^2} + \frac{\eta_r}{q_r w_r} \|g^r\|_{L^2}, \tag{40}$$

where $g(x, t_r) = g^r(x)$, and $h_l = \left(\frac{1}{q_l} - \frac{1}{q_{l+1}} \right) \eta_l,\ l = 1, 2, \ldots r - 1$.

Proof. The proof of this theorem is similar to the Theorem 1 of [32]. In [32], the authors proved the stability for the time-fractional KdV equation. Here, we extend the proof for the time-fractional diffusion equation. To prove Theorem 4, we first rewrite Equation (26) over the summation up to time step t_{r-1} in discrete form. Thus, we have

$$\frac{1}{\eta_r} q_r w_r v(x, t_r) = \frac{1}{\eta_r} \left[\sum_{l=1}^{r-1} (q_{l+1} - q_l) w_l v(x, t_l) + q_1 w_0 v(x, t_0) \right] + v''(x, t_r) + g(x, t_r), \tag{41}$$

where $r = 1, 2, ..\mathcal{M}$, and $\eta_l = w_l \Gamma(2 - \gamma)$.

Let $w^{\alpha,\beta}(x) = x^{\beta}(1 - x)^{\alpha}$, and $u_{\mathcal{N}-2}^r(x) = u_{\mathcal{N}-2}(x, t_r)$ is the polynomial of $\mathcal{N} - 2$ degree satisfying $v_{\mathcal{N}}^r(x) = u_{\mathcal{N}-2}^r(x) w^{\alpha,\beta}(x)$. Multiplying both sides of Equation (41) by $u_{\mathcal{N}-2}(x_i, t_r) w^{\alpha,\beta}(x_i)$, and taking the summation on i from 0 to $\mathcal{N}$, we have

$$\sum_{i=0}^{N} \left[\frac{1}{\eta_r} q_r w_r v(x_i, t_r) \right] u_{\mathcal{N}-2}(x_i, t_r) w^{\alpha,\beta}(x_i) = \sum_{i=0}^{N} \frac{1}{\eta_r} \left[\sum_{l=1}^{r-1} (q_{l+1} - q_l) w_l v(x_i, t_l) + q_1 w_0 v(x_i, t_0) \right] \tag{42}$$

$$+ \frac{1}{\eta_r} \left[v''(x_i, t_r) + g(x_i, t_r) u_{\mathcal{N}-2}(x_i, t_r) w^{\alpha,\beta}(x_i) \right],$$

where $w^{\alpha,\beta}(x_i)$ is the corresponding weight function. Since the degree of $v_{\mathcal{N}}^r(x)$ does not exceed $\mathcal{N} + 1$, then from Equation (11),

$$(v_{\mathcal{N}}^r, u_{\mathcal{N}-2}^r)_{w^{\alpha,\beta}(x)} = (v_{\mathcal{N}}^r, v_{\mathcal{N}}^r). \tag{43}$$

It can be easily shown that

$$\int_0^1 v''(x,t_r)u_{\mathcal{N}-2}(x,t_r)w^{\alpha,\beta}(x,t_r)\,dx = \int_0^1 v''(x,t_r)v(x,t_r)w^{\alpha,\beta}(x,t_r)\,dx = 0. \tag{44}$$

Now, the discrete form of the Equation (42) at the nodes x_i can be rewritten as

$$q_r w_r \|v_{\mathcal{N}}^r(x)\|_{L^2} \leq \sum_{l=1}^{r-1}(q_{l+1}-q_l)w_l\|v_{\mathcal{N}}^l(x)\|_{L^2} + q_1 w_0\|v_{\mathcal{N}}^0(x)\|_{L^2} + \eta_r\|g^r(x)\|_{L^2}, \tag{45}$$

by using the Cauchy–Schwartz inequality and Lemma (1). The remaining part of the proof can be completed following similar steps to those shown in Theorem 1 of [32]. $\square$

5. Numerical Results

In this section, we provide two numerical examples to validate the presented finite difference–collocation method. In the given examples, we calculate the maximum absolute error (MAE), absolute error (AE), and the order of convergence (CO) for each example. With the help of the MAE and CO, we analyze the error and convergence analysis numerically. Furthermore, we have plotted the graphs of the numerical solutions by changing the various parameters of γ, α, β, and scale function $z(t)$. For the numerical simulations, we take the weight function $w(t) = 1$. All numerical simulations were performed with Mathematica software.

The MAE at time t is given by

$$\mathcal{E}_n(t) = \max_{0 \leq x \leq 1}|v(x,t) - v_{\mathcal{N}}(x,t)|, \tag{46}$$

and the order of convergence is defined by

$$CO = \frac{\log\left(\frac{\mathcal{E}_{n_1}(t)}{\mathcal{E}_{n_2}(t)}\right)}{\log\left(\frac{n_2}{n_1}\right)}, \tag{47}$$

where $v(x,t)$ and $v_{\mathcal{N}}(x,t)$ are the exact and approximate solutions, respectively. $\mathcal{E}_{n_1}(t)$ and $\mathcal{E}_{n_2}(t)$ are the MAEs for two consecutive values n_1 and n_2.

Example 1. *Here, we consider the generalized version of the problem given in [34] as*

$$* \mathcal{D}_t^\gamma v(x,t) - \frac{x^2}{2}\frac{\partial^2 v(x,t)}{\partial x^2} = 0, \ t \in [0,\tau], \tag{48}$$

the initial and boundary conditions are given by

$$\begin{cases} v(x,0) = x^2,\ 0 \leq x \leq 1, \\ v(0,t) = 0,\ v(1,t) = e^{t+\gamma},\ 0 \leq t \leq \tau. \end{cases} \tag{49}$$

The exact solution of Example (1) is $x^2 e^{t+\gamma}$. This problem is solved for various values of $\mathcal{N}$, $\gamma, z(t)$, and t. In Tables 1 and 2, we compare the results obtained by our technique to the given methods in [35–37] at $\gamma = 0$. We observed that the results obtained by the present method (PM) provided a better approximation for this problem. In Table 3, we have discussed the MAE and CO for various values of γ and $\mathcal{M}$. Further, in Figure 1, we plotted the AE comparison graphs for different values of γ and observe that the numerical approximation showed good agreement with the exact solution. Figure 2 shows the behavior of the AEs for various values of $\mathcal{N}$ with a fixed $\gamma = 0.2$. We observe from Figure 2 that the numerical solution at $\mathcal{N} = 6, 8$ showed good agreement with the exact solution at

$\gamma = 0.2$. Finally, in Figure 3, we plotted the numerical solutions for the various values of $\gamma = 0.1, 0.3, 0.5, 0.7, 0.9$ for a fixed value of $\mathcal{N} = 5$.

Table 1. Comparison of MAE for Example 1 with $\gamma = 0.75$, $z(t) = t$.

t	x	Method of [35]	Method of [36]	Method of [37]	Present Method
0.25	0.3	1.312×10^{-1}	1.346×10^{-1}	1.293×10^{-1}	$\mathbf{8.009} \times 10^{-2}$
.	0.6	4.957×10^{-1}	5.385×10^{-1}	5.175×10^{-1}	$\mathbf{1.313} \times 10^{-2}$
.	0.9	1.055×10^{-1}	1.211×10^{0}	1.164×10^{0}	$\mathbf{6.041} \times 10^{-2}$
0.5	0.3	1.685×10^{-1}	1.795×10^{-1}	1.695×10^{-1}	$\mathbf{6.342} \times 10^{-2}$
.	0.6	6.303×10^{-1}	7.183×10^{-1}	6.780×10^{-1}	$\mathbf{9.543} \times 10^{-2}$
.	0.9	1.352×10^{0}	1.616×10^{-1}	1.525×10^{-1}	$\mathbf{5.315} \times 10^{-2}$
0.75	0.3	2.118×10^{-1}	2.313×10^{-1}	2.154×10^{-1}	$\mathbf{5.242} \times 10^{-2}$
.	0.6	7.962×10^{-1}	9.255×10^{-1}	8.618×10^{-1}	$\mathbf{6.875} \times 10^{-2}$
.	0.9	1.733×10^{0}	2.082×10^{0}	1.939×10^{0}	$\mathbf{4.891} \times 10^{-2}$
1	0.3	2.645×10^{-1}	2.909×10^{-1}	2.687×10^{-1}	$\mathbf{4.349} \times 10^{-2}$
.	0.6	9.745×10^{-1}	1.163×10^{0}	1.075×10^{0}	$\mathbf{4.802} \times 10^{-2}$
.	0.9	2.014×10^{0}	2.618×10^{0}	2.419×10^{0}	$\mathbf{1.048} \times 10^{-2}$

Table 2. Comparison of MAE for Example 1 with $\gamma = 0.9$, $z(t) = t$.

t	x	Method of [35]	Method of [36]	Method of [37]	Present Method
0.25	0.3	1.122×10^{-1}	1.218×10^{-1}	1.210×10^{-1}	$\mathbf{9.795} \times 10^{-2}$
.	0.6	4.762×10^{-1}	4.872×10^{-1}	4.841×10^{-1}	$\mathbf{1.543} \times 10^{-1}$
.	0.9	1.046×10^{0}	1.096×10^{0}	1.089×10^{0}	$\mathbf{5.303} \times 10^{-3}$
0.5	0.3	1.564×10^{-1}	1.588×10^{-1}	1.567×10^{-1}	$\mathbf{7.852} \times 10^{-2}$
.	0.6	6.086×10^{-1}	6.355×10^{-1}	6.268×10^{-1}	$\mathbf{1.129} \times 10^{-1}$
.	0.9	1.342×10^{0}	1.429×10^{0}	1.410×10^{0}	$\mathbf{9.122} \times 10^{-2}$
0.75	0.3	1.9948×10^{-1}	2.041×10^{-1}	1.998×10^{-1}	$\mathbf{6.677} \times 10^{-2}$
.	0.6	7.761×10^{-1}	8.165×10^{-1}	7.992×10^{-1}	$\mathbf{9.517} \times 10^{-2}$
.	0.9	1.722×10^{0}	1.837×10^{0}	1.798×10^{0}	$\mathbf{5.475} \times 10^{-2}$
1	0.3	2.529×10^{-1}	2.588×10^{-1}	2.517×10^{-1}	$\mathbf{5.719} \times 10^{-2}$
.	0.6	9.938×10^{-1}	1.035×10^{0}	1.007×10^{0}	$\mathbf{6.471} \times 10^{-2}$
.	0.9	2.144×10^{0}	2.329×10^{0}	2.265×10^{0}	$\mathbf{1.964} \times 10^{-2}$

Table 3. The CO and MAE for Example 1 with various values of γ and $z(t) = t^2$.

	$\gamma = 0.3$		$\gamma = 0.5$		$\gamma = 0.7$	
$\mathcal{M}$	MAE	CO	MAE of PM	CO	MAE of PM	CO
50	2.637×10^{-1}	$\ldots$	4.921×10^{-1}	$\ldots$	7.994×10^{-1}	$\ldots$
100	9.246×10^{-2}	1.617	1.839×10^{-1}	1.419	3.310×10^{-1}	1.263
200	2.919×10^{-2}	1.663	6.782×10^{-2}	1.439	1.380×10^{-1}	1.270
400	9.186×10^{-3}	1.667	2.464×10^{-2}	1.460	5.677×10^{-2}	1.282
800	2.850×10^{-3}	1.688	8.878×10^{-3}	1.473	2.344×10^{-2}	1.275

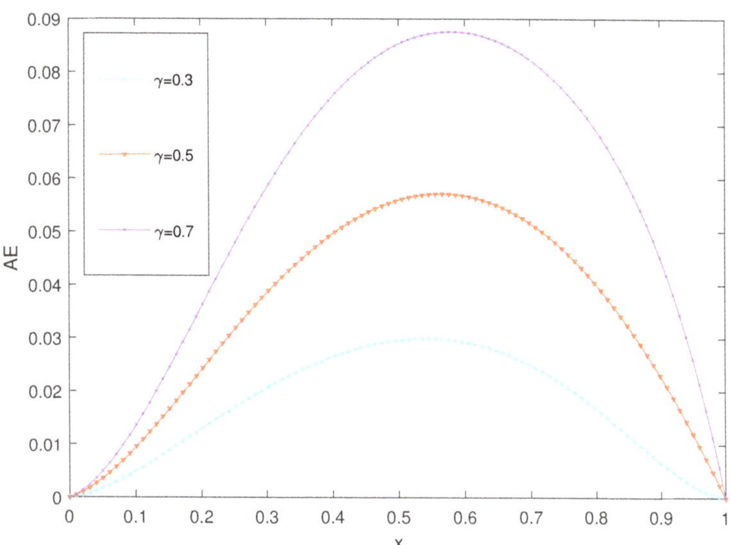

Figure 1. Comparison of AE at $t = 0.5$, $\mathcal{N} = 5$, $z(t) = t$, and different values of γ for Example 1.

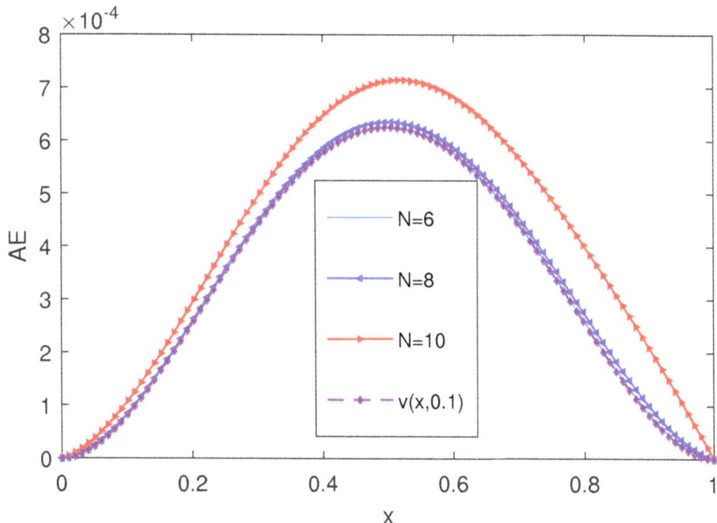

Figure 2. Plot of AE for different values of $\mathcal{N}$ at $t = 0.1$ and $z(t) = t$ for Example 1.

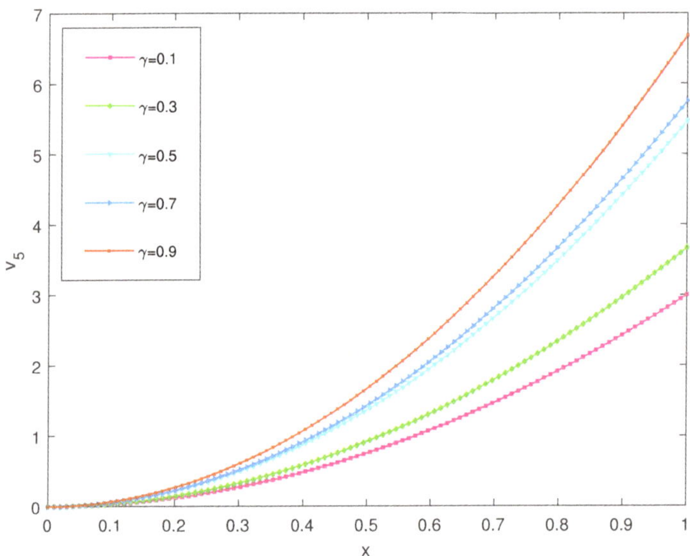

Figure 3. Comparison of the numerical solution for different values of γ at $t = 1$ and $z(t) = t^2$ for Example 1.

Example 2. *Consider the following Example [34],*

$$* \mathcal{D}_t^\gamma v(x,t) = \frac{\partial^2 v(x,t)}{\partial x^2} + g(x,t), \quad t \in [0,\tau], \tag{50}$$

where

$$g(x,t) = 4\pi^2 t^2 \sin(2\pi x) + \frac{2t^{2-\gamma} \sin(2\pi x)}{(2 - 3\gamma + \gamma^2)\Gamma(1 - \gamma)},$$

with the initial and boundary conditions given by,

$$\begin{cases} v(x,0) = 0, \\ v(0,t) = 0, \ v(1,t) = 0. \end{cases} \tag{51}$$

The exact solution for this Example (2) is $t^2 \sin(2\pi x)$.

We shall apply the numerical scheme (26) to solve this problem (2) for different values of $\mathcal{N}$ with $z(t) = t$ varying the fractional order $\gamma = 0.1, 0.3, 0.5, 0.7$. We obtain the AEs at the grid points in the given domain, which are shown in Tables 4 and 5, respectively. The results presented in Tables 4 and 5 establish the convergence of the proposed method for different values of γ. In Table 6, we show the MAE by varying the different values of $\gamma = 0.3, 0.5, 0.7$ and $\mathcal{M}$. Further, we show the CO for each value of γ, which proves the accuracy of the present method. In Figure 4, we compared the numerical solutions for various choices of γ with the exact solution known at $\gamma = 0.2$. In Figure 5, the solution graphs for different values of $\mathcal{N}$ and plot of the exact solution (for $\gamma = 0.2$) are shown. From Figures 4 and 5, we conclude that the numerical solution obtained by the proposed method converges to the exact solution. Finally, we compared our results with the existing method [34] in Table 7. We see that the proposed method gives better accuracy in approximating the numerical solutions.

Table 4. Comparison of AE for Example 2 at $\gamma = 0.1, 0.3$ and various values of $\mathcal{N}$.

	$\gamma = 0.1$			$\gamma = 0.3$		
x	$\mathcal{N} = 5$	$\mathcal{N} = 7$	$\mathcal{N} = 9$	$\mathcal{N} = 5$	$\mathcal{N} = 7$	$\mathcal{N} = 9$
0.1	5.899×10^{-4}	4.786×10^{-5}	2.407×10^{-6}	5.801×10^{-4}	4.707×10^{-5}	2.352×10^{-6}
0.2	3.927×10^{-4}	3.174×10^{-5}	1.694×10^{-6}	3.801×10^{-4}	3.074×10^{-5}	1.618×10^{-4}
0.3	2.403×10^{-4}	2.173×10^{-5}	1.126×10^{-6}	2.296×10^{-4}	2.087×10^{-5}	1.058×10^{-6}
0.4	1.376×10^{-4}	1.057×10^{-5}	5.571×10^{-6}	1.315×10^{-4}	1.008×10^{-5}	5.174×10^{-7}
0.5	1.051×10^{-18}	7.952×10^{-20}	3.281×10^{-19}	1.233×10^{-4}	1.067×10^{-18}	5.225×10^{-19}
0.6	1.436×10^{-4}	1.045×10^{-5}	5.560×10^{-7}	1.246×10^{-4}	1.000×10^{-5}	5.186×10^{-7}
0.7	2.313×10^{-4}	2.343×10^{-5}	1.344×10^{-6}	2.378×10^{-4}	2.066×10^{-4}	1.077×10^{-6}
0.8	3.567×10^{-4}	3.164×10^{-5}	1.678×10^{-6}	3.875×10^{-4}	3.099×10^{-5}	1.623×10^{-6}
0.9	5.679×10^{-4}	4.567×10^{-5}	2.457×10^{-6}	5.112×10^{-4}	4.888×10^{-5}	2.399×10^{-6}

Table 5. Comparison of AE for Example 2 at $\gamma = 0.5, 0.7$ and various values of $\mathcal{N}$.

	$\gamma = 0.5$			$\gamma = 0.7$		
x	$\mathcal{N} = 5$	$\mathcal{N} = 7$	$\mathcal{N} = 9$	$\mathcal{N} = 5$	$\mathcal{N} = 7$	$\mathcal{N} = 9$
0.1	3.386×10^{-5}	4.579×10^{-5}	2.194×10^{-6}	5.350×10^{-4}	4.407×10^{-5}	2.406×10^{-6}
0.2	6.886×10^{-4}	2.914×10^{-5}	1.387×10^{-6}	3.230×10^{-4}	2.717×10^{-5}	1.694×10^{-6}
0.3	8.617×10^{-4}	1.953×10^{-5}	8.399×10^{-7}	1.815×10^{-4}	1.800×10^{-5}	1.126×10^{-6}
0.4	5.455×10^{-4}	9.335×10^{-6}	3.863×10^{-7}	1.047×10^{-4}	8.528×10^{-6}	5.570×10^{-7}
0.5	1.191×10^{-18}	1.087×10^{-18}	5.186×10^{-18}	6.447×10^{-19}	3.683×10^{-18}	3.280×10^{-19}
0.6	5.575×10^{-4}	9.435×10^{-6}	3.789×10^{-7}	1.046×10^{-4}	8.546×10^{-6}	5.580×10^{-7}
0.7	8.637×10^{-4}	1.944×10^{-6}	8.457×10^{-7}	1.824×10^{-4}	1.890×10^{-5}	1.136×10^{-6}
0.8	6.745×10^{-4}	2.879×10^{-5}	1.478×10^{-6}	3.320×10^{-4}	2.654×10^{-5}	1.794×10^{-6}
0.9	3.378×10^{-4}	4.543×10^{-5}	2.148×10^{-6}	5.458×10^{-4}	4.568×10^{-5}	2.451×10^{-6}

Table 6. The MAE and CO for Example 2 for various values of γ and $\mathcal{M}$.

	$\gamma = 0.3$		$\gamma = 0.5$		$\gamma = 0.7$	
$\mathcal{M}$	MAE	CO	MAE of PM	CO	MAE of PM	CO
50	1.475×10^{-6}	$\ldots$	1.718×10^{-6}	$\ldots$	3.746×10^{-6}	$\ldots$
100	4.636×10^{-6}	1.667	6.336×10^{-6}	1.439	1.553×10^{-6}	1.271
200	1.425×10^{-6}	1.701	2.300×10^{-6}	1.461	6.385×10^{-7}	1.282
400	4.459×10^{-7}	1.676	8.228×10^{-7}	1.483	2.613×10^{-7}	1.289
800	1.387×10^{-7}	1.685	2.938×10^{-7}	1.485	1.065×10^{-7}	1.296

Table 7. Comparison of the MAE of Example 2 with $\gamma = 0.5$.

$\mathcal{M}$	Present Method	Method [34]
2	1.061×10^{-4}	1.627×10^{-1}
4	3.879×10^{-5}	3.101×10^{-2}
6	2.103×10^{-5}	1.140×10^{-2}
8	1.342×10^{-5}	5.378×10^{-3}
10	3.925×10^{-6}	2.726×10^{-3}
12	7.953×10^{-7}	1.461×10^{-3}
14	1.047×10^{-6}	8.187×10^{-4}
16	1.252×10^{-6}	4.755×10^{-4}

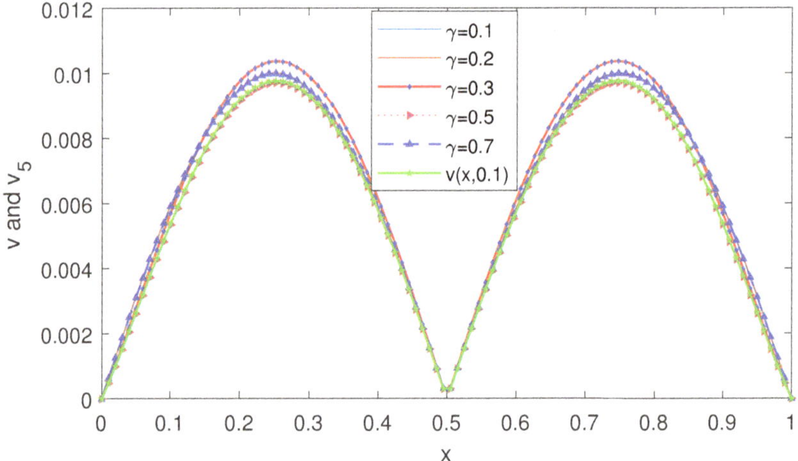

Figure 4. Comparison of the numerical and the exact solution at $t = 0.1$ and different values of γ for Example 2.

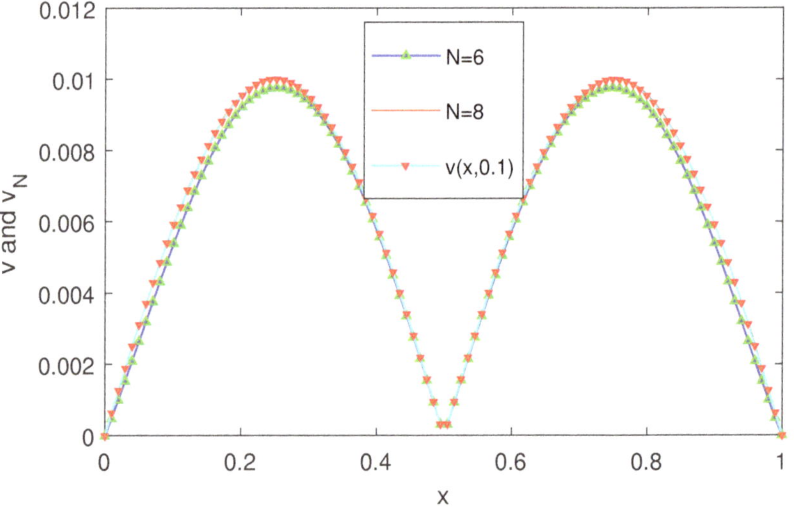

Figure 5. Plot of the numerical solutions for different values of $\mathcal{N}$ at $t = 0.1$ for Example 2.

6. Conclusions

A numerical scheme for a new class of fractional diffusion equation was studied in this paper in which the time derivative was considered as the generalized fractional derivative. The scheme used the finite difference and collocation methods to find the numerical solution. The theoretical error and convergence analysis were also validated numerically. The numerical examples showed that the proposed method achieved high accuracy in comparison to other methods [34,36–38] presented recently.

Author Contributions: Conceptualization, S.K. (Sandeep Kumar) and R.K.P.; methodology, S.K. (Sandeep Kumar), K.K. and R.K.P.; software, S.K. (Sandeep Kumar) and R.K.P.; validation, S.K. (Sandeep Kumar), K.K. and R.K.P., writing—original draft preparation, S.K. (Sandeep Kumar), K.K., R.K.P., S.K. (Shyam Kamal) and T.N.D.; writing—review and editing, R.K.P., S.K. (Shyam Kamal) and T.N.D.; supervision, R.K.P.; funding acquisition, R.K.P., S.K. (Shyam Kamal), and T.N.D. All authors have read and agreed to the published version of the manuscript.

Funding: This research received no external funding.

Data Availability Statement: Not applicable.

Acknowledgments: The authors sincerely thank the reviewers for their constructive comments to improve the manuscript.

Conflicts of Interest: The authors declare no conflict of interest.

References

1. Podlubny, I. *Fractional Differential Equations: An Introduction to Fractional Derivatives, Fractional Differential Equations, to Methods of Their Solution and Some of Their Applications*; Elsevier: Amsterdam, The Netherlands, 1998.
2. McBride, A. V. Kiryakova Generalized fractional calculus and applications (Pitman Research Notes in Mathematics Vol. 301, Longman1994), 388 pp., 0 582 21977 9,£ 39. *Proc. Edinb. Math. Soc.* **1995**, *38*, 189–190. [CrossRef]
3. Podlubny, I. An introduction to fractional derivatives, fractional differential equations, methods of their solution and some of their applications. *Math. Sci. Eng.* **1999**, *198*, 340.
4. Miller, K.S.; Ross, B. *An Introduction to the Fractional Calculus and Fractional Differential Equations*; Wiley: Hoboken, NJ, USA, 1993.
5. Cattani, C.; Srivastava, H.M.; Yang, X.J. *Fractional Dynamics*; De Gruyter Open: Warsaw, Poland, 2016.
6. Hilfer, R. *Applications of Fractional Calculus in Physics*; World Scientific: Singapore, 2000.
7. Hassani, H.; Avazzadeh, Z.; Machado, J.A.T. Solving two-dimensional variable-order fractional optimal control problems with transcendental Bernstein series. *J. Comput. Nonlinear Dyn.* **2019**, *14*, 061001. [CrossRef]
8. Li, X.; Chen, X.D.; Chen, N. A third-order approximate solution of the reaction-diffusion process in an immobilized biocatalyst particle. *Biochem. Eng. J.* **2004**, *17*, 65–69. [CrossRef]
9. Magin, R. Fractional calculus in bioengineering, part 1. *Crit. Rev. Biomed. Eng.* **2004**, *32*, 1–104. [CrossRef] [PubMed]
10. Baleanu, D.; Sajjadi, S.S.; Jajarmi, A.; Asad, J.H. New features of the fractional Euler–Lagrange equations for a physical system within a non-singular derivative operator. *Eur. Phys. J. Plus* **2019**, *134*, 181. [CrossRef]
11. Bagley, R.L.; Torvik, P. A theoretical basis for the application of fractional calculus to viscoelasticity. *J. Rheol.* **1983**, *27*, 201–210. [CrossRef]
12. Carpinteri, A.; Mainardi, F. *Fractals and Fractional Calculus in Continuum Mechanics*; Springer: Berlin/Heidelberg, Germany, 2014; Volume 378.
13. Shukla, A.K.; Pandey, R.K.; Pachori, R.B. A fractional filter based efficient algorithm for retinal blood vessel segmentation. *Biomed. Signal Process. Control* **2020**, *59*, 101883. [CrossRef]
14. Agrawal, O.P. Some generalized fractional calculus operators and their applications in integral equations. *Fract. Calc. Appl. Anal.* **2012**, *15*, 700–711. [CrossRef]
15. Xu, Y.; Agrawal, O.P. Numerical solutions and analysis of diffusion for new generalized fractional Burgers equation. *Fract. Calc. Appl. Anal.* **2013**, *16*, 709–736. [CrossRef]
16. Xu, Y.; He, Z.; Xu, Q. Numerical solutions of fractional advection–diffusion equations with a kind of new generalized fractional derivative. *Int. J. Comput. Math.* **2014**, *91*, 588–600. [CrossRef]
17. Kumar, K.; Pandey, R.K.; Sharma, S.; Xu, Y. Numerical scheme with convergence for a generalized time-fractional Telegraph-type equation. *Numer. Methods Partial Differ. Equ.* **2019**, *35*, 1164–1183. [CrossRef]
18. Yadav, S.; Pandey, R.K.; Shukla, A.K.; Kumar, K. High-order approximation for generalized fractional derivative and its application. *Int. J. Numer. Methods Heat Fluid Flow* **2019**, *29*, 3515–3534. [CrossRef]
19. Kumar, K.; Pandey, R.K.; Sultana, F. Numerical schemes with convergence for generalized fractional integro-differential equations. *J. Comput. Appl. Math.* **2021**, *388*, 113318. [CrossRef]
20. Kumari, S.; Pandey, R.K. High-order approximation to generalized Caputo derivatives and generalized fractional advection-diffusion equations. *arXiv* **2022**, arXiv:2206.04033.
21. Li, X.; Wong, P.J. A gWSGL numerical scheme for generalized fractional sub-diffusion problems. *Commun. Nonlinear Sci. Numer. Simul.* **2020**, *82*, 104991. [CrossRef]
22. Heydari, M.; Atangana, A.; Avazzadeh, Z.; Mahmoudi, M. An operational matrix method for nonlinear variable-order time fractional reaction-diffusion equation involving Mittag–Leffler kernel. *Eur. Phys. J. Plus* **2020**, *135*, 237. [CrossRef]
23. Duan, X.; Leng, J.; Cattani, C.; Li, C. A shannon-runge-kutta-gill method for convection-diffusion equations. *Math. Probl. Eng.* **2013**, *2013*, 163734. [CrossRef]
24. Bhrawy, A.H.; Baleanu, D.; Mallawi, F. A new numerical technique for solving fractional sub-diffusion and reaction sub-diffusion equations with a nonlinear source term. *Therm. Sci.* **2015**, *19*, S25–S34. [CrossRef]

25. Lin, Y.; Xu, C. Finite difference/spectral approximations for the time-fractional diffusion equation. *J. Comput. Phys.* **2007**, *225*, 1533–1552. [CrossRef]
26. Murio, D.A. Implicit finite difference approximation for time-fractional diffusion equations. *Comput. Math. Appl.* **2008**, *56*, 1138–1145. [CrossRef]
27. Sweilam, N.; Khader, M.; Mahdy, A. Crank-Nicolson finite difference method for solving time-fractional diffusion equation. *J. Fract. Calc. Appl.* **2012**, *2*, 1–9.
28. Luchko, Y. Initial-boundary-value problems for the generalized multi-term time-fractional diffusion equation. *J. Math. Anal. Appl.* **2011**, *374*, 538–548. [CrossRef]
29. Dubey, V.P.; Kumar, R.; Kumar, D.; Khan, I.; Singh, J. An efficient computational scheme for nonlinear time-fractional systems of partial differential equations arising in physical sciences. *Adv. Differ. Equ.* **2020**, *2020*, 46. [CrossRef]
30. Oldham, K.; Spanier, J. *The Fractional Calculus Theory and Applications of Differentiation and Integration to Arbitrary Order*; Elsevier: Amsterdam, The Netherlands, 1974.
31. Askey, R.; Wilson, J.A. *Some Basic Hypergeometric Orthogonal Polynomials that Generalize Jacobi Polynomials*; American Mathematical Soc.: Providence, RI, USA, 1985; Volume 319.
32. Cao, W.; Xu, Y.; Zheng, Z. Finite difference/collocation method for a generalized time-fractional KDV equation. *Appl. Sci.* **2018**, *8*, 42. [CrossRef]
33. Bellomo, N.; Lods, B.; Revelli, R.; Ridolfi, L. *Generalized Collocation Methods: Solutions to Nonlinear Problems*; Springer Science & Business Media: Berlin/Heidelberg, Germany, 2007.
34. Pirkhedri, A.; Javadi, H.H.S. Solving the time-fractional diffusion equation via Sinc–Haar collocation method. *Appl. Math. Comput.* **2015**, *257*, 317–326. [CrossRef]
35. Saadatmandi, A.; Dehghan, M.; Azizi, M.R. The Sinc–Legendre collocation method for a class of fractional convection-diffusion equations with variable coefficients. *Commun. Nonlinear Sci. Numer. Simul.* **2012**, *17*, 4125–4136. [CrossRef]
36. Momani, S. An analytical approximate solution for fractional heat-like and wave-like equations with variable coefficients using the decomposition method. *Appl. Math. Comput.* **2005**, *165*, 459–472. [CrossRef]
37. Molliq, Y.; Noorani, M.S.M.; Hashim, I. Variational iteration method for fractional heat-and wave-like equations. *Nonlinear Anal. Real World Appl.* **2009**, *10*, 1854–1869. [CrossRef]
38. Chen, R.; Liu, F.; Anh, V. Numerical methods and analysis for a multi-term time-space variable-order fractional advection–diffusion equations and applications. *J. Comput. Appl. Math.* **2019**, *352*, 437–452. [CrossRef]

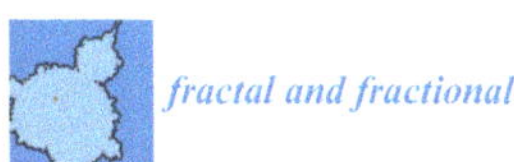
fractal and fractional

Article

Multiweighted-Type Fractional Fourier Transform: Unitarity

Tieyu Zhao * and Yingying Chi

Information Science Teaching and Research Section, School of Mathematics and Statistics, Northeastern
University at Qinhuangdao, Qinhuangdao 066004, China; chiyingying@neuq.edu.cn
* Correspondence: zhaotieyu@neuq.edu.cn

Abstract: The definition of the discrete fractional Fourier transform (DFRFT) varies, and the multiweighted-
type fractional Fourier transform (M-WFRFT) is its extended definition. It is not easy to prove
its unitarity. We use the weighted-type fractional Fourier transform, fractional-order matrix and
eigendecomposition-type fractional Fourier transform as basic functions to prove and discuss the
unitarity. Thanks to the growing body of research, we found that the effective weighting term of the
M-WFRFT is only four terms, none of which are extended to M terms, as described in the definition.
Furthermore, the program code is analyzed, and the result shows that the previous work (Digit
Signal Process 2020: 104: 18) based on MATLAB for unitary verification is inaccurate.

Keywords: fractional fourier transform; weighted-type fractional Fourier transform; multiweighted-type
fractional fourier transform; unitarity

Citation: Zhao, T.; Chi, Y.
Multiweighted-Type Fractional
Fourier Transform: Unitarity. *Fractal
Fract.* **2021**, *5*, 205. https://doi.org/
10.3390/fractalfract5040205

Academic Editors: Thach
Ngoc Dinh, Shyam Kamal
and Rajesh Kumar Pandey

Received: 3 October 2021
Accepted: 1 November 2021
Published: 8 November 2021

Publisher's Note: MDPI stays neutral
with regard to jurisdictional claims in
published maps and institutional affiliations.

1. Introduction

The multiweighted-type fractional Fourier transform (M-WFRFT) is the extended
definition of the weighted-type fractional Fourier transform (WFRFT), and its application
has been described in detail in our previous research [1]. Here, we focus on summarizing
and analyzing the theory of the M-WFRFT. In [2], Zhu et al. proposed the definition of
the multifraction Fourier transform, i.e., the M-WFRFT. Researchers have applied this
definition to image encryption but have not discussed the properties of the definition itself.
Early research work [3–5] laid a solid foundation for the proposal of the M-WFRFT. In
1995, Shih proposed a new type of fractional-order Fourier transform (FRFT), which is
called WFRFT because it is a linear summation [3]. This definition has period 4, so it is
also called the 4-WFRFT. Subsequently, Liu et al. extended the definition of the WFRFT,
and the generalized definition has period M = 4l, where l = 1,2, ... [4,5]. Zhu's M-WFRFT is
proposed on this basis, and its period is any integer M > 4 [2]. However, little is known
about the properties of these definitions. Ran et al. sought to present a unified framework
with the help of a generalized permutation matrix group and discussed its properties [6].
This research greatly promotes the theoretical development of WFRFTs. Unfortunately,
there is no proof of unitarity, and the focus of the previous studies has been the generality
of weighted coefficients. Recently, some new definitions based on the M-WFRFT have
been proposed [7–11]. For example, Tao et al. proposed multiple-parameter fractional
Fourier transforms (MPFRFTs) [8], Ran et al. proposed modified multiple-parameter
fractional Fourier transforms (m-MPFRFTs) [9], and Zhao et al. proposed vector power
multiple-parameter fractional Fourier transforms (VPMPFRFTs) [10,11]. Unfortunately, the
properties of these definitions have not been discussed.

First, Santhanam et al. demonstrated the properties of the WFRFT and proved its
unitarity using weighted coefficients [12]. However, this work ignores that the basis
function is also a part of the definition. For the M-WFRFT, its basis function is the fractional
power of the Fourier transform, so it is not easy to prove its properties. Some recent research
results have also failed to prove its properties [13–18]. We proposed a new reformulation
of the M-WFRFT to prove its periodicity, additivity and boundary [1]. Unfortunately,

unitarity is only discussed by means of numerical simulation. This paper is a follow-up of previous research work and mainly seeks to prove and discuss the unitarity of the M-WFRFT. However, the most recent studies have also enlightened our research [19,20].

The remainder of this paper is organized as follows. Section 2 proposes a new reformulation of the M-WFRFT. The unitarity of the M-WFRFT is proven in Section 3. The deviation caused by the numerical simulation is discussed in Section 4. Finally, the conclusions are presented in Section 5.

2. Reformulation of M-WFRFT

Shih proposed the WFRFT [3], and its definition can be expressed as

$$F^{\alpha}[f(t)] = \sum_{l=0}^{3} A_l^{\alpha} f_l(t),\tag{1}$$

with

$$A_l^{\alpha} = \cos\left(\frac{(\alpha-l)\pi}{4}\right)\cos\left(\frac{2(\alpha-l)\pi}{4}\right)\exp\left(\frac{3(\alpha-l)i\pi}{4}\right),\tag{2}$$

where $f_l(t) = F^l[f(t)]; l = 0,1,2,3$, (F denotes Fourier transform). Shih's WFRFT with period 4 is also called the 4-weighted-type fractional Fourier transform (4-WFRFT).

We further improve the weighted coefficient A_l^{α}, as shown in Equation (3).

$$
\begin{aligned}
A_l^{\alpha} &= \cos\left(\frac{(\alpha-l)\pi}{4}\right)\cos\left(\frac{2(\alpha-l)\pi}{4}\right)\exp\left(\frac{3(\alpha-l)i\pi}{4}\right)\\
&= \tfrac{1}{2}\times\left[\exp\left(\frac{(\alpha-l)\pi i}{4}\right)+\exp\left(\frac{-(\alpha-l)\pi i}{4}\right)\right]\\
&\quad \times\tfrac{1}{2}\times\left[\exp\left(\frac{2(\alpha-l)\pi i}{4}\right)+\exp\left(\frac{-2(\alpha-l)\pi i}{4}\right)\right]\times\exp\left(\frac{3(\alpha-l)i\pi}{4}\right)\\
&= \tfrac{1}{4}\left(1+\exp\left(\frac{2(\alpha-l)\pi i}{4}\right)+\exp\left(\frac{4(\alpha-l)\pi i}{4}\right)+\exp\left(\frac{6(\alpha-l)\pi i}{4}\right)\right)\\
&= \tfrac{1}{4}\sum_{k=0}^{3}\exp\left(\frac{2\pi i k(\alpha-l)}{4}\right).
\end{aligned}\tag{3}
$$

Then, we can obtain Equation (4) as

$$
\begin{pmatrix} A_0^{\alpha} \\ A_1^{\alpha} \\ A_2^{\alpha} \\ A_3^{\alpha} \end{pmatrix} = \frac{1}{4}\begin{pmatrix} 1 & 1 & 1 & 1 \\ 1 & -i & -1 & i \\ 1 & -1 & 1 & -1 \\ 1 & i & -1 & -i \end{pmatrix}\begin{pmatrix} B_0^{\alpha} \\ B_1^{\alpha} \\ B_2^{\alpha} \\ B_3^{\alpha} \end{pmatrix},\tag{4}
$$

where $B_k^{\alpha} = \exp\left(\frac{2\pi i k\alpha}{4}\right)$, $k = 0,1,2,3$. Equation (4) provides ideas for expanding the definition in the future.

Liu et al., generalize Shih's definition, and the generalized definition is shown to have M-periodic eigenvalues with respect to the order of Hermite–Gaussian functions ($M = 4l$, where $l = 1,2,3, ...$) [4,5].

Subsequently, Zhu et al. proposed a multifractional Fourier transform whose period can be any integer ($M > 4$), so this definition is also called the M-WFRFT [2]. Zhu et al., extended the weighting coefficient A_l^{α}, which is more general; the result is shown in Equation (5).

$$
\begin{pmatrix} A_0^{\alpha} \\ A_1^{\alpha} \\ \vdots \\ A_{M-1}^{\alpha} \end{pmatrix} = \frac{1}{M}\begin{pmatrix} u^{0\times0} & u^{0\times1} & \cdots & u^{0\times(M-1)} \\ u^{1\times0} & u^{1\times1} & \cdots & u^{1\times(M-1)} \\ \vdots & \vdots & \ddots & \vdots \\ u^{(M-1)\times0} & u^{(M-1)\times1} & \cdots & u^{(M-1)\times(M-1)} \end{pmatrix}\begin{pmatrix} B_0^{\alpha} \\ B_1^{\alpha} \\ \vdots \\ B_{M-1}^{\alpha} \end{pmatrix},\tag{5}
$$

where $u = \exp(-2\pi i/M)$ and $B_k^\alpha = \exp\left(\frac{2\pi i k\alpha}{M}\right)$, $k = 0, 1, \cdots, M-1$. Then,

$$A_l^\alpha = \frac{1}{M} \sum_{k=0}^{M-1} \exp\left[\frac{2\pi i k(\alpha - l)}{M}\right];$$
$$l = 0, 1, \cdots, M-1. \tag{6}$$

The M-WFRFT is defined as

$$F_M^\alpha[f(t)] = \sum_{l=0}^{M-1} A_l^\alpha f_l(t), \tag{7}$$

where the basic functions are $f_l(t) = F^{4l/M}[f(t)]$; $l = 0, 1, \cdots, M-1$ (F denotes the Fourier transform).

At present, the M-WFRFT is widely used in image encryption and signal processing [7–11,21–25]. Unfortunately, few researchers have discussed its properties, and the proponents of the definition have not explained the properties. We find that it is not easy to prove the properties of the M-WFRFT (Equation (7)). Some researchers have discussed the properties using the weighted coefficient A_l^α but ignore that the basis function is also a part of the definition [6,12]. Therefore, we present a new reformulation of the M-WFRFT. As such, Equation (7) can be expressed as

$$
\begin{aligned}
F_M^\alpha[f(t)] &= A_0^\alpha f_0(t) + A_1^\alpha f_1(t) + \cdots + A_{M-1}^\alpha f_{M-1}(t) \\
&= A_0^\alpha F^{\frac{0}{M}}[f(t)] + A_1^\alpha F^{\frac{4}{M}}[f(t)] + \cdots + A_{M-1}^\alpha F^{\frac{4(M-1)}{M}}[f(t)] \\
&= \left(A_0^\alpha I + A_1^\alpha F^{\frac{4}{M}} + \cdots + A_{M-1}^\alpha F^{\frac{4(M-1)}{M}}\right) f(t) \\
&= \left(I, F^{\frac{4}{M}}, \cdots, F^{\frac{4(M-1)}{M}}\right)
\begin{pmatrix} A_0^\alpha \\ A_1^\alpha \\ \vdots \\ A_{M-1}^\alpha \end{pmatrix} f(t).
\end{aligned}
\tag{8}
$$

By Equations (5) and (8), Equation (9) is obtained as

$$
F_M^\alpha[f(t)] = \frac{1}{M}\left(I, F^{\frac{4}{M}}, \cdots, F^{\frac{4(M-1)}{M}}\right)
\begin{pmatrix}
u^{0\times 0} & u^{0\times 1} & \cdots & u^{0\times(M-1)} \\
u^{1\times 0} & u^{1\times 1} & \cdots & u^{1\times(M-1)} \\
\vdots & \vdots & \ddots & \vdots \\
u^{(M-1)\times 0} & u^{(M-1)\times 1} & \cdots & u^{(M-1)\times(M-1)}
\end{pmatrix}
\begin{pmatrix} B_0^\alpha \\ B_1^\alpha \\ \vdots \\ B_{M-1}^\alpha \end{pmatrix} f(t), \tag{9}
$$

where $u = \exp(-2\pi i/M)$, $B_k^\alpha = \exp\left(\frac{2\pi i k\alpha}{M}\right)$, $k = 0, 1, \ldots, M-1$ and F denotes the Fourier transform. Here, let

$$
\begin{cases}
Y_0 = u^{0\times 0} I + u^{1\times 0} F^{\frac{4}{M}} + \cdots + u^{(M-1)\times 0} F^{\frac{4(M-1)}{M}}, \\[2mm]
Y_1 = u^{0\times 1} I + u^{1\times 1} F^{\frac{4}{M}} + \cdots + u^{(M-1)\times 1} F^{\frac{4(M-1)}{M}}, \\[2mm]
Y_2 = u^{0\times 2} I + u^{1\times 2} F^{\frac{4}{M}} + \cdots + u^{(M-1)\times 2} F^{\frac{4(M-1)}{M}}, \\[2mm]
\quad\quad \vdots \\[2mm]
Y_{M-1} = u^{0\times(M-1)} I + u^{1\times(M-1)} F^{\frac{4}{M}} + \cdots + u^{(M-1)\times(M-1)} F^{\frac{4(M-1)}{M}}.
\end{cases}
\tag{10}
$$

Definition 1. A new reformulation of the M-WFRFT as

$$
T^\alpha_{MW}[f(t)] = \frac{1}{M}(Y_0, Y_1, \cdots, Y_{M-1})
\begin{pmatrix}
B^\alpha_0 \\
B^\alpha_1 \\
\vdots \\
B^\alpha_{M-1}
\end{pmatrix} f(t)
$$

$$
= \frac{1}{M}\sum_{k=0}^{M-1} Y_k B^\alpha_k f(t),
$$

(11)

where $B^\alpha_k = \exp\left(\frac{2\pi i k\alpha}{M}\right)$; $k = 0, 1, \cdots, M-1$.

Remark 1. *Our previous work [1] discussed that the new reformulation helps to prove the properties. Unitarity is often used in signal processing. Unfortunately, previous research work only presents simulation verification. This paper will seek to prove and discuss the unitarity.*

3. Unitarity

A complex matrix U satisfies

$$
UU^H = U^H U = I,
$$

(12)

where H denotes the conjugate transpose and I is the identity matrix. Then, U is called a unitary matrix. The greatest difficulty in proving the unitarity of the M-WFRFT is considering the basis function $F^{4l/M}$, $l = 0, 1, \cdots, M-1$. The basis function is related to the discrete fractional Fourier transform (DFRFT), and the definition of the DFRFT varies. Therefore, we seek to use different types of DFRFT as the basis function to verify the unitarity of M-WFRFT.

3.1. 4-WFRFT as the Basis Function

Proposition 1. *4-WFRFT is used as the basis function, so the M-WFRFT has unitarity.*

Proof. The definition of the 4-WFRFT is shown in Equation (1), and Equation (13) can be obtained as

$$
F^\alpha[f(t)] = \left(A^\alpha_0 \cdot I + A^\alpha_1 \cdot F + A^\alpha_2 \cdot F^2 + A^\alpha_3 \cdot F^3\right)f(t)
$$

$$
= (I, F, F^2, F^3)
\begin{pmatrix}
A^\alpha_0 \\
A^\alpha_1 \\
A^\alpha_2 \\
A^\alpha_3
\end{pmatrix} f(t).
$$

(13)

From Equations (4) and (13), we obtain

$$
F^\alpha[f(t)] = \frac{1}{4}\left(I, F, F^2, F^3\right)
\begin{pmatrix}
1 & 1 & 1 & 1 \\
1 & -i & -1 & i \\
1 & -1 & 1 & -1 \\
1 & i & -1 & -i
\end{pmatrix}
\begin{pmatrix}
B^\alpha_0 \\
B^\alpha_1 \\
B^\alpha_2 \\
B^\alpha_3
\end{pmatrix} f(t),
$$

(14)

where $B^\alpha_k = \exp\left(\frac{2\pi i k\alpha}{4}\right)$, $k = 0, 1, 2, 3$. Here, let

$$
\begin{cases}
P_0 = I + F + F^2 + F^3 \\
P_1 = I - F * i - F^2 + F^3 * i \\
P_2 = I - F + F^2 - F^3 \\
P_3 = I + F * i - F^2 - F^3 * i.
\end{cases}
$$

(15)

Then, the 4-WFRFT can be re-expressed as

$$T_{4W}^{\alpha}[f(t)] = \frac{1}{4}(P_0, P_1, P_2, P_3)\begin{pmatrix} B_0^{\alpha} \\ B_1^{\alpha} \\ B_2^{\alpha} \\ B_3^{\alpha} \end{pmatrix} f(t). \tag{16}$$

Thus, the discrete 4-WFRFT can be expressed as

$$T_{4W}^{\alpha} = \frac{1}{4}(P_0, P_1, P_2, P_3)\begin{pmatrix} B_0^{\alpha} \\ B_1^{\alpha} \\ B_2^{\alpha} \\ B_3^{\alpha} \end{pmatrix}. \tag{17}$$

From Equation (10), Y_k can be expressed as

$$Y_k = u^{0\times k} \times I + u^{1\times k} \times F^{\frac{4}{M}} + \cdots + u^{(M-1)\times k} \times F^{\frac{4(M-1)}{M}}; \\ k = 0, 1, \cdots, M-1, \tag{18}$$

The 4-WFRFT as the basis function is

$$Y_k = u^{0\times k} \times T_{4M}^0 + u^{1\times k} \times T_{4M}^{\frac{4}{M}} + \cdots + u^{(M-1)\times k} \times T_{4M}^{\frac{4(M-1)}{M}}. \tag{19}$$

From Equations (17) and (19), we can obtain

$$
\begin{aligned}
Y_k &= \frac{1}{4}(P_0, P_1, P_2, P_3)\left(u^{0\times k} \times \begin{pmatrix} B_0^0 \\ B_1^0 \\ B_2^0 \\ B_3^0 \end{pmatrix} + u^{1\times k} \times \begin{pmatrix} B_0^{\frac{4}{M}} \\ B_1^{\frac{4}{M}} \\ B_2^{\frac{4}{M}} \\ B_3^{\frac{4}{M}} \end{pmatrix} + \cdots + u^{(M-1)\times k} \times \begin{pmatrix} B_0^{\frac{4(M-1)}{M}} \\ B_1^{\frac{4(M-1)}{M}} \\ B_2^{\frac{4(M-1)}{M}} \\ B_3^{\frac{4(M-1)}{M}} \end{pmatrix} \right) \\
&= \frac{1}{4}(P_0, P_1, P_2, P_3)\begin{pmatrix} u^{0\times k} \times B_0^0 + u^{1\times k} \times B_0^{\frac{4}{M}} + \cdots + u^{(M-1)\times k} \times B_0^{\frac{4(M-1)}{M}} \\ u^{0\times k} \times B_1^0 + u^{1\times k} \times B_1^{\frac{4}{M}} + \cdots + u^{(M-1)\times k} \times B_1^{\frac{4(M-1)}{M}} \\ u^{0\times k} \times B_2^0 + u^{1\times k} \times B_2^{\frac{4}{M}} + \cdots + u^{(M-1)\times k} \times B_2^{\frac{4(M-1)}{M}} \\ u^{0\times k} \times B_3^0 + u^{1\times k} \times B_3^{\frac{4}{M}} + \cdots + u^{(M-1)\times k} \times B_3^{\frac{4(M-1)}{M}} \end{pmatrix},
\end{aligned}
\tag{20}
$$

where $k = 0, 1, \cdots, M-1$ and $u = \exp(-2\pi i/M)$. Therefore, we obtain

$$
\begin{aligned}
Y_k &= \frac{1}{4}(P_0, P_1, P_2, P_3)\begin{pmatrix} 1 + \exp\left(\frac{-2\pi i 1 k}{M}\right) + \exp\left(\frac{-2\pi i 2 k}{M}\right) + \cdots + \exp\left(\frac{-2\pi i (M-1) k}{M}\right) \\ 1 + \exp\left(\frac{-2\pi i 1 (k-1)}{M}\right) + \exp\left(\frac{-2\pi i 2 (k-1)}{M}\right) + \cdots + \exp\left(\frac{-2\pi i (M-1)(k-1)}{M}\right) \\ 1 + \exp\left(\frac{-2\pi i 1 (k-2)}{M}\right) + \exp\left(\frac{-2\pi i 2 (k-2)}{M}\right) + \cdots + \exp\left(\frac{-2\pi i (M-1)(k-2)}{M}\right) \\ 1 + \exp\left(\frac{-2\pi i 1 (k-3)}{M}\right) + \exp\left(\frac{-2\pi i 2 (k-3)}{M}\right) + \cdots + \exp\left(\frac{-2\pi i (M-1)(k-3)}{M}\right) \end{pmatrix} \\
&= \frac{1}{4}(P_0, P_1, P_2, P_3)\begin{pmatrix} S_0(k) \\ S_1(k) \\ S_2(k) \\ S_3(k) \end{pmatrix}.
\end{aligned}
\tag{21}
$$

For sequence $S_0(k)$, it can be expressed as

$$S_0(k) = \frac{a_1(1 - q^M)}{1 - q} = \frac{1 - \exp\left(\frac{-2\pi i k}{M}\right)^M}{1 - \exp\left(\frac{-2\pi i k}{M}\right)}. \tag{22}$$

where $a_1 = 1$. Then, we obtain

$$S_0(k) = \begin{cases} M, & k \equiv 0 \bmod M \\ 0, & k \not\equiv 0 \bmod M. \end{cases} \tag{23}$$

For sequence $S_1(k)$,

$$S_1(k) = \frac{1 - \left(e^{-2\pi i(k-1)/M} \right)^M}{1 - e^{-2\pi i(k-1)/M}}, \tag{24}$$

we obtain

$$S_1(k) = \begin{cases} M, & k \equiv 1 \bmod M \\ 0, & k \not\equiv \ \bmod M. \end{cases} \tag{25}$$

For sequence $S_2(k)$,

$$S_2(k) = \frac{1 - \left(e^{-2\pi i(k-2)/M} \right)^M}{1 - e^{-2\pi i(k-2)/M}}, \tag{26}$$

we obtain

$$S_2(k) = \begin{cases} M, & k \equiv 2 \bmod M \\ 0, & k \not\equiv 2 \bmod M. \end{cases} \tag{27}$$

For sequence $S_3(k)$,

$$S_3(k) = \frac{1 - \left(e^{-2\pi i(k-3)/M} \right)^M}{1 - e^{-2\pi i(k-3)/M}}, \tag{28}$$

we obtain

$$S_3(k) = \begin{cases} M, & k \equiv 3 \bmod M \\ 0, & k \not\equiv 3 \bmod M. \end{cases} \tag{29}$$

Then, Equation (21) can be expressed as

$$Y_k = \begin{cases} \frac{M}{4} P_k, & k = 0, 1, 2, 3 \\ 0, & k = 4, 5, \cdots, M - 1. \end{cases} \tag{30}$$

Therefore, the M-WFRFT Equation (11) is written as

$$
\begin{aligned}
T_{MW}^{\alpha} &= \frac{1}{M} (Y_0, Y_1, \cdots, Y_{M-1}) \begin{pmatrix} B_0^{\alpha} \\ B_1^{\alpha} \\ \vdots \\ B_{M-1}^{\alpha} \end{pmatrix} \\
&= \frac{1}{4} (P_0, P_1, P_2, P_3, 0, \cdots, 0) \begin{pmatrix} B_0^{\alpha} \\ B_1^{\alpha} \\ \vdots \\ B_{M-1}^{\alpha} \end{pmatrix} \\
&= \frac{1}{4} (P_0, P_1, P_2, P_3) \begin{pmatrix} B_0^{\alpha} \\ B_1^{\alpha} \\ B_2^{\alpha} \\ B_3^{\alpha} \end{pmatrix}.
\end{aligned} \tag{31}
$$

From the expressions, we notice that Equations (17) and (31) are the same, but in fact they are different. The difference is that for Equation (31), $B_k^{\alpha} = \exp\left(\frac{2\pi i k \alpha}{M} \right)$; $k = 0, 1, \cdots, M - 1$. However, this does not affect the proof of unitarity. $\square$

Remark 2. *In our previous work [1], we proved the unitarity of Equation (17). When the 4-WFRFT is selected as the basis function, the M-WFRFT has unitarity. From Equation (31), we notice that the weighted sum of the M-WFRFT is only four terms.*

3.2. Fractional-Order Matrix as the Basis Function

In our previous numerical simulation, a fractional-order matrix was used to verify the unitarity of the M-WFRFT [1]. In this section, we present the theoretical analysis to improve the previous work.

Proposition 2. *Fractional-order matrix is used as the basis function, so the M-WFRFT has unitarity.*

Proof. The calculation of the fractional power of the matrix is applied to the eigenvalues, so eigenvalue decomposition of the matrix is required. Therefore, the eigendecomposition of the matrix can be expressed as

$$F = VDV^H, \tag{32}$$

where F is the DFT matrix, V is the eigenvector, and D is the eigenvalue matrix.

In [26,27], the eigenvalues of the DFT can be expressed as $\lambda_n = e^{n\pi i/2}$. Then, the possible values of the eigenvalue are $\lambda_r = \{1, -1, i, -i\}; r = 1, 2, \cdots, n$. In this way, the eigenvalue matrix D can be expressed as

$$D = \begin{pmatrix} \lambda_1 & 0 & \cdots & 0 \\ 0 & \lambda_2 & \cdots & 0 \\ \vdots & \vdots & \ddots & \vdots \\ 0 & 0 & \cdots & \lambda_n \end{pmatrix}. \tag{33}$$

Then, the fractional power operation of matrix F can be expressed as

$$F^{4l/M} = VD^{4l/M}V^H. \tag{34}$$

where $l = 0, 1, \cdots, M - 1$. For Equation (10), Y_k can be expressed as

$$\begin{aligned} Y_k &= u^{0 \times k}I + u^{1 \times k} \times F^{\frac{4}{M}} + \cdots + u^{(M-1) \times k} \times F^{\frac{4(M-1)}{M}} \\ &= u^{0 \times k}VD^0V^H + u^{1 \times k}VD^{4/M}V^H + \cdots + u^{(M-1) \times k}VD^{4(M-1)/M}V^H. \end{aligned} \tag{35}$$

where $k = 0, 1, \cdots, M - 1$. Therefore, we can obtain

$$\begin{aligned} Y_k &= V\left(u^{0 \times k} \times D^0 + u^{1 \times k} \times D^{4/M} + \cdots + u^{(M-1) \times k} \times D^{4(M-1)/M}\right)V^H \\ &= V\begin{pmatrix} u^{0 \times k}\lambda_1^0 + u^{1 \times k}\lambda_1^{4/M} + \cdots + u^{(M-1) \times k}\lambda_1^{4(M-1)/M} & 0 & \cdots & 0 \\ 0 & u^{0 \times k}\lambda_2^0 + u^{1 \times k}\lambda_2^{4/M} + \cdots + u^{(M-1) \times k}\lambda_2^{4(M-1)/M} & \cdots & 0 \\ \vdots & \vdots & \ddots & \vdots \\ 0 & 0 & \cdots & u^{0 \times k}\lambda_n^0 + u^{1 \times k}\lambda_n^{4/M} + \cdots + u^{(M-1) \times k}\lambda_n^{4(M-1)/M} \end{pmatrix}V^H \\ &= V\begin{pmatrix} Q_1(k) & 0 & \cdots & 0 \\ 0 & Q_2(k) & \cdots & 0 \\ \vdots & \vdots & \ddots & \vdots \\ 0 & 0 & 0 & Q_n(k) \end{pmatrix}V^H. \end{aligned} \tag{36}$$

Here, let

$$\begin{cases} Q_1(k) = u^{0 \times k}\lambda_1^0 + u^{1 \times k}\lambda_1^{4/M} + \cdots + u^{(M-1) \times k}\lambda_1^{4(M-1)/M} \\ Q_2(k) = u^{0 \times k}\lambda_2^0 + u^{1 \times k}\lambda_2^{4/M} + \cdots + u^{(M-1) \times k}\lambda_2^{4(M-1)/M} \\ \vdots \\ Q_n(k) = u^{0 \times k}\lambda_n^0 + u^{1 \times k}\lambda_n^{4/M} + \cdots + u^{(M-1) \times k}\lambda_n^{4(M-1)/M}. \end{cases} \tag{37}$$

The multiplicities of the DFT eigenvalues [26,27] are shown in Table 1. Therefore, there is

$$
\begin{aligned}
\lambda_r &= \{1, i, -1, -i\} \\
&= \left\{ e^{4n\pi i/2}, e^{(4n+1)\pi i/2}, e^{(4n+2)\pi i/2}, e^{(4n+3)\pi i/2} \right\} \\
&= \left\{ e^{2n\pi i} e^{0\pi i/2}, e^{2n\pi i} e^{\pi i/2}, e^{2n\pi i} e^{2\pi i/2}, e^{2n\pi i} e^{3\pi i/2} \right\} \\
&= \left\{ e^{0\pi i/2}, e^{\pi i/2}, e^{2\pi i/2}, e^{3\pi i/2} \right\}.
\end{aligned}
\tag{38}
$$

Table 1. Multiplicities of the DFT eigenvalues.

N	1	-1	$-i$	i
$4n$	$n+1$	n	n	$n-1$
$4n+1$	$n+1$	n	n	n
$4n+2$	$n+1$	$n+1$	n	n
$4n+3$	$n+1$	$n+1$	$n+1$	n

For Equation (37), $Q_r(k), r = 1, 2, \cdots, n$ can be expressed as

$$
Q_r(k) = u^{0\times k}\lambda_r^0 + u^{1\times k}\lambda_r^{4/M} + \cdots + u^{(M-1)\times k}\lambda_r^{4(M-1)/M}.
\tag{39}
$$

When the eigenvalues $\lambda_r = e^{0\pi i/2} = 1$ and $u = e^{-2\pi i/M}$, $Q_r^{(1)}(k)$ can be expressed using Equation (39), as

$$
\begin{aligned}
Q_r^{(1)}(k) &= u^{0\times k}\lambda_r^0 + u^{1\times k}\lambda_r^{4/M} + \cdots + u^{(M-1)\times k}\lambda_r^{4(M-1)/M} \\
&= 1 + e^{-2\pi i 1(k-0)/M} + e^{-2\pi i 2(k-0)/M} + \cdots + e^{-2\pi i(M-1)(k-0)/M} \\
&= \frac{1 - \left(e^{-2\pi i(k-0)/M}\right)^M}{1 - e^{-2\pi i(k-0)/M}}.
\end{aligned}
\tag{40}
$$

Therefore, we obtain

$$
Q_r^{(1)}(k) = \begin{cases} 0, & k \equiv 0 \bmod M \\ M, & k \not\equiv 0 \bmod M. \end{cases}
\tag{41}
$$

When the eigenvalue $\lambda_r = e^{\pi i/2} = i$, $Q_r^{(i)}(k)$ can be expressed using Equation (39), as

$$
\begin{aligned}
Q_r^{(i)}(k) &= u^{0\times k}\lambda_r^0 + u^{1\times k}\lambda_r^{4/M} + \cdots + u^{(M-1)\times k}\lambda_r^{4(M-1)/M} \\
&= 1 + e^{-2\pi i 1(k-1)/M} + e^{-2\pi i 2(k-1)/M} + \cdots + e^{-2\pi i(M-1)(k-1)/M} \\
&= \frac{1 - \left(e^{-2\pi i(k-1)/M}\right)^M}{1 - e^{-2\pi i(k-1)/M}}.
\end{aligned}
\tag{42}
$$

Therefore, there is

$$
Q_r^{(i)}(k) = \begin{cases} 0, & k \equiv 1 \bmod M \\ M, & k \not\equiv 1 \bmod M. \end{cases}
\tag{43}
$$

When the eigenvalue $\lambda_r = e^{2\pi i/2} = -1$, $Q_r^{(-1)}(k)$ can be expressed using Equation (39), as

$$
\begin{aligned}
Q_r^{(-1)}(k) &= u^{0\times k}\lambda_r^0 + u^{1\times k}\lambda_r^{4/M} + \cdots + u^{(M-1)\times k}\lambda_r^{4(M-1)/M} \\
&= 1 + e^{-2\pi i 1(k-2)/M} + e^{-2\pi i 2(k-2)/M} + \cdots + e^{-2\pi i(M-1)(k-2)/M} \\
&= \frac{1 - \left(e^{-2\pi i(k-2)/M}\right)^M}{1 - e^{-2\pi i(k-2)/M}}.
\end{aligned}
\tag{44}
$$

Then, we can obtain

$$Q_r^{(-1)}(k) = \begin{cases} 0, & k \equiv 2 \bmod M \\ M, & k \not\equiv 2 \bmod M. \end{cases} \tag{45}$$

When the eigenvalue $\lambda_r = e^{3\pi i/2} = -i$, $Q_r^{(-i)}(k)$ can be expressed using Equation (39), as

$$\begin{aligned} Q_r^{(-i)}(k) &= u^{0 \times k}\lambda_r^0 + u^{1 \times k}\lambda_r^{4/M} + \cdots + u^{(M-1) \times k}\lambda_r^{4(M-1)/M} \\ &= 1 + e^{-2\pi i 1(k-3)/M} + e^{-2\pi i 2(k-3)/M} + \cdots + e^{-2\pi i(M-1)(k-3)/M} \\ &= \frac{1 - \left(e^{-2\pi i(k-3)/M}\right)^M}{1 - e^{-2\pi i(k-3)/M}}. \end{aligned} \tag{46}$$

Therefore, there is

$$Q_r^{(-i)}(k) = \begin{cases} 0, & k \equiv 3 \bmod M \\ M, & k \not\equiv 3 \bmod M. \end{cases} \tag{47}$$

Using Equations (41), (43), (45) and (47), we can formulate Equation (36) as

$$Y_k = \begin{cases} Y_k, & k = 0,1,2,3 \\ 0, & k = 4,5,\cdots, M-1. \end{cases} \tag{48}$$

In this way, the M-WFRFT of Equation (11) can also be expressed as Equation (49).

$$\begin{aligned} T_{MW}^{\alpha} &= \tfrac{1}{M}(Y_0, Y_1, \cdots, Y_{M-1}) \begin{pmatrix} B_0^{\alpha} \\ B_1^{\alpha} \\ \vdots \\ B_{M-1}^{\alpha} \end{pmatrix} \\ &= \tfrac{1}{M}(Y_0, Y_1, Y_2, Y_3, 0, \cdots, 0) \begin{pmatrix} B_0^{\alpha} \\ B_1^{\alpha} \\ \vdots \\ B_{M-1}^{\alpha} \end{pmatrix} \\ &= \tfrac{1}{M}(Y_0, Y_1, Y_2, Y_3) \begin{pmatrix} B_0^{\alpha} \\ B_1^{\alpha} \\ B_2^{\alpha} \\ B_3^{\alpha} \end{pmatrix}, \end{aligned} \tag{49}$$

where $B_k^{\alpha} = \exp\left(\frac{2\pi i k\alpha}{M}\right); k = 0, 1, \cdots, M-1$.

The effective weighted sum of the M-WFRFT based on the fractional-order matrix is also four terms. In order to prove its unitarity, we denote

$$(T_{MW}^{\alpha})^H = \frac{1}{M}\left(Y_0^H B_0^{-\alpha} + Y_1^H B_1^{-\alpha} + Y_2^H B_2^{-\alpha} + Y_3^H B_3^{-\alpha}\right), \tag{50}$$

Therefore, there is

$$T_{MW}^{\alpha}(T_{MW}^{\alpha})^H = \frac{1}{M^2}\left(\sum_{k=0}^{3}\sum_{l=0}^{3} Y_k Y_l^H B_k^{\alpha} B_l^{-\alpha}\right). \tag{51}$$

From Equation (36), we can obtain

$$Y_k Y_l^H = V \begin{pmatrix} Q_1(k) & 0 & \cdots & 0 \\ 0 & Q_2(k) & \cdots & 0 \\ \vdots & \vdots & \ddots & \vdots \\ 0 & 0 & 0 & Q_n(k) \end{pmatrix} V^H \left[V \begin{pmatrix} Q_1(l) & 0 & \cdots & 0 \\ 0 & Q_2(l) & \cdots & 0 \\ \vdots & \vdots & \ddots & \vdots \\ 0 & 0 & 0 & Q_n(l) \end{pmatrix} V^H \right]^H. \tag{52}$$

The eigenvector V of the DFT can be defined as a real symmetric matrix [27–29]; and through Equations (41), (43), (45) and (47), we know that the value of $Q_r(k)$ is 0 or M (M is an integer greater than 4). Therefore, $Y_l^H = Y_l$. Then, Equation (52) can be expressed as

$$Y_k Y_l^H = Y_k Y_l = V \begin{pmatrix} Q_1(k)Q_1(l) & 0 & \cdots & 0 \\ 0 & Q_2(k)Q_2(l) & \cdots & 0 \\ \vdots & \vdots & \ddots & \vdots \\ 0 & 0 & 0 & Q_n(k)Q_n(l) \end{pmatrix} V^H, \tag{53}$$

and

$$Q_r(k)Q_r(l) = \begin{cases} M^2, & k = l \\ 0, & k \neq l. \end{cases} \tag{54}$$

Therefore, we can obtain

$$Y_k Y_l^H = \begin{cases} M Y_k, & k = l \\ 0, & k \neq l. \end{cases} \tag{55}$$

Then, the result of Equation (51) is

$$\begin{aligned} T_{MW}^\alpha \left(T_{MW}^\alpha\right)^H &= \frac{1}{M^2} \left(\sum_{k=0}^{3} \sum_{l=0}^{3} Y_k Y_l^H B_k^\alpha B_l^{-\alpha} \right) \\ &= \frac{1}{M}(Y_0 + Y_1 + Y_2 + Y_3) \\ &= \frac{1}{M} \left(V \begin{pmatrix} M & 0 & \cdots & 0 \\ 0 & M & \cdots & 0 \\ \vdots & \vdots & \ddots & \vdots \\ 0 & 0 & \cdots & M \end{pmatrix} V^H \right) \\ &= I. \end{aligned} \tag{56}$$

$\square$

Remark 3. *With the help of theoretical analysis, we can confirm that the M-WFRFT based on the fractional-order matrix has unitarity. However, we find that the theoretical analysis deviates from the previous numerical simulation [1], which we will discuss further in Section 4.*

3.3. Eigendecomposition-Type FRFT as the Basis Function

Proposition 3. *Eigendecomposition-type FRFT is used as the basis function, so the M-WFRFT has unitarity.*

Proof. In [2], Zhu et al. proposed the M-WFRFT and stated that the basis function is the FRFT, as shown in Equation (57).

$$F^\alpha[f(t)] = \int_{-\infty}^{\infty} K_\alpha(u,t)f(t)dt, \tag{57}$$

where the transform kernel is given by

$$K_\alpha(u,t) = \begin{cases} A_\alpha e^{i\frac{u^2+t^2}{2}\cot\phi - iut\csc\phi} & \alpha \neq k\pi \\ \delta(u-t) & \alpha = 2k\pi \\ \delta(u+t) & \alpha = (2k+1)\pi \end{cases}, \tag{58}$$

where $\phi = \alpha\pi/2$ is interpreted as a rotation angle in the phase plane and $A_\alpha = \sqrt{(1 - i\cot\alpha)/2\pi}$.

As we know, Equation (57) is a continuous FRFT, and a discrete FRFT is used for numerical simulation. At present, the discrete definition [29] closest to the continuous FRFT is

$$F^\alpha(m,n) = \sum_{k=0}^{N-1} v_k(m)e^{-i\frac{\pi}{2}k\alpha}v_k(n),\tag{59}$$

where $v_k(n)$ is an arbitrary orthonormal eigenvector set of the $N \times N$ DFT matrix. Equation (59) can be written as

$$F^\alpha = VD^\alpha V^H,\tag{60}$$

where $V = (v_0, v_1, \cdots, v_{N-1})$, v_k is the kth-order DFT Hermite eigenvector, and D^α is a diagonal matrix, defined as

$$D^\alpha = \mathrm{diag}\left(1, e^{-i\frac{\pi}{2}\alpha}, \cdots, e^{-i\frac{\pi}{2}(N-2)\alpha}, e^{-i\frac{\pi}{2}(N-1)\alpha}\right), \text{ when } N \text{ is odd},\tag{61}$$

and

$$D^\alpha = \mathrm{diag}\left(1, e^{-i\frac{\pi}{2}\alpha}, \cdots, e^{-i\frac{\pi}{2}(N-2)\alpha}, e^{-i\frac{\pi}{2}(N)\alpha}\right), \text{ when } N \text{ is even}.\tag{62}$$

We only prove that N is odd (when N is even, the proof process is the same). Therefore, there is

$$D^\alpha = \mathrm{diag}\left((1)^\alpha, (-i)^\alpha, (-1)^\alpha, (i)^\alpha, (1)^\alpha, (-i)^\alpha, (-1)^\alpha, (i)^\alpha, \cdots\cdots, (1 \ or -1)^\alpha\right).\tag{63}$$

Then, Equation (10) can be written as

$$\begin{aligned}
Y_k &= u^{0\times k} \times F^0 + u^{1\times k} \times F^{\frac{4}{M}} + \cdots + u^{(M-1)\times k} \times F^{\frac{4(M-1)}{M}}\\[4pt]
&= u^{0\times k}V\begin{pmatrix} 1 & 0 & \cdots & 0 \\ 0 & (-i)^0 & \cdots & 0 \\ \vdots & \vdots & \ddots & \vdots \\ 0 & 0 & \cdots & (1 \ or -1)^0 \end{pmatrix}V^H + u^{1\times k}V\begin{pmatrix} 1 & 0 & \cdots & 0 \\ 0 & (-i)^{\frac{4}{M}} & \cdots & 0 \\ \vdots & \vdots & \ddots & \vdots \\ 0 & 0 & \cdots & (1 \ or -1)^{\frac{4}{M}} \end{pmatrix}V^H + \cdots + u^{(M-1)\times k}V\begin{pmatrix} 1 & 0 & \cdots & 0 \\ 0 & (-i)^{\frac{4(M-1)}{M}} & \cdots & 0 \\ \vdots & \vdots & \ddots & \vdots \\ 0 & 0 & \cdots & (1 \ or -1)^{\frac{4(M-1)}{M}} \end{pmatrix}V^H.
\end{aligned}\tag{64}$$

We can further obtain Equation (65) as

$$Y_k = V\begin{pmatrix} Q^{(1)}(k) & 0 & \cdots & 0 \\ 0 & Q^{(-i)}(k) & \cdots & 0 \\ \vdots & \vdots & \ddots & \vdots \\ 0 & 0 & \cdots & Q^{(1 \ or -1)}(k) \end{pmatrix}V^H.\tag{65}$$

The diagonal matrix of Equation (65) can be expressed as

$$\mathrm{diag}\left(Q^{(1)}(k), Q^{(-i)}(k), Q^{(-1)}(k), Q^{(i)}(k), Q^{(1)}(k), Q^{(-i)}(k), \cdots\cdots, Q^{(1 \ or -1)}(k)\right).\tag{66}$$

Then, $Q^{(1)}(k)$ is the same as Equation (40), $Q^{(-i)}(k)$ is the same as Equation (46), $Q^{(-1)}(k)$ is the same as Equation (44), and $Q^{(i)}(k)$ is the same as Equation (42). Thus, Y_k can be obtained as

$$Y_k = \begin{cases} Y_k, & k = 0,1,2,3 \\ 0, & k = 4,5,\cdots, M-1. \end{cases}\tag{67}$$

All the following proofs are the same as Section 3.2. In other words, the M-WFRFT has unitarity. $\qquad\square$

Remark 4. *From Equation (67), it is not difficult to find that there are only four weighted terms of the M-WFRFT based on the eigendecomposition-type FRFT.*

3.4. Other Types of FRFTs

There are three types of discrete definitions of the FRFT. In Section 3.1, the linear WFRFT is used. The fractional-order matrix is used in Section 3.2. The discrete FRFT, which is called the eigendecomposition type, is used in Section 3.3. Then, there is a sampling-type FRFT.

In [30], a sampling-type FRFT is proposed, and its process can be written as follows:

(a)　Chirp multiplication

$$g(x_0) = \exp\left[-ipx_0^2 \tan(f/2)\right] f(x_0);\tag{68}$$

(b)　Chirp convolution

$$g'(x) = A_\phi \int_{-\infty}^{\infty} \exp[i\pi \csc(\phi)(x - x_0)^2] g(x_0) dx_0;\tag{69}$$

(c)　Chirp multiplication

$$f_\alpha(x) = \exp\left[-i\pi x^2 \tan(\phi/2)\right] g'(x).\tag{70}$$

The definition of the sampling type is the numerical simulation of a continuous FRFT.

The discretization of the FRFT has been extensively studied [12], and the three main types of DFRFTs are compared, as shown in Table 2. We noticed that the sampling-type FRFT did not satisfy additivity and unitarity.

Table 2. Comparison of the three main types of DFRFT.

	Linear Weighted Type	Eigendecomposition Type	Sampling Type
Unitarity	$\checkmark$	$\checkmark$	$\times$
Additivity	$\checkmark$	$\checkmark$	$\times$
Approximation	$\times$	$\checkmark$	$\checkmark$
Closed-form	$\checkmark$	$\times$	$\checkmark$
Complexity	$O(N\log N)$	$O(N^2)$	$O(N\log N)$

Remark 5. *The M-WFRFT is an extended definition, and its basis function can be expressed as shown in Figure 1. The sampling type FRFT does not satisfy the additivity and unitarity, so it cannot be used as a basis function.*

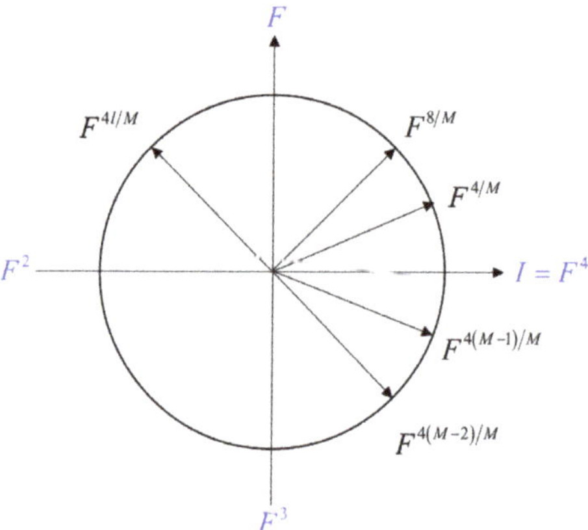

Figure 1. Time-frequency denotation of the M-WFRFT operator.

4. Discussion

Our previous research only verified the unitarity of the M-WFRFT via numerical simulation [1], but the simulation results are different from the theoretical proof in Section 3.2. Next, we will analyze and discuss this issue. Equation (10) can be verified using MATLAB, and its program is shown in Code 1.

Code 1. The program of Equation (10).

```
1.    function Yk = Yk(N,M)
2.    % M is the resulting weighting term, for example: M = 4(4-WFRFT); M = 5(5-WFRFT)
3.    % N is the length of the signal;
4.    F = zeros(N);
5.    for k = 1:N
6.        for h = 1:N
7.            F(h,k) = exp(2*pi*i*(h−1)*(k−1)/N)/sqrt(N); % IDFT
8.        end
9.    end
10.   F = fftshift(F);
11.   for k = 0:M−1
12.       yy = F^(4*k/M);     % Fractional power of Fourier transform
13.       y{k + 1} = yy;
14.   end
15.   % celldisp(y);
16.   u = zeros(M);
17.   for k = 1:M
18.       for h = 1:M
19.           u(h,k) = exp(−2*pi*i*(h−1)*(k−1)/M); %DFT
20.       end
21.   end
22.   for k = 1:M
23.       YY = zeros(N);
24.       for h = 1:M
25.           YY = YY + u(h,k)*y{h};
26.       end
27.           Y{k} = YY;     % Yk in the paper is obtained
28.   end
29.   Celldisp(y)
```

We tested from 2 to 1000 dimension and found that Y_k was a real matrix only when the dimensions were

$$N = 2, 3, 4, 5, 6, 7, 8, 9, 10, 11, 12, 13, 16, 18, 21, 28, 29, 32, 33, 44. \tag{71}$$

Therefore, the unitarity of the M-WFRFT is only available in the aforementioned cases. Y_k has the following rules:

$$\begin{cases} 5\text{-}WFRFT \Rightarrow & Y_0 \quad Y_1 \quad Y_2 \quad Y_3 \quad Y_4 \\ 6\text{-}WFRFT \Rightarrow & Y_0 \quad Y_1 \quad Y_2 \quad Y_3 \quad Y_4 \quad Y_5 \\ 7\text{-}WFRFT \Rightarrow & Y_0 \quad Y_1 \quad Y_2 \quad Y_3 \quad Y_4 \quad Y_5 \quad Y_6 \\ 8\text{-}WFRFT \Rightarrow & Y_0 \quad Y_1 \quad Y_2 \quad Y_3 \quad Y_4 \quad Y_5 \quad Y_6 \quad Y_7 \\ \quad\vdots & \quad\quad\quad\quad\vdots \\ M\text{-}WFRFT \Rightarrow & Y_0 \quad Y_1 \quad Y_2 \quad Y_3 \quad Y_4 \quad \cdots \quad Y_{M-3} \quad Y_{M-2} \quad Y_{M-1} \end{cases} \tag{72}$$

where the blue Y_k indicates that the result is zero.

For other dimensions, the M-WFRFT does not have unitarity, and Y_k is as follows:

$$\begin{cases} 5\text{-}WFRFT \Rightarrow & Y_0 \quad Y_1 \quad Y_2 \quad Y_3 \quad Y_4 \\ 6\text{-}WFRFT \Rightarrow & Y_0 \quad Y_1 \quad Y_2 \quad Y_3 \quad Y_4 \quad Y_5 \\ 7\text{-}WFRFT \Rightarrow & Y_0 \quad Y_1 \quad Y_2 \quad Y_3 \quad Y_4 \quad Y_5 \quad Y_6 \\ 8\text{-}WFRFT \Rightarrow & Y_0 \quad Y_1 \quad Y_2 \quad Y_3 \quad Y_4 \quad Y_5 \quad Y_6 \quad Y_7 \\ \quad\vdots & \qquad\qquad \vdots \\ M\text{-}WFRFT \Rightarrow & Y_0 \quad Y_1 \quad Y_2 \quad Y_3 \quad Y_4 \quad \cdots \quad Y_{M-3} \quad Y_{M-2} \quad Y_{M-1} \end{cases} \tag{73}$$

where the blue Y_k indicates that the result is zero.

The numerical simulation results show that the M-WFRFT has unitarity only in certain dimensions. Following the theory of Section 3.2, the program is shown in Code 2. Our purpose is to compare the results of Code 2 with the results of Code 1.

Code 2. The program of Equation (36).

```
1.    function Yk = Yk1(N.M)
2.    % M is the resulting weighting term, for example: M = 4(4-WFRFT); M = 5(5-WFRFT)
3.    % N is the length of the signal;
4.    F = zeros(N);
5.    for k = 1:N
6.            for h = 1:N
7.                F(h,k) = exp(2*pi*i*(h−1)*(k−1)/N)/sqrt(N); % IDFT
8.            end
9.    end
10.   F = fftshift(F);
11.   [V,D] = eig(F);
12.   for k = 0:M−1
13.            YY = zeros(N);
14.            for l = 0:M−1
15.                YY = YY + exp(−2*pi*i*k*l/M)*D^(4*l/M);
16.            end
17.            YY = V*YY*inv(V);
18.            Y{k + 1} = YY;          % Yk in the paper is obtained
19.   end
20.   celldisp(Y)
```

After verification, we found that the results of Codes 1 and 2 are the same. Therefore, the numerical simulation shows that the unitarity of the M-WFRFT is related to signal length. However, our theoretical analysis shows that the unitarity of the M-WFRFT does not depend on signal length. Therefore, there is a problem insofar as the simulation verification is inconsistent with the theoretical analysis. In order to solve this problem, we will analyze it with a specific numerical value. For Code 2, when $M = 7$ and $N = 13$, we can obtain the eigenvalue of the DFT in *line* 11 of Code 2. Therefore, the eigenvalue matrix D is

$$D = \begin{pmatrix} i & & & & & & & & & & & & \\ & i & & & & & & & & & & & \\ & & i & & & & & & & & & & \\ & & & -i & & & & & & & & & \\ & & & & -i & & & & & & & & \\ & & & & & -i & & & & & & & \\ & & & & & & -1 & & & & & & \\ & & & & & & & -1 & & & & & \\ & & & & & & & & -1 & & & & \\ & & & & & & & & & 1 & & & \\ & & & & & & & & & & 1 & & \\ & & & & & & & & & & & 1 & \\ & & & & & & & & & & & & 1 \end{pmatrix}_{13\times13} \tag{74}$$

Then, for Equation (36), the calculated values of Y_k ($k = 0, 1, \cdots, 6$) are

$$Y_0 = V \begin{pmatrix} 0 & & & & & & \\ & 0 & & & & & \\ & & 0 & & & & \\ & & & 0 & & & \\ & & & & 0 & & \\ & & & & & 7 & \\ & & & & & & 7 \\ & & & & & & & 7 \end{pmatrix} V^{-1}; \tag{75}$$

$$Y_1 = V \begin{pmatrix} 7 & & & & & & \\ & 7 & & & & & \\ & & 7 & & & & \\ & & & 0 & & & \\ & & & & 0 & & \\ & & & & & 0 & \\ & & & & & & 0 \end{pmatrix} V^{-1}; \tag{76}$$

$$Y_2 = V \begin{pmatrix} 0 & & & & & & \\ & 0 & & & & & \\ & & 0 & & & & \\ & & & 7 & & & \\ & & & & 7 & & \\ & & & & & 7 & \\ & & & & & & 0 \end{pmatrix} V^{-1}; \tag{77}$$

$$Y_3 = V \begin{pmatrix} 0 & & 0 \\ & 0 & \\ & & \ddots & \\ 0 & & 0 \end{pmatrix} V^{-1}; \tag{78}$$

and $Y_4 = Y_5 = Y_3$;

$$Y_6 = V \begin{pmatrix} 0 & & & & & & \\ & 0 & & & & & \\ & & 0 & & & & \\ & & 7 & & & & \\ & & & 7 & & & \\ & & & & 0 & & \\ & & & & & 0 \end{pmatrix} V^{-1}. \tag{79}$$

In the results obtained, the values of Y_3, Y_4 and Y_5 are zero; Equation (72) is verified.

If $M = 7$ and $N = 14$, we can obtain the eigenvalue of the DFT in *line* 11 of Code 2. Therefore, the eigenvalue matrix D is

$$
D = \begin{pmatrix}
1 & & & & & & & & & & & & & \\
& 1 & & & & & & & & & & & & \\
& & 1 & & & & & & & & & \mathbf{0} & & \\
& & & 1 & & & & & & & & & & \\
& & & & i & & & & & & & & & \\
& & & & & i & & & & & & & & \\
& & & & & & -i & & & & & & & \\
& & & & & & & i & & & & & & \\
& & & & & & & & -i & & & & & \\
& & & & & & & & & -i & & & & \\
& & & & & & & & & & -1 & & & \\
& \mathbf{0} & & & & & & & & & & -1 & & \\
& & & & & & & & & & & & -1 & \\
& & & & & & & & & & & & & -1
\end{pmatrix}_{14 \times 14}
\tag{80}
$$

Then, using Equation (36), the calculated values of Y_k $(k = 0, 1, \cdots, 6)$ are

$$
Y_0 = V \begin{pmatrix}
7 & & & & & & & & & & & & & \\
& 7 & & & & & & & & & & & & \\
& & 7 & & & & & & & & \mathbf{0} & & & \\
& & & 7 & & & & & & & & & & \\
& & & & 0 & & & & & & & & & \\
& & & & & 0 & & & & & & & & \\
& & & & & & 0 & & & & & & & \\
& & & & & & & 0 & & & & & & \\
& & & & & & & & 0 & & & & & \\
& & & & & & & & & 0 & & & & \\
& \mathbf{0} & & & & & & & & & 0 & & & \\
& & & & & & & & & & & 0 & & \\
& & & & & & & & & & & & 0 &
\end{pmatrix} V^{-1};
\tag{81}
$$

$$
Y_1 = V \begin{pmatrix}
0 & & & & & & & & & & & & & \\
& 0 & & & & & & & & & & & & \\
& & 0 & & & & & & & & \mathbf{0} & & & \\
& & & 0 & & & & & & & & & & \\
& & & & 7 & & & & & & & & & \\
& & & & & 7 & & & & & & & & \\
& & & & & & 0 & & & & & & & \\
& & & & & & & 7 & & & & & & \\
& & & & & & & & 0 & & & & & \\
& & & & & & & & & 0 & & & & \\
& \mathbf{0} & & & & & & & & & 0 & & & \\
& & & & & & & & & & & 0 & & \\
& & & & & & & & & & & & 0 &
\end{pmatrix} V^{-1};
\tag{82}
$$

$$
Y_2 = V \begin{pmatrix}
0 & & & & & & & & & & & & & \\
& 0 & & & & & & & & & & & & \\
& & 0 & & & & & & & & \mathbf{0} & & & \\
& & & 0 & & & & & & & & & & \\
& & & & 0 & & & & & & & & & \\
& & & & & 0 & & & & & & & & \\
& & & & & & 0 & & & & & & & \\
& & & & & & & 0 & & & & & & \\
& & & & & & & & 0 & & & & & \\
& & & & & & & & & 7 & & & & \\
& \mathbf{0} & & & & & & & & & 0 & & & \\
& & & & & & & & & & & 7 & & \\
& & & & & & & & & & & & 7 &
\end{pmatrix} V^{-1};
\tag{83}
$$

$$
Y_3 = V \begin{pmatrix}
0 & & & 0 \\
& 0 & & \\
& & \ddots & \\
0 & & & 0
\end{pmatrix} V^{-1};
\tag{84}
$$

and $Y_4 = Y_3$;

$$Y_5 = V \begin{pmatrix} 0 & & & & & & & & & & & \\ & 0 & & & & & & & & & & \\ & & 0 & & & & & & & & & \\ & & & 0 & & & & & & & & \\ & & & & 0 & & & & & & & \\ & & & & & 0 & & & & & & \\ & & & & & & 0 & & & & & \\ & & & & & & & 0 & & & & \\ & & & & & & & & 0 & & & \\ & & & & & & & & & 7 & & \\ & & & & & & & & & & 0 & \\ & & & & & & & & & & & 0 \end{pmatrix} V^{-1}; \tag{85}$$

$$Y_6 = V \begin{pmatrix} 0 & & & & & & & & & & & \\ & 0 & & & & & & & & & & \\ & & 0 & & & & & & & & & \\ & & & 0 & & & & & & & & \\ & & & & 0 & & & & & & & \\ & & & & & 7 & & & & & & \\ & & & & & & 0 & & & & & \\ & & & & & & & 7 & & & & \\ & & & & & & & & 7 & & & \\ & & & & & & & & & 0 & & \\ & & & & & & & & & & 0 & \\ & & & & & & & & & & & 0 \end{pmatrix} V^{-1}; \tag{86}$$

In the results obtained, the values of Y_3 and Y_4 are zero, and Equation (73) is verified.

When $N = 13$, Equation (72) is obtained by means of Code 1. However, in the theoretical analysis, the nonzero terms of Y_k are Y_0, Y_1, Y_2 and Y_3, which are different from the simulation results presented by Equation (72). This problem is generated by fractional power operation, based on MATLAB, mainly in *line* 15 of Code 2 (*line* 12 of Code 1), and its operation $D^{4l/M}$ ($F^{4l/M}$).

According to the deMoivre theorem, we know that

$$[r(\cos\theta + i\sin\theta)]^n = r^n(\cos n\theta + i\sin n\theta). \tag{87}$$

where n is a positive integer. Therefore, for Equation (87),

$$x^n = r(\cos\theta + i\sin\theta), \tag{88}$$

the results have n roots

$$x_k = \sqrt[n]{r}(\cos((\theta + 2k\pi)/n) + i\sin((\theta + 2k\pi)/n)). \tag{89}$$

where $k = 0, 1, \cdots, n - 1$. However, in the numerical simulation, we only obtained one of the roots. For example, $-i = \cos(3\pi/2) + i\sin(3\pi/2)$. Using MATLAB to calculate $(-i)^{1/2}$, we obtain $0.7071 - 0.7071i$. The actual results should be that the two roots are $0.7071 - 0.7071i$ and $-0.7071 + 0.7071i$, respectively. This leads to the deviation between the simulation results (Equation (72)) and the theory (Section 3.2).

For $N = 14$, the simulation results (Equation (73)) show that the M-WFRFT does not have unitarity. However, the theoretical (Section 3.2) explanation has unitarity. This problem is caused by fractional exponentiation operation based on MATLAB. In Equation (80), we notice the position of the eigenvalue (-1), but after the fractional power operation based on MATLAB, Equations (83) and (85) appear. Therefore, the final numerical simulation results show that the M-WFRFT does not have unitarity. In fact, the correct result is the sum of Equations (83) and (85).

Using the above analysis, we explain the error of the operation based on MATLAB, which is also the clarification of our previous research work. The final conclusion is that the M-WFRFT has unitarity. The M-WFRFT code is shown in Appendix A, and interested researchers can verify it.

5. Conclusions

In this paper, we present a new reformulation of the M-WFRFT to prove its unitarity. The M-WFRFT uses the DFRFT as the basis function, and the diversity of the DFRFT leads to different definitions of the M-WFRFT. We use the linear weighted-type, fractional-order matrix and eigendecomposition-type FRFT as the basis functions and prove the unitarity of the M-WFRFT. The results show that M-WFRFTs based on these three definitions have unitarity. However, with greater research, the results also show that the effective weighted sum of the M-WFRFT is only four terms. That is to say, as an extended definition of the WFRFT, the M-WFRFT shows no increase in its weighting term. It has great reference value for the application of the M-WFRFT. Furthermore, we note the deviation between the numerical simulation and the theoretical analysis, which reveals that the unitary verification based on MATLAB is inaccurate for the previous work. Finally, we analyze two examples and establish the reasons for the deviation. In other words, the fractional power operation directly based on MATLAB can only obtain one root at a time.

Author Contributions: Methodology, T.Z.; software, T.Z.; validation, T.Z. and Y.C.; investigation, Y.C.; writing—original draft preparation, T.Z.; writing—review and editing, T.Z.; supervision, T.Z.; funding acquisition, T.Z. All authors have read and agreed to the published version of the manuscript.

Funding: This study was supported by the Fundamental Research Funds for the Central Universities (N2123016); and the Scientific Research Projects of Hebei colleges and universities (QN2020511).

Acknowledgments: We would like to express our great appreciation to the editors and reviewers.

Conflicts of Interest: The authors declare no conflict of interest.

Appendix A

M-WFRFT code is written; its basis function is the WFRFT. By calling "celldisp (Y)", Y_k is verified in Section 3.1.

```
%% M-WFRFT (multi-weighted type fractional Fourier transform)
% The basis function F^(4*l/M) is WFRFT
function F = mwfrft(alpha,M,N)
% This code is written by Tieyu Zhao, E-mail: zhaotieyu@neuq.edu.cn;
% alpha is the transform order;
% M is the resulting weighting term, for example: M = 4(4-WFRFT); M = 5(5-WFRFT)
% N is the length of the signal;
for l = 0:M−1
    yy = wfrft(N,4*l/M); % WFRFT
    y{l + 1} = yy;
end
% celldisp(y);
D = zeros(M);
for k = 1:M
    for h = 1:M
        D(h,k) = exp(−2*pi*i*(h−1)*(k−1)/M); % DFT
    end
end
for k = 1:M
    YY = zeros(N);
    for h = 1:M
        YY = YY + D(h,k)*y{h};
    end
    Y{k} = YY; % Yk is obtained in Section 3.1
end
```

```
% celldisp(Y)
B = zeros(1,M);
for k = 0:M−1
      B(k + 1) = B(k + 1) + exp(2*pi*i*k*alpha/M); % B_alpha
end
F = zeros(N);
for k = 0:M−1
      F = F + B(k + 1)*Y{k + 1}/M; % M-WFRFT
end

function F = wfrft(N,beta) % WFRFT
Y = eye(N);
y1 = fftshift(fft(Y))/(sqrt(N));
y2 = y1*y1;
y3 = conj(y1);
pl = zeros(1,4);
for k = 0:3
        pl(k + 1) = pl(k + 1) + exp(i*3*pi*(beta−k)/4)*cos(pi*(beta−k)/2)*cos(pi*(beta−k)/4);
end
F = pl(1)*Y + pl(2)*y1 + pl(3)*y2 + pl(4)*y3;
```

References

1. Zhao, T.; Yuan, L.; Li, M.; Chi, Y. A reformulation of weighted fractional Fourier transform. *Digit. Signal Process.* **2020**, *104*, 102807. [CrossRef]
2. Zhu, B.; Liu, S.; Ran, Q. Optical image encryption based on multifractional Fourier transforms. *Opt. Lett.* **2000**, *25*, 1159–1161. [CrossRef] [PubMed]
3. Shih, C.C. Fractionalization of Fourier transform. *Opt. Commun.* **1995**, *118*, 495–498. [CrossRef]
4. Liu, S.; Jiang, J.; Zhang, Y.; Zhang, J. Generalized fractional Fourier transforms. *J. Phys. A Math. Gen.* **1997**, *30*, 973. [CrossRef]
5. Liu, S.T.; Zhang, J.D.; Zhang, Y. Properties of the fractionalization of a Fourier transform. *Opt. Commun.* **1997**, *133*, 50–54. [CrossRef]
6. Ran, Q.; Yeung, D.S.; Tsang, E.C.; Wang, Q. General multifractional Fourier transform method based on the generalized permutation matrix group. *IEEE Trans. Signal Process.* **2004**, *53*, 83–98.
7. Zhao, T.Y.; Ran, Q.W. The Weighted Fractional Fourier Transform and Its Application in Image Encryption. *Math. Probl. Eng.* **2019**, *2019*, 1–10. [CrossRef]
8. Tao, R.; Lang, J.; Wang, Y. Optical image encryption based on the multiple-parameter fractional Fourier transform. *Opt. Lett.* **2008**, *33*, 581–583. [CrossRef]
9. Ran, Q.W.; Zhang, H.Y.; Zhang, J.; Tan, L.Y.; Ma, J. Deficiencies of the cryptography based on multiple-parameter fractional Fourier transform. *Opt. Lett.* **2009**, *34*, 1729–1731. [CrossRef]
10. Ran, Q.W.; Zhao, T.Y.; Yuan, L.; Wang, J.; Xu, L. Vector power multiple-parameter fractional Fourier transform of image encryption algorithm. *Opt. Laser Eng.* **2014**, *62*, 80–86. [CrossRef]
11. Zhao, T.; Ran, Q.; Yuan, L.; Chi, Y.; Ma, J. Security of image encryption scheme based on multi-parameter fractional Fourier transform. *Opt. Commun.* **2016**, *376*, 47–51. [CrossRef]
12. Santhanam, B.; McClellan, J.H. The discrete rotational Fourier transform. *IEEE Trans. Signal Process.* **1996**, *44*, 994–998. [CrossRef]
13. Saxsena, R.; Singh, K. Fractional Fourier Transform: A novel tool for signal processing. *J. Indian Inst. Sci.* **2005**, *85*, 11–26.
14. Tao, R.; Zhang, F.; Wang, Y. Research progress on discretization of fractional Fourier transform. *Sci. China Ser. F Inf. Sci.* **2008**, *51*, 859. [CrossRef]
15. Kang, X.; Tao, R.; Zhang, F. Multiple-parameter discrete fractional transform and its applications. *IEEE Trans. Signal Process.* **2016**, *64*, 3402–3417. [CrossRef]
16. Zhang, Y.; Wang, S.; Yang, J.F.; Zhang, Z.; Phillips, P.; Sun, P.; Yan, J. A comprehensive survey on fractional Fourier transform. *Fund. Inf.* **2017**, *151*, 1–48. [CrossRef]
17. Zayed, A. Two-dimensional fractional Fourier transform and some of its properties. *Integral Transforms Spec. Funct.* **2018**, *29*, 553–570. [CrossRef]
18. Gómez-Echavarría, A.; Ugarte, J.P.; Tobón, C. The Fractional Fourier transform as a biomedical signal and image processing tool: A review. *Biocybern. Biomed. Eng.* **2020**, *40*, 1081–1093. [CrossRef]
19. Liu, Y.; Zhang, F.; Miao, H.; Tao, R. The hopping discrete fractional Fourier transform. *Signal Process.* **2021**, *178*, 7763. [CrossRef]
20. Hosny, K.M.; Darwish, M.M.; Aboelenen, T. New fractional-order Legendre-Fourier moments for pattern recognition applications. *Commun. Comput. Inf. Sci.* **2020**, *103*, 7324. [CrossRef]

21. Li, Y.; Song, Z.Q.; Sha, X.J. The multi-weighted type fractional fourier transform scheme and its application over wireless communications. *EURASIP J. Wirel. Commun.* **2018**, *2018*, 1–10. [CrossRef]
22. Fang, X.; Zhang, N.; Zhang, S.; Chen, D.; Sha, X.; Shen, X. On physical layer security: Weighted fractional Fourier transform based user cooperation. *IEEE Trans. Wirel. Commun.* **2017**, *16*, 5498–5510. [CrossRef]
23. Zhang, Y.D.; Chen, S.; Wang, S.H.; Yang, J.F.; Phillips, P. Magnetic resonance brain image classification based on weighted-type fractional Fourier transform and nonparallel support vector machine. *Int. J. Imaging Syst. Technol.* **2015**, *25*, 317–327.
24. Liang, Y.; Da, X.; Wang, S. On narrowband interference suppression in OFDM-based systems with CDMA and weighted-type fractional Fourier transform domain preprocessing. *KSII Trans. Internet Inf. Syst. (TIIS)* **2017**, *11*, 5377–5391.
25. Sha, X.j.; Qiu, X.; Mei, L. Hybrid carrier CDMA communication system based on weighted-type fractional Fourier transform. *IEEE Commun. Lett.* **2012**, *16*, 432–435.
26. McClellan, J.; Parks, T. Eigenvalue and eigenvector decomposition of the discrete Fourier transform. *IEEE Trans. Audio Electroacoust.* **1972**, *20*, 66–74. [CrossRef]
27. Dickinson, B.; Steiglitz, K. Eigenvectors and functions of the discrete Fourier transform. *IEEE Trans. Acoust. Speech Signal Process.* **1982**, *30*, 25–31.
28. Bose, N. Eigenvectors and eigenvalues of 1-D and nD DFT matrices. *AEU-Int. J. Electron. Commun.* **2001**, *55*, 131–133.
29. Candan, C.; Kutay, M.A.; Ozaktas, H.M. The discrete fractional Fourier transform. *IEEE Trans. Signal Process.* **2000**, *48*, 1329–1337.
30. Ozaktas, H.M.; Arikan, O.; Kutay, M.A.; Bozdagt, G. Digital computation of the fractional Fourier transform. *IEEE Trans. Signal Process.* **1996**, *44*, 2141–2150.

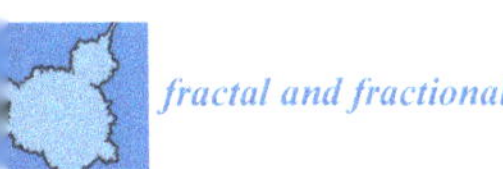
fractal and fractional

Article

Spiral Dive Control of Underactuated AUV Based on a Single-Input Fractional-Order Fuzzy Logic Controller

Zhiyu Cui [1,2], Lu Liu [1,2,*], Boyu Zhu [1], Lichuan Zhang [1,2], Yang Yu [1,2], Zhexuan Zhao [1], Shiyuan Li [1] and Mingwei Liu [1]

[1] Research & Development Institute, Northwestern Polytechnical University, Shenzhen 518057, China
[2] School of Marine Science and Technology, Northwestern Polytechnical University, Xi'an 710072, China
* Correspondence: liulu12201220@nwpu.edu.cn

Abstract: Autonomous underwater vehicles (AUVs) have broad applications owing to their ability to undertake long voyages, strong concealment, high level of intelligence and ability to replace humans in dangerous operations. AUV motion control systems can ensure stable operation in the complex ocean environment and have attracted significant research attention. In this paper, we propose a single-input fractional-order fuzzy logic controller (SIFOFLC) as an AUV motion control system. First, a single-input fuzzy logic controller (SIFLC) was proposed based on the signed distance method, whose control input is the linear combination of the error signal and its derivative. The SIFLC offers a significant reduction in the controller design and calculation process. Then, a SIFOFLC was obtained with the derivative of the error signal extending to a fractional order and offering greater flexibility and adaptability. Finally, to verify the superiority of the proposed control algorithm, comparative numerical simulations in terms of spiral dive motion control were conducted. Meanwhile, the parameters of different controllers were optimized according to the hybrid particle swarm optimization (HPSO) algorithm. The simulation results illustrate the superior stability and transient performance of the proposed control algorithm.

Keywords: fractional calculus; autonomous underwater vehicle; fuzzy logic control; particle swarm optimization algorithm

Citation: Cui, Z.; Liu, L.; Zhu, B.; Zhang, L.; Yu, Y.; Zhao, Z.; Li, S.; Liu, M. Spiral Dive Control of Underactuated AUV Based on a Single-Input Fractional-Order Fuzzy Logic Controller. *Fractal Fract.* **2022**, *6*, 519. https://doi.org/10.3390/fractalfract6090519

Academic Editors: Thach Ngoc Dinh, Shyam Kamal and Rajesh Kumar Pandey

Received: 24 July 2022
Accepted: 29 August 2022
Published: 14 September 2022

Publisher's Note: MDPI stays neutral with regard to jurisdictional claims in published maps and institutional affiliations.

1. Introduction

Autonomous underwater vehicles (AUVs) are a crucial technical platform for ocean information acquisition and autonomous operation. They have extensive application prospects, such as marine environment observation, marine resources exploration and security defense. Nevertheless, motion control systems for AUV have become very challenging due to their high nonlinearity, strong coupling, model parameter uncertainties and external disturbances. In addition, an AUV system is usually designed to be underactuated to save cost and improve propulsion efficiency.

With regard to the motion control of underactuated AUV, a variety of control algorithms are available, including proportional-integral-derivative (PID) control, backstepping control, fuzzy logic control, and sliding mode control [1–7]. In [4], a single-input fuzzy logic controller (SIFLC) was proposed for AUV depth control. Simulation results show that the SIFLC gives an identical response as Mamdani and T-S type FLC to the same input sets, while its execution time is more than two orders of magnitude faster than the conventional FLC. In [6], a switching control algorithm based on active disturbance rejection control (ADRC) and fuzzy logic control was applied to the depth control of a self-developed AUV. Numerical simulations showed that the proposed method is more efficient in suppressing external disturbances and inner signal transmission disturbance than PID controller.

Fractional calculus is an extension of traditional integral calculus, it describes the fractal dimension of space. In recent years, its applications in the control field, such as fractional-order model [8,9], fractional-order control algorithm [10–12], fractional-order optimization algorithm [13,14], have attracted significant research attention. Furthermore, stability analysis for fractional-order control systems have been proposed in several studies [15–17]. A fractional-order, proportional–integral–derivative (FOPID) controller has been proposed by Podlubny. Their proposed controller has two additional parameters: integral order and differential order compared with a PID controller [18]. In [19], an optimized FOPID controller for improved transient control performance was applied to an AUV yaw control system. In addition, a fractional-order Mamdani fuzzy logic controller has been proposed for vehicle nonlinear active suspension, which effectively improves ride comfort and handling stability [20]. However, there has been no report on the application of fractional-order fuzzy logic control in AUV motion.

In this paper, a single-input fractional-order fuzzy logic controller (SIFOFLC) is proposed and applied to an AUV motion control system. Its control input was simplified to a single variable known as distance variable by applying the signed distance method [21], which aims to reduce the computation burden and complex parameter tuning process. Furthermore, a fractional calculus operator was applied to the enhanced FLC due to its recognized ability to increase the controller's flexibility and adaptability. With respect to the controller parameters, we developed and applied a hybrid particle swarm optimization (HPSO) algorithm to obtain optimal control performance. Unlike a conventional PSO algorithm, this includes the local optimal particle term to avoid falling into local optimal region, and the fitness value function includes both steady-state performance and transient performance of an AUV motion control system. To verify the effectiveness of SIFOFLC, we conducted comparative numerical simulations of spiral dive motion control. The object of study, REMUS-100 AUV, was developed by Woods Hole Oceanographic Institution [22], while the simulation was performed using the marine systems simulator (MSS) by Fossen and Perez [23]. Simulation results show that, compared with a FOPID controller and conventional T-S FLC, the SIFOFLC is more efficient in reducing angular velocity oscillations, shortening settling time and improving control accuracy.

The remainder of this paper is organized as follows. Section 2 discusses the six degrees of freedom nonlinear motion equations of AUV. The SIFOFLC design is introduced in Section 3, along with its advantages compared with traditional T-S FLC. In Section 4, the HPSO algorithm is described and is applied to various control systems to obtain optimal parameters. To verify the effectiveness of the proposed method, simulations and numerical comparisons are carried out in Section 5. Finally, some concluding remarks are presented in Section 6.

2. Kinematic and Dynamic Modeling of AUV

Six degrees of freedom motion equations of AUV can be described using the earth-fixed coordinate system and body-fixed coordinate system shown in Figure 1, both of which are right-handed. The earth-fixed coordinate system $O - xyz$ has its origin O fixed to the earth, and the body-fixed coordinate system $O_b - x_b y_b z_b$ is a moving reference frame with its origin O_b fixed to AUV center of buoyancy.

The general motion of a vehicle in six degrees of freedom can be described with the following vectors:

$$\eta_1 = \begin{bmatrix} x & y & z \end{bmatrix}^T \ \eta_2 = \begin{bmatrix} \phi & \theta & \psi \end{bmatrix}^T$$
$$v_1 = \begin{bmatrix} u & v & w \end{bmatrix}^T \ v_2 = \begin{bmatrix} p & q & r \end{bmatrix}^T$$
$$\tau_1 = \begin{bmatrix} X & Y & Z \end{bmatrix}^T \ \tau_2 = \begin{bmatrix} K & M & N \end{bmatrix}^T$$

where η describes the position and orientation of the vehicle with respect to the earth-fixed reference frame, v denotes the linear and angular velocities with respect to the body-fixed reference frame, and τ describes the total forces and moments acting on the vehicle in the body-fixed reference frame.

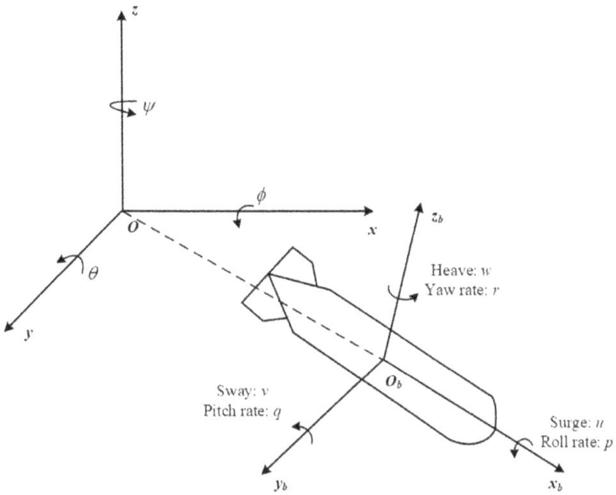

Figure 1. Coordinate systems of AUV.

The coordinate transformation of the translational velocity between earth-fixed and body-fixed coordinate systems can be expressed as

$$
\begin{bmatrix} \dot{x} \\ \dot{y} \\ \dot{z} \end{bmatrix} = J_1 \begin{bmatrix} u \\ v \\ w \end{bmatrix}
\tag{1}
$$

where

$$
J_1 = \begin{bmatrix}
\cos\psi\cos\theta & -\sin\psi\cos\phi + \cos\psi\sin\theta\sin\phi & \sin\psi\sin\phi + \cos\psi\sin\theta\cos\phi \\
\sin\psi\cos\theta & \cos\psi\cos\phi + \sin\psi\sin\theta\sin\phi & -\cos\psi\sin\phi + \sin\psi\sin\theta\cos\phi \\
-\sin\theta & \cos\theta\sin\phi & \cos\theta\cos\phi
\end{bmatrix}
$$

The coordinate transformation relates rotational velocity between two coordinate systems and can be described as

$$
\begin{bmatrix} \dot{\phi} \\ \dot{\theta} \\ \dot{\psi} \end{bmatrix} = J_2 \begin{bmatrix} p \\ q \\ r \end{bmatrix}
\tag{2}
$$

where

$$
J_2 = \begin{bmatrix}
1 & \sin\phi\tan\theta & \cos\phi\tan\theta \\
0 & \cos\phi & -\sin\phi \\
0 & \sin\phi/\cos\theta & \cos\phi/\cos\theta
\end{bmatrix}
$$

The locations of the AUV centers of gravity and buoyancy are defined in the body-fixed coordinate system as follows:

$$
r_G = \begin{bmatrix} x_g & y_g & z_g \end{bmatrix}^T \quad r_B = \begin{bmatrix} x_b & y_b & z_b \end{bmatrix}^T
$$

Based on the theory of rigid body dynamics and the analysis of total forces and moments acting on AUV, the nonlinear motion equations for the REMUS vehicle in six degrees of freedom can be expressed as follows [24]:

$$m[\dot{u} - vr + wq - x_g(q^2 + r^2) + y_g(pq - \dot{r}) + z_g(pr + \dot{q})] = X_{HS} + X_{u|u|}u|u| +$$
$$X_{\dot{u}}\dot{u} + X_{wq}wq + X_{qq}qq + X_{vr}vr + X_{rr}rr + X_{prop}$$
$$m[\dot{v} - wp + ur - y_g(r^2 + p^2) + z_g(qr - \dot{p}) + x_g(pq + \dot{r})] = Y_{HS} + Y_{v|v|}v|v| + Y_{r|r|}r|r| +$$
$$Y_{\dot{v}}\dot{v} + Y_{\dot{r}}\dot{r} + Y_{ur}ur + Y_{wp}wp + Y_{pq}pq + Y_{uv}uv + Y_{uu\delta_r}u^2\delta_r$$
$$m[\dot{w} - uq + vp - z_g(p^2 + q^2) + x_g(rp - \dot{q}) + y_g(rq + \dot{p})] = Z_{HS} + Z_{w|w|}w|w| + Z_{q|q|}q|q| +$$
$$Z_{\dot{w}}\dot{w} + Z_{\dot{q}}\dot{q} + Z_{uq}uq + Z_{vp}vp + Z_{rp}rp + Z_{uw}uw + Z_{uu\delta_s}u^2\delta_s$$
$$I_{xx}\dot{p} + (I_{zz} - I_{yy})qr + m[y_g(\dot{w} - uq + vp) - z_g(\dot{v} - wp + ur)] = K_{HS} + K_{p|p|}p|p| + K_{\dot{p}}\dot{p} + K_{prop}$$
$$I_{yy}\dot{q} + (I_{xx} - I_{zz})rp + m[z_g(\dot{u} - vr + wq) - x_g(\dot{w} - uq + vp)] = M_{HS} + M_{w|w|}w|w| + M_{q|q|}q|q| +$$
$$M_{\dot{w}}\dot{w} + M_{\dot{q}}\dot{q} + M_{uq}uq + M_{vp}vp + M_{rp}rp + M_{uw}uw + M_{uu\delta_s}u^2\delta_s$$
$$I_{zz}\dot{r} + (I_{yy} - I_{xx})pq + m[x_g(\dot{v} - wp + ur) - y_g(\dot{u} - vr + wq)] = N_{HS} + N_{v|v|}v|v| + N_{r|r|}r|r| +$$
$$N_{\dot{v}}\dot{v} + N_{\dot{r}}\dot{r} + N_{ur}ur + N_{wp}wp + N_{pq}pq + N_{uv}uv + N_{uu\delta_r}u^2\delta_r$$

$$(3)$$

where m is AUV's mass, I_{xx}, I_{yy}, I_{zz} are the moments of inertia of AUV to three coordinate axes, $X_{HS}, Y_{HS}, Z_{HS}, K_{HS}, M_{HS}, N_{HS}$ are hydrostatics, $X_{u|u|}, Y_{v|v|}, Y_{r|r|}, Z_{w|w|}, Z_{q|q|}, K_{p|p|}, M_{w|w|}, M_{q|q|}, N_{v|v|}, N_{r|r|}$ are hydrodynamic drag coefficients, $Y_{uv}, Y_{uu\delta_r}, Z_{uw}, Z_{uu\delta_s}, M_{uw}, M_{uu\delta_s}, N_{uv}, N_{uu\delta_r}$ are lift coefficients and lift moment coefficients of body and control fins, respectively, X_{prop}, K_{prop} are propeller thrust and torque, respectively, and δ_r, δ_s are rudder angle and stern plane angle, respectively. The remaining coefficients are added mass coefficients.

Separate the acceleration terms from the other terms in the equations of AUV motion so that the equations can be summarized in matrix form as follows:

$$\begin{bmatrix} \dot{u} \\ \dot{v} \\ \dot{w} \\ \dot{p} \\ \dot{q} \\ \dot{r} \end{bmatrix} = \begin{bmatrix} m - X_{\dot{u}} & 0 & 0 & 0 & mz_g & -my_g \\ 0 & m - Y_{\dot{v}} & 0 & -mz_g & 0 & mx_g - Y_{\dot{r}} \\ 0 & 0 & m - Z_{\dot{w}} & my_g & -mx_g - Z_{\dot{q}} & 0 \\ 0 & -mz_g & my_g & I_{xx} - K_{\dot{p}} & 0 & 0 \\ mz_g & 0 & -mx_g - M_{\dot{w}} & 0 & I_{yy} - M_{\dot{q}} & 0 \\ -my_g & mx_g - N_{\dot{v}} & 0 & 0 & 0 & I_{zz} - N_{\dot{r}} \end{bmatrix}^{-1} \begin{bmatrix} \sum X \\ \sum Y \\ \sum Z \\ \sum K \\ \sum M \\ \sum N \end{bmatrix}$$

$$(4)$$

where $\sum X \cdots \sum N$ refer to the sum of terms without acceleration. So far, six degrees of freedom nonlinear motion equations of AUV can be obtained by combining (4) with (1) and (2).

3. Design of SIFOFLC

3.1. Fundamentals of Fractional Calculus

Fractional calculus, essentially the non-integer order calculus, has the same history as integer order calculus. The three frequently used definitions of fractional calculus are the Grunwald–Letnikov definition, the Riemann–Liouville definition and the Caputo definition [25].

We consider the Caputo definition in this study because of its wide applications in engineering problems, the fractional integral and derivative by Caputo definition are as follows:

$$_{t_0}D_t^{-\alpha}f(t) = \frac{1}{\Gamma(\alpha)}\int_{t_0}^{t}\frac{f(\tau)}{(t-\tau)^{1-\alpha}}d\tau \tag{5}$$

$$_{t_0}D_t^{\alpha}f(t) = \frac{1}{\Gamma(m-\alpha)}\int_{t_0}^{t}\frac{f^{(m)}(\tau)}{(t-\tau)^{1+\alpha-m}}d\tau \tag{6}$$

where $_{t_0}D_t^{\alpha}$ represents the fractional calculus operator, $f(t)$ is a continuous function and t_0 denotes the initial time. α represents the fractional-order, $m = \lceil \alpha \rceil$. $\Gamma(\cdot)$ denotes the Gamma function as in (7).

$$\Gamma(z) = \int_{0}^{\infty} e^{-t}t^{z-1}dl \tag{7}$$

In the numerical simulations, we adopt the standard Oustaloup approximation method to obtain the consistent frequency characteristics as fractional differential operator. A rational transfer function in the form of zero-pole type is described according to the Oustaloup method, that N is the order of filter, $[\omega_b, \omega_h]$ is the selected frequency bound. The zero and pole are defined as ω'_k and ω_k, which divide the frequency band into $2N + 1$ intervals.

The Oustaloup rational approximation is described as

$$s^\alpha \approx G(s) = K \prod_{k=1}^{N} \frac{s + \omega'_k}{s + \omega_k} \tag{8}$$

where $\omega'_k = \omega_b \omega_u^{(2k-1-\alpha)/N}$, $\omega_u = \sqrt{\omega_h / \omega_b}$, $\omega_k = \omega_b \omega_u^{(2k-1+\alpha)/N}$, $K = \omega_h^\alpha$.

3.2. Structure Design of SIFOFLC

A fuzzy logic controller has four components: knowledge base, inference engine, fuzzification interface, and defuzzification interface [26]. The basic structure of a two-dimensional fuzzy logic controller is described as in Figure 2, where the continuous input signal, e and $\dot{e}$, convert to the membership degree vector of the fuzzy variables through a fuzzification interface, the inference engine carries out rule inference and actual output signal is obtained through defuzzification interface. The data base denotes membership functions of the total input and output variables and the rule base is performed using a collection of fuzzy if–then rules by expert experience, both of which make up the knowledge base.

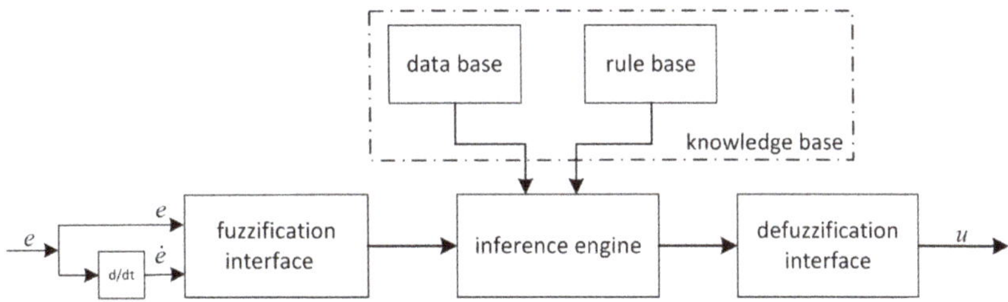

Figure 2. Block diagram of two-dimensional FLC.

Compared with the linear controller, fuzzy logic control method is more robust and suitable for complex control requirements, while its complex decision-making process brings a challenge for real-time operation. Thus, we aim to adjust the controller structure to reduce computation burden and achieve better performance.

Typically, a fuzzy logic controller has two control inputs, namely error (e) and its derivative ($\dot{e}$). It is common for its rule table to have the same output membership in a diagonal direction, something known as the Toeplitz structure, as shown in Table 1 [4]. In addition, each position on a diagonal line has the same distance from the main diagonal line of rule table. Thus, instead of using two-variable input sets $(e, \dot{e})$, the corresponding control output can be obtained using the distance between input signal and the main diagonal line. This finding was first proposed by Choi et al. and is known as the signed distance method [21]. To derive the distance, d, a two-dimensional space of e and $\dot{e}$ is established as shown in Figure 3.

Table 1. Rule table with the Toeplitz structure.

$\dot{e}$ \ e	PL	PM	PS	Z	NS	NM	NL
NL	Z	NS	NM	NL	NL	NL	NL
NM	PS	Z	NS	NM	NL	NL	NL
NS	PM	PS	Z	NS	NM	NL	NL
Z	PL	PM	PS	Z	NS	NM	NL
PS	PL	PL	PM	PS	Z	NS	NM
PM	PL	PL	PL	PM	PS	Z	NS
PL	PL	PL	PL	PL	PM	PS	Z

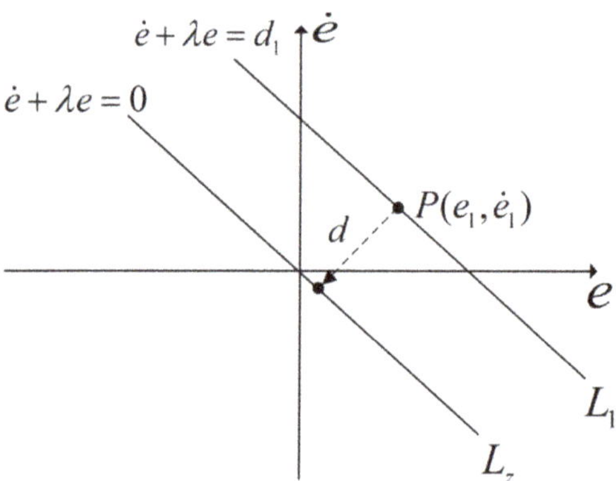

Figure 3. Distance variable.

The main diagonal line of the rule table is presented as a straight line crossing over the origin, whose function is $L_z : \dot{e} + \lambda e = 0$. In this case, the distance from point $P(e_1, \dot{e}_1)$ to L_z can be obtained as $d = \frac{\dot{e}_1 + \lambda e_1}{\sqrt{1+\lambda^2}}$.

Furthermore, in order to achieve better control performance, we extend the derivative of error signal to fractional order, thus the distance variable is described as

$$d = \frac{D_t^{\alpha} e + \lambda e}{\sqrt{1 + \lambda^2}} \tag{9}$$

where α is the fractional order of error signal and it further increases degrees of freedom and complexity of the controller.

Based on the above analysis, the overall structure of SIFLC can be depicted as in Figure 4. The distance variable is obtained through linear combination of error signal and its fractional derivative, the inner fuzzy logic controller is single input single output (SISO) and the final output u is obtained by multiplying u_0 with the scale factor, which is denoted as r. We set λ to 1. Thus, in addition to the membership functions, there are two adjustable parameters in the SIFOFLC, namely α and r.

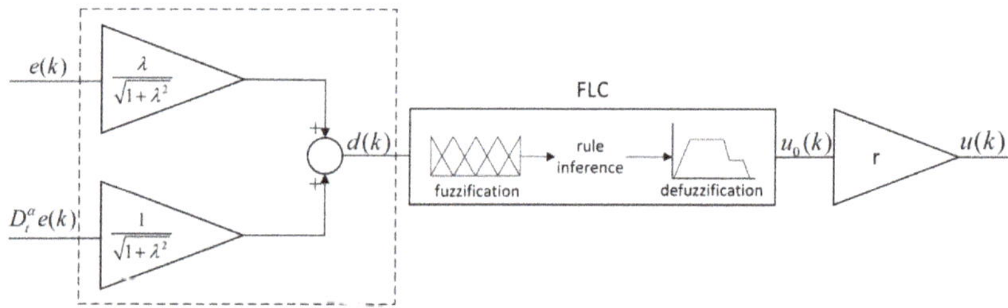

Figure 4. The SIFLC control structure.

3.3. Characteristics of SIFOFLC

In this section, the SIFOFLC and T-S FLC used in the simulation research of Section 4 are proposed and the superiority of SIFOFLC is presented through comparison.

To the SIFOFLC, the fuzzy sets of d and u_0 are both {NL, NM, NS, Z, PS, PM, PL} and the membership functions are shown in Figure 5. The membership functions of d include both S shape and triangular shape membership functions, with a singleton value membership function for u_0. In this case, the rule table is reduced to the one-dimensional vector as shown in Table 2.

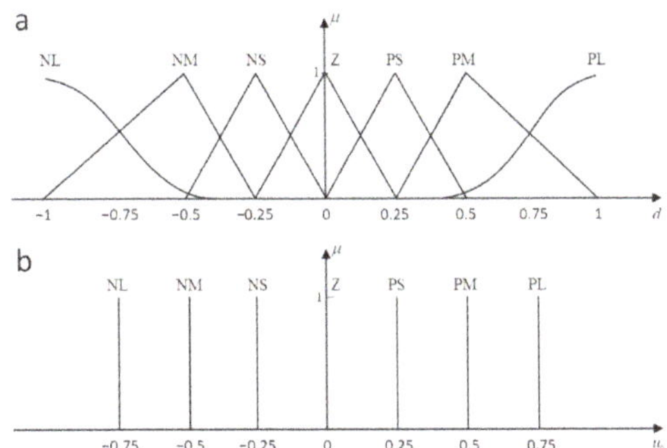

Figure 5. (**a**) Input membership function. (**b**) Output membership function.

Table 2. Rule table for the above SIFOFLC.

d	NL	NM	NS	Z	PS	PM	PL
u_0	NL	NM	NS	Z	PS	PM	PL

Based on the work mentioned above, the proposed SIFOFLC has the following advantages compared to conventional FLC:

(1)   Simplified design process.

For the two-dimensional FLC, the membership functions for error input and its derivative are required simultaneously, which means a lengthy complex tuning process. With the only input, d, the parameter tuning process for SIFOFLC is significantly reduced. Further, in a two-dimensional FLC, the number of fuzzy rules to be inferred is the square of n, which is the size of the fuzzy set. The distinguishing feature of SIFLC is that it requires only n rules, which is an exponential decay. Typically, better performance can be obtained with more a complicated control algorithm, such as the increase of fuzzy sets and rules, which further reveals the superiority of SIFOFLC.

(2)   Reduced computation burden.

The control surfaces for the SIFOFLC and conventional T-S FLC are respectively shown in Figures 6 and 7. Compared with the complex curved surface of T-S FLC, the control surface of SIFOFLC has been simplified to a linear with different slopes. Thus, the computation burden of controller operation, which includes fuzzification, fuzzy inference and defuzzification, has been significantly reduced.

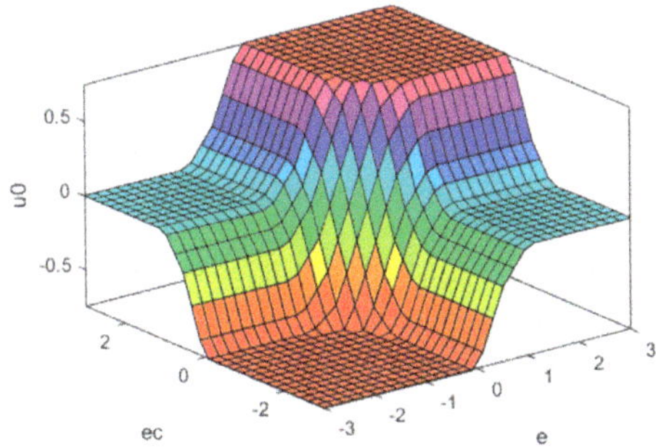

Figure 6. T-S FLC control surface.

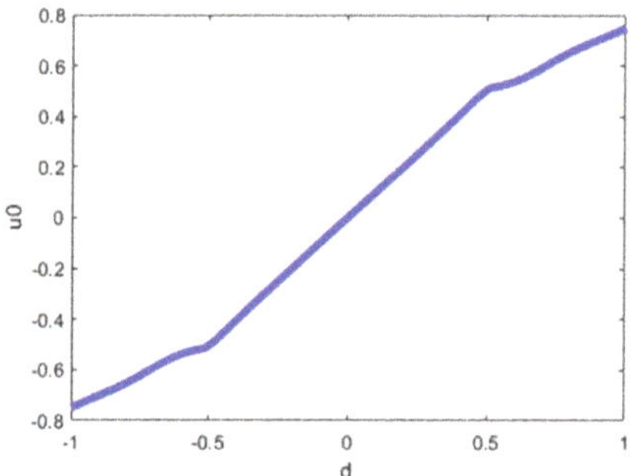

Figure 7. SIFLC control surface.

4. Parameter Optimization for Three Controllers with HPSO Algorithm

In this section, the HPSO algorithm is developed and applied to optimize the controller parameters. Except for the SIFOFLC, the T-S FLC and FOPID control systems are optimized for further comparison in Section 5, as the SIFOFLC algorithm essentially originates from the T-S FLC, and because the FOPID controller is the most extensively used fractional order control algorithm.

It is necessary to mention that our optimization and research scenario is the spiral dive motion control of REMUS-100 AUV. It is performed using Marine Systems Simulator (MSS), which is a MATLAB toolbox and offers a set of tools for marine engineering researchers [23]. Some research details are as follows: The target depth of AUV motion linearly increases from 0 m to 30 m and the target yaw angle from $0°$ to $720°$. The simulation time is 300 s and the input constraint of fins is set to $13.8°$ in accordance with [24]. Furthermore, the external disturbances are ignored.

Figure 8 illustrates the block diagram of an AUV motion control system. The propeller speed is fixed at 1500 r/m so that the cruise speed maintains a constant value. In this case, the target trajectory is obtained with heading controller and depth controller.

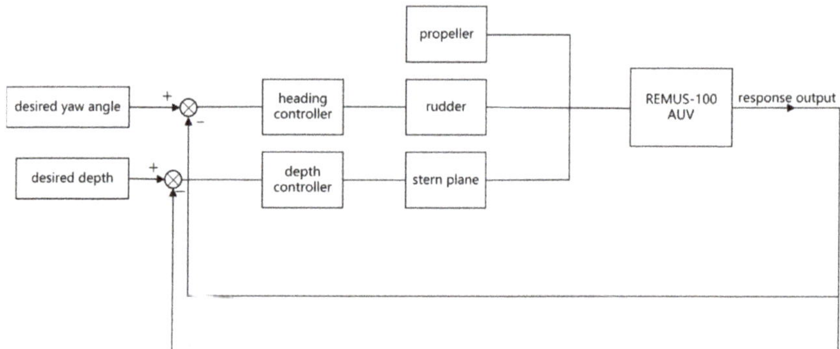

Figure 8. REMUS-100 AUV control system.

4.1. HPSO Algorithm

Because of the uncertainty and nonlinearity of the control system, adjusting parameters of the controller by manual experience is usually hard, thus optimization algorithms are applied to obtain the optimal or suboptimal solution. PSO is a typical swarm intelligence algorithm developed in 1995, stemming from research on the foraging behavior of birds [27]. Studies in [28] have shown that it lacks the capability to achieve sustainable development and the swarm becomes stagnant after a certain number of iterations. To improve this, the term of local optimal particle is introduced in the HPSO algorithm and the velocity and position of particles are updated according to the following two equations:

$$V_i^{K+1} = \omega \cdot V_i^K + c_1 rand(p_i^K - X_i^K) + c_2(q \cdot rand(S_i^K - X_i^K) + (1 - q) \cdot rand(p_{iL}^K - X_i^K)) \tag{10}$$

$$X_i^{K+1} = X_i^K + V_i^{K+1} \cdot dt \tag{11}$$

where K is the current number of evolution and at the K-th evolution, V_i^K denotes the velocity of the i-th particle; X_i^K represents the position of the i-th particle; p_i^K depicts the best position of i-th particle; S_i^K is the best position of the particle swarm; p_{iL}^K is the position of the local optimal particle for the i-th particle, and the introduction of p_{iL}^K avoids falling into the local optimal region. ω is inertia weight, which decreases as evolution unfolds and aids in global search in the early stage and local optimization in the late stage. c_1 and c_2 are learning factors, q represents social factor, and dt denotes time interval coefficient.

Instead of single-object optimization, we consider multiple objects in our study, including the steady-state performance and transient performance of AUV motion control system. And the objective function is defined as follows:

$$\begin{aligned} F = {} & \log(1 + \tfrac{p\_vibration}{p\_vibration0}) + \log(1 + \tfrac{q\_vibration}{q\_vibration0}) + \log(1 + \tfrac{r\_vibration}{r\_vibration0}) + \log(1 + \tfrac{p\_ts}{p\_ts0}) \\ & + \log(1 + \tfrac{q\_ts}{q\_ts0}) + \log(1 + \tfrac{r\_ts}{r\_ts0}) + \log(1 + \tfrac{z\_ITAE}{z\_ITAE0}) + \log(1 + \tfrac{\psi\_ITAE}{\psi\_ITAE0}) \end{aligned} \tag{12}$$

where $(\cdot)\_vibration$ represents oscillation times of AUV angular velocity; $(\cdot)\_ts$ is the settling time in which the angular velocity is kept within a $\pm 5\%$ range of the steady state value; $(\cdot)\_ITAE$ describes the ITAE index of track error; $p\_(vibration0)$ which denotes the expected value and is set artificially, all of the expected values are presented in Appendix A.

The flowchart of optimization is shown in Figure 9 and the implementation procedure of the HPSO algorithm is summarized as follows:

Step 1: Specify the population size and the maximum number of evolutions, as well as other coefficients that can be noted from Equations (10)–(12). Determine the value range of controller parameters and initialize a population of particles with random positions.

Step 2: Evaluate the fitness value of all particles according to Equation (12), let p_i^K of each particle equal to its current position and let S_i^K equal the position of the best initial particle. After that, the local optimal position of each particle, p_{iL}^K, is obtained in accordance with Appendix A.

Step 3: Update the particles' velocities and positions in terms of Equations (10) and (11). Compare the fitness value of each particle with its own best fitness value, if the current one is better, update the p_i^K. Similarly, compare the best fitness value of the new generation with the fitness value of the global best position S_i^K and decide whether it has to be updated. The local optimal position of each particle is also updated.

Step 4: The optimization is terminated when it reaches the maximum number of evolutions. Then the global best position is output, which is the target parameters of the controller. The convergence curve of optimal fitness value is also printed.

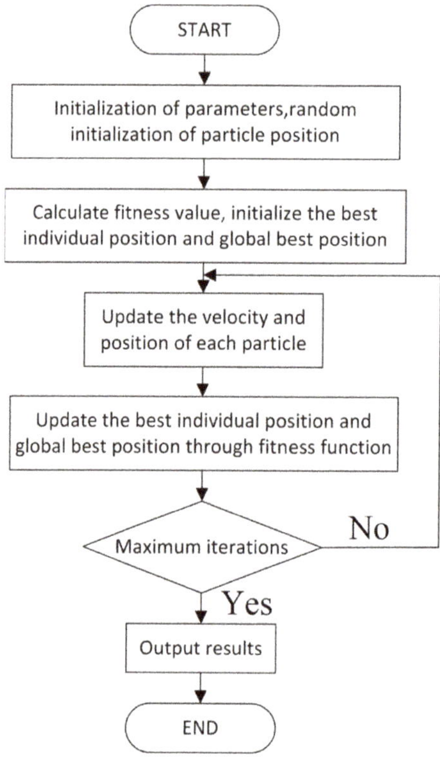

Figure 9. Flowchart of HPSO algorithm.

4.2. Optimization Experiments and Result Analysis

Here we conduct optimization of three control systems by using the HPSO algorithm mentioned above. To the SIFOFLC system, the particle dimension is four with the fractional order of the error signal and the scale factor of the depth controller and heading controller. We adopted the standard Oustaloup filter module deriving from FOTF toolbox [29] to perform fractional derivative operator, the frequency band was set to [0.001,1000] and the filter order was 4. As for the FOPID control system, the dimension of particle is 10, so that each FOPID controller has five unknown parameters. For the T-S FLC control system, only two scale factors are to be optimized. The membership functions and fuzzy rules of T-S FLC were determined through empirical approach, and its control surface is shown in Figure 6.

The coefficients of the HPSO algorithm are set as follows: the particle size is 10 and the maximum number of evolution is 20. The limit of the inertia weight, ω, is set to between 1 and 0.7. The learning factors c_1 and c_2 are set as value 2, the social factor q is set as value 0.7 and the time interval coefficient dt is set as value 0.5. The searching range for parameters are presented in Appendix A. Furthermore, we implement multiple optimizations and adopt the optimal result to solve randomness.

Figures 10–12 respectively show the convergence curve of optimal fitness of three control systems and the obtained target controller parameters are illustrated in Table 3. It can be observed that all the curves tend to decline through evolution, which illustrates the effectiveness of optimization. Actually, the optimal fitness of three control systems respectively decreases by 19.8%, 37.9% and 9.6%. The FOPID controller clearly outperforms the other two controllers as it has the highest degrees of freedom. The optimization effect of the SIFOFLC system is twice as good as the T-S FLC system and its ultimate fitness is significantly less than the others. These results demonstrate that the introduction of a fractional differential operator not only increases degrees of freedom and the flexibility of the controller but also improves the control performance, as is particularly demonstrated in the next section.

Table 3. Controller parameters optimized by HPSO algorithm.

Control System	Parameter	Heading Controller	Depth Controller
SIFOFLC	α	1.08	1.21
	r	131.02	-102.50
FOPID	K_p	19.88	-15.50
	K_i	18.69	0
	K_d	2.96	-50
	λ	0.10	1.5
	μ	1.11	1.12
T-S FLC	r	33.20	-140.43

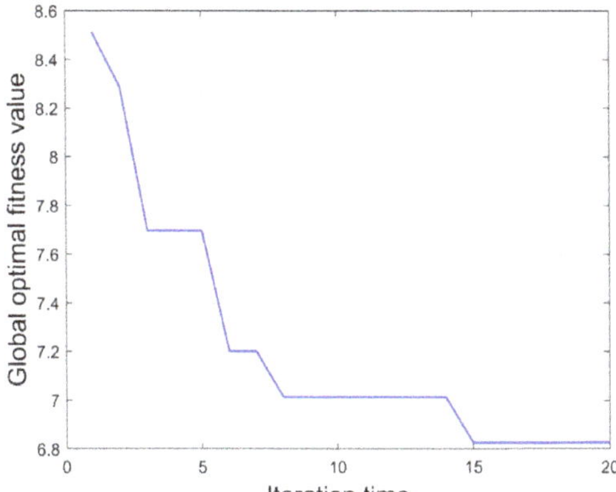

Figure 10. Convergence curve of SIFOFLC system.

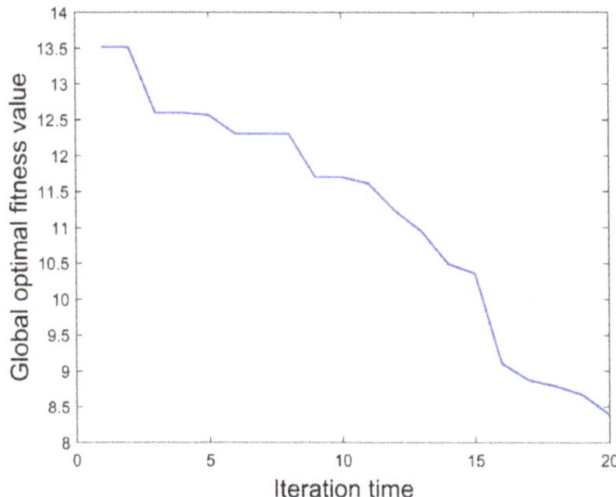

Figure 11. Convergence curve of FOPID control system.

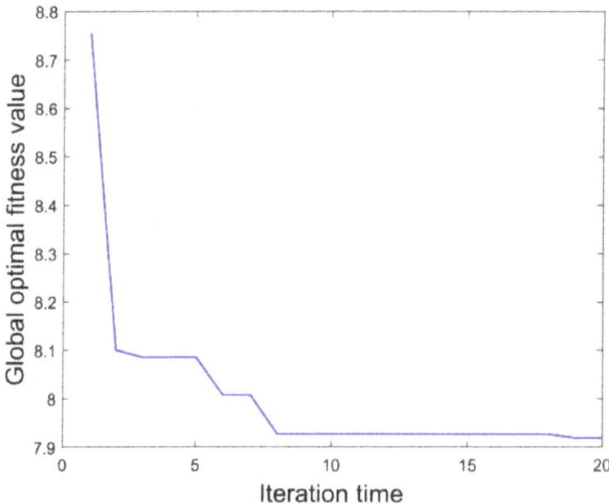

Figure 12. Convergence curve of T-S FLC system.

5. Simulation and Analysis

In this section, we conduct comparative simulations to verify the effectiveness of the proposed SIFOFLC algorithm for AUV motion control. The optimized controllers expressed in Table 3 are applied and the target trajectory of AUV does not change.

Figure 13 shows the trajectory of AUV under SIFOFLC and the state of the response is depicted in Figure 14. Simulation results reveal that the trajectory tracking is accurate and rapid with SIFOFLC. Actually, the steady state error of depth is within 0.01 m in 6.5 s and gradually decreases to approximately 0.002 m in 7 s. Correspondingly, the maximum error of yaw angle is 2.5° and it gradually decreases to 0.01° in 13 s. The cruise speed maintains 1.54 m/s and the peak overshoot of pitch angle is 10°.

Figures 15 and 16 respectively illustrate the angular velocity of an AUV using FOPID controller and T-S FLC. Furthermore, the oscillation and settling times of angular velocity, the ITAE index of track error and the optimal fitness value are all presented in Table 4. The responses with SIFOFLC algorithm have much shorter settling times and steady state errors. Compared with the SIFOFLC, the control performance obtained via the FOPID controller and T-S FLC requires a much longer settling time, and they also oscillate considerably in the beginning, which may lead to an unstable performance of the controlled system. Simulation results clearly reveal the superior stability and transient control performance of the proposed SIFOFLC.

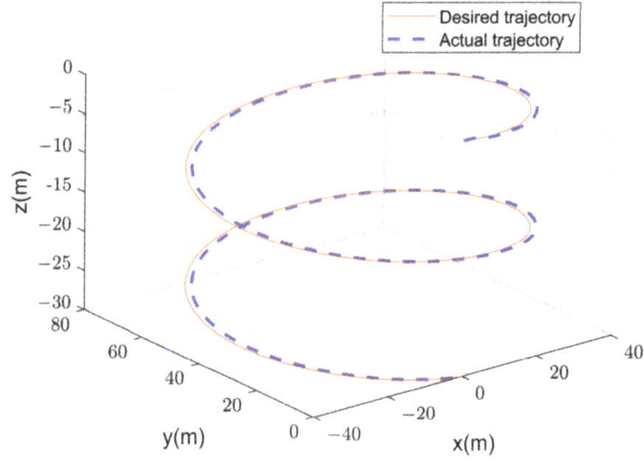

Figure 13. Comparison of desired trajectory and actual trajectory with SIFOFLC.

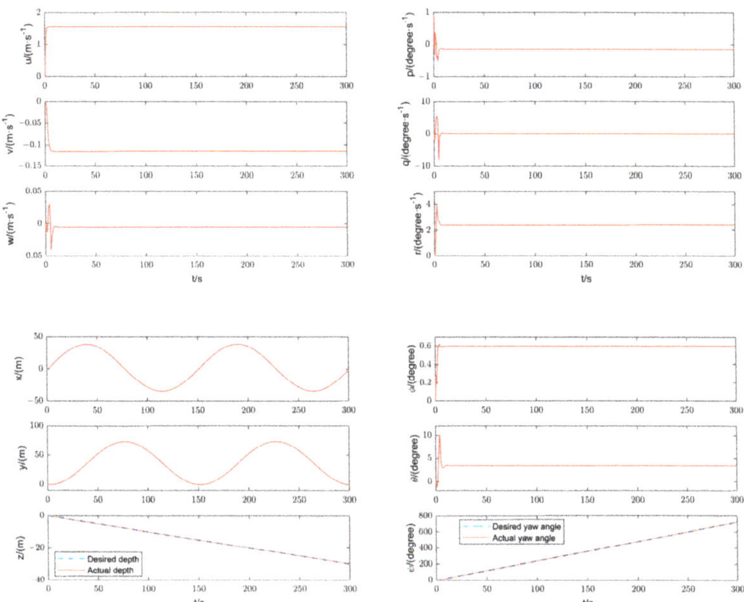

Figure 14. State response of AUV with SIFOFLC.

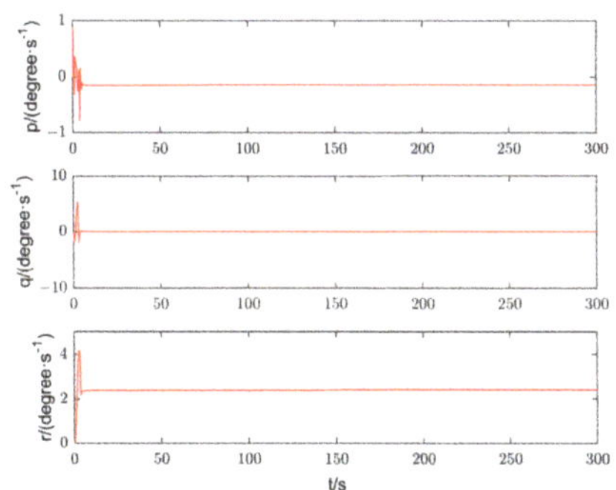

Figure 15. Angular velocity curves with FOPID.

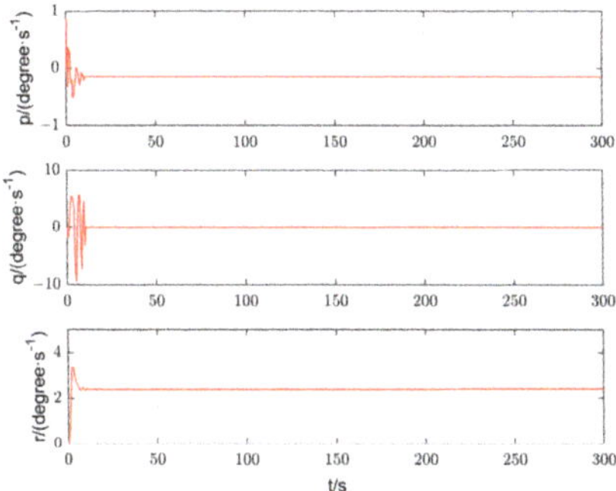

Figure 16. Angular velocity curves with T-S FLC.

Table 4. Quantitative analysis of system performance.

Characteristic	SIFOFLC	FOPID	T-S FLC
p_vibration	9	12	12
q_vibration	5	5	9
r_vibration	1	3	1
p_ts	8.15	10.62	10.17
q_ts	14.71	18.44	13.53
r_ts	4.97	3.87	5.43
z_ITAE	2.69	7.70	2.90
ψ_ITAE	12.21	14.84	20.38
Fitness value	6.83	8.39	7.92

6. Conclusions

In this study, we proposed a SIFOFLC algorithm for an AUV motion control system. Unlike a conventional FLC algorithm, this is reduced to a SISO controller by using the signed distance method, which provides a significant reduction to parameter tuning and computation burden. In addition, a fractional derivative operator was applied to increase degrees of freedom of the controller, hence the proposed control algorithm is more flexible and adaptive to the AUV motion control system. Furthermore, we developed an HPSO algorithm and applied it to optimize the controller parameters. The simulation results show that the proposed controller enhances the stability and transient performance of the controlled AUV motion system, which manifests in less oscillations of angular velocity, shorter dynamic settling time, and higher control accuracy. In future studies, we will perform experiments using the proposed controller to verify its practicability.

Author Contributions: Conceptualization, Z.C. and B.Z.; methodology and software, Z.C. and Z.Z.; validation, S.L. and M.L.; writing—original draft preparation, Z.C. and L.L.; writing—review and editing, L.L. and B.Z.; supervision, L.L., L.Z. and Y.Y. All authors have read and agreed to the published version of the manuscript.

Funding: This research was funded by the Shenzhen Science and Technology Program under Grant JCYJ20210324122010027; the Science and Development Program of Local Lead by Central Government, Shenzhen Science and Technology Innovation Committee under Grant 2021Szvup111; the Guangdong Basic and Applied Basic Research Foundation under Grant 2019A1515111073; the National Natural Science Foundation of China under Grant 52001259, 11902252; the Young Talent fund of University Association for Science and Technology in Shaanxi, China under Grant 20200502; Science and Technology on Avionics Integration Laboratory and Aeronautical Science Foundation of China under Grant 201955053003; Maritime Defense Technology Innovation Center Innovation Fund under Grant JJ-2021-702-09; National Research and Development Project under Grant 2021YFC2803000.

Institutional Review Board Statement: Not applicable.

Informed Consent Statement: Not applicable.

Data Availability Statement: The data presented in this study are available in the article.

Conflicts of Interest: The authors declare no conflict of interest.

Appendix A

Table A1 shows the searching area of parameters, they are established through trial and error. The target values of characteristics of AUV motion control system, which are defined in Equation (12), are depicted in Table A2. The MATLAB function code for particle distance is given as follows:

```
function DM = Distance_Matrix (X,xmin,xmax)
n = size (X,1);
DM = zeros (n,n);
Normal = xmax-xmin;
for t = 1:n
Dif = (X (t,:)-X)./Normal;
 Dis = sum (Dif.^2,2);
DM (t,:) = Dis;
end
```

where X represents the parameters of particle swarm, xmin and xmax denote the searching bound of parameters. DM is a square matrix and DM (i,j) is the distance between the i-th particle and j-th particle. By the way, if the distance is less than 1, we consider them as local particles.

Table A1. Searching range of parameters for three control systems.

Control System	Parameter	Heading Controller	Depth Controller
SIFOFLC	α	$[0.1, 1.5]$	$[0.1, 1.5]$
	r	$[5, 150]$	$[-150, -5]$
FOPID	K_p	$[0, 20]$	$[-20, 0]$
	K_i	$[0, 20]$	$[-20, 0]$
	K_d	$[0, 10]$	$[-50, 0]$
	λ	$[0.1, 1.5]$	$[0.1, 1.5]$
	μ	$[0.1, 1.5]$	$[0.1, 1.5]$
T-S FLC	r	$[0, 150]$	$[-150, 0]$

Table A2. Target performance of control system.

Characteristics of Control System	Target Value
p_vibration0	6
q_vibration0	4
r_vibration0	2
p_ts0	5
q_ts0	5
r_ts0	5
z_ITAE0	5
ψ_ITAE0	5

References

1. Ma, Y.T.; Zheng, R. Autonomous underwater vehicle deepening control based on transiting target value nonlinear PID. *Control Theory Appl.* **2018**, *35*, 1120–1125.
2. Kumar, A.; Sharma, R. A Genetic Algorithm based Fractional Fuzzy PID Controller for Integer and Fractional order Systems. *Int. J. Intell. Syst. Appl.* **2018**, *10*, 23–32. [CrossRef]
3. Xu, J.; Wang, M.; Qiao, L. Backstepping-based controller for three-dimentional trajectory of underactuated unmanned underwater vehicles. *Control Theory Appl.* **2014**, *31*, 1589–1596.
4. Ishaque, K.; Abdullah, S.S.; Ayob, S.M.; Salam, Z. A simplified approach to design fuzzy logic controller for an underwater vehicle. *Ocean Eng.* **2011**, *38*, 271–284. [CrossRef]
5. Qiao, L.; Zhang, W. Adaptive Second-Order Fast Nonsingular Terminal Sliding Mode Tracking Control for Fully Actuated Autonomous Underwater Vehicles. *IEEE J. Ocean. Eng.* **2019**, *44*, 363–385. [CrossRef]
6. Wan, L.; Zhang, Y.H. Depth Control of Underactuated AUV Under Complex Environment. *J. Shanghai Jiao Tong Univ.* **2015**, *49*, 1849–1854.

7. Fan, Q.Y.; Wang, D.S.; Xu, B. H∞ Codesign for Uncertain Nonlinear Control Systems Based on Policy Iteration Method. *IEEE Trans. Cybern.* **2021**, 1–10. [CrossRef]

8. Fatmawati; Yuliani, E.; Alfiniyah, C. On the Modeling of COVID-19 Transmission Dynamics with Two Strains: Insight through Caputo Fractional Derivative. *Fractal Fract.* **2022**, *6*, 346. [CrossRef]

9. Monika, S.; Bharat, S.R. Development of Fractional Order Modeling of Voltage Source Converters. *IEEE Access* **2020**, *8*, 131750–131759.

10. Liu, L.; Zhang, L.C.; Pan, G. Robust yaw control of autonomous underwater vehicle based on fractional-order PID controller. *Ocean Eng.* **2022**, *257*, 111493. [CrossRef]

11. Ivo, P. Novel Fractional-Order Model Predictive Control: State-Space Approach. *IEEE Access* **2021**, *9*, 92769–92775.

12. Zheng, W.J.; Huang, R.Q.; Luo, Y. A Look-Up Table Based Fractional Order Composite Controller Synthesis Method for the PMSM Speed Servo System. *Fractal Fract.* **2022**, *6*, 47. [CrossRef]

13. Liu, L.; Zhang, S.; Zhang, L.C. Multi-Auv Dynamic Maneuver Decision-Making Based on Intuitionistic Fuzzy Counter-Game and Fractional-Order Particle Swarm Optimization. *Fractals-Complex Geom. Patterns Scaling Nat. Soc.* **2021**, *29*, 2140039. [CrossRef]

14. Liu, L.; Wang, J.; Zhang, L.C. Multi-AUV Dynamic Maneuver Countermeasure Algorithm Based on Interval Information Game and Fractional-Order DE. *Fractal Fract.* **2022**, *6*, 235. [CrossRef]

15. Zhang, S.; Liu, L.; Xue, D.Y. Nyquist-based stability analysis of non-commensurate fractional-order delay systems. *Appl. Math. Comput.* **2020**, *377*, 125111. [CrossRef]

16. Zhang, S.; Liu, L.; Xue, D.Y. Stability and resonance analysis of a general non-commensurate elementary fraction-al-order system. *Fract. Calc. Appl. Anal.* **2020**, *23*, 183–210. [CrossRef]

17. Dai, C.P.; Ma, W.Y. Lyapunov Direct Method for Nonlinear Hadamard-Type Fractional Order Systems. *Fractal Fract.* **2022**, *6*, 405. [CrossRef]

18. Podlubny, I. Fractional-order systems and PI/sup/spl lambda//D/sup /spl mu//-controllers. *IEEE Trans. Autom. Control* **2002**, *44*, 208–214. [CrossRef]

19. Zhao, R.; Xu, J.; Wang, M. Heading control of AUV based on GA and fractional order technology. *Chin. J. Ship Res.* **2018**, *13*, 87–93.

20. Gao, Y.; Fan, J.W. Fractional-order Fuzzy Control Method for Vehicle Nonlinear Active Suspension. *China Mech. Eng.* **2015**, *26*, 1403–1408.

21. Choi, B.J.; Kwak, S.W.; Kim, B.K. Design and stability analysis of single-input fuzzy logic controller. *IEEE Trans. Syst. Man Cybern. Part B (Cybern.)* **2000**, *30*, 303–309. [CrossRef] [PubMed]

22. Allen, B.; Stokey, R. REMUS: A small, low cost AUV.; System Description, Field Trials and Performance Results. In Proceedings of the Ocean '97. MTS/IEEE Conference Proceedings, Halifax, NS, Canada, 6–9 October 1997.

23. Marine Systems Simulator (MSS). Available online: https://github.com/cybergalactic/MSS (accessed on 8 January 2022).

24. Prestero, T. Verification of a six-degree of freedom simulation model for the REMUS autonomous underwater vehicle. Master's Thesis, Massachusetts Institute of Technology, Cambridge, UK, 2001.

25. Xue, D.Y. *Advanced Applied Mathematical Problem Solutions with MATLAB*, 4th ed.; Tsinghua University Publishing House: Beijing, China, 2018; pp. 436–444.

26. Johnston, R. Fuzzy logic control. *Microelectron. J.* **1995**, *26*, 481–495. [CrossRef]

27. Kennedy, J.; Eberhart, R. Particle Swarm Optimization. In Proceedings of the Icnn95-international Conference on Neural Networks, Perth, WA, Australia, 27 November–1 December 1995.

28. Xie, X.F.; Zhang, W.J.; Yang, Z.L. Dissipative particle swarm optimization. In Proceedings of the 2002 Congress on Evolutionary Computation, Honolulu, HI, USA, 12–17 May 2002.

29. FOTF Toolbox. Available online: https://ww2.mathworks.cn/matlabcentral/fileexchange/60874-fotf-toolbox (accessed on 30 December 2021).

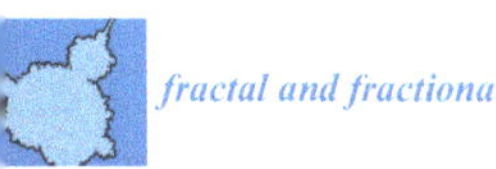

fractal and fractional

Review

A Review of Recent Developments in Autotuning Methods for Fractional-Order Controllers

Cristina I. Muresan [1], Isabela Birs [1,2,3,*], Clara Ionescu [1,2,3], Eva H. Dulf [1] and Robin De Keyser [2,3]

1   Automation Department, Technical University of Cluj-Napoca, 400114 Cluj-Napoca, Romania; Cristina.Muresan@aut.utcluj.ro (C.I.M.); ClaraMihaela.Ionescu@UGent.Be (C.I.); Eva.Dulf@aut.utcluj.ro (E.H.D.)
2   DySC Research Group on Dynamical Systems and Control, Faculty of Engineering and Architecture, Ghent University, Tech Lane Science Park 125, B-9052 Ghent, Belgium; Robain.DeKeyser@ugent.be
3   EEDT Decision & Control, Flanders Make Consortium, Tech Lane Science Park 131, B-9052 Ghent, Belgium
*   Correspondence: Isabela.Birs@aut.utcluj.ro

Abstract: The scientific community has recently seen a fast-growing number of publications tackling the topic of fractional-order controllers in general, with a focus on the fractional order PID. Several versions of this controller have been proposed, including different tuning methods and implementation possibilities. Quite a few recent papers discuss the practical use of such controllers. However, the industrial acceptance of these controllers is still far from being reached. Autotuning methods for such fractional order PIDs could possibly make them more appealing to industrial applications, as well. In this paper, the current autotuning methods for fractional order PIDs are reviewed. The focus is on the most recent findings. A comparison between several autotuning approaches is considered for various types of processes. Numerical examples are given to highlight the practicality of the methods that could be extended to simple industrial processes.

Keywords: direct autotuning methods; indirect autotuning methods; fractional-order controllers; simulation results

Citation: Muresan, C.I.; Birs, I.; Ionescu, C.; Dulf, E.H.; De Keyser, R. A Review of Recent Developments in Autotuning Methods for Fractional-Order Controllers. *Fractal Fract.* **2022**, *6*, 37. https://doi.org/10.3390/fractalfract6010037

Academic Editors: Thach Ngoc Dinh, Shyam Kamal and Rajesh Kumar Pandey

Received: 13 December 2021
Accepted: 5 January 2022
Published: 11 January 2022

Publisher's Note: MDPI stays neutral with regard to jurisdictional claims in published maps and institutional affiliations.

1. Introduction

Despite the abundance of research in advanced control strategies, the PID (proportional-integrative-derivative) controller remains the preferred control algorithm in industrial applications [1,2]. To produce the desired effects, PIDs need to be adequately tuned. A mathematical model is usually needed in order to properly tune the controller. However, large industrial plants are characterized by numerous sub-systems and obtaining an accurate process model is not cost effective as it can be difficult and/or time consuming. To overcome this issue, two different approaches for autotuning PIDs were developed, as indicated in Figure 1.

Both approaches use step or sinusoidal input data and collect the process output response. For a direct autotuner the PID parameters are determined directly from process input/output data, while for the indirect PID autotuner, simple process models are first determined and then the PID parameters are computed according to some tuning rules based on the model parameters. The majority of indirect methods use either first-order plus dead time (FOPDT) or second-order plus dead time (SOPDT) models.

Two of the most popular autotuning methods have been developed by Ziegler and Nichols [3]. One of these methods is a direct approach, based on the relay experiment, as indicated in Figure 2. Once the relay test is performed on a process, it will lead to a sinusoidal output signal which is used to estimate the process critical frequency and the corresponding critical gain. Tuning rules based on the process critical frequency and gain are employed to compute the PID controller parameters. The Ziegler-Nichols direct autotuning method is highly popular because of its simplicity and good performance results.

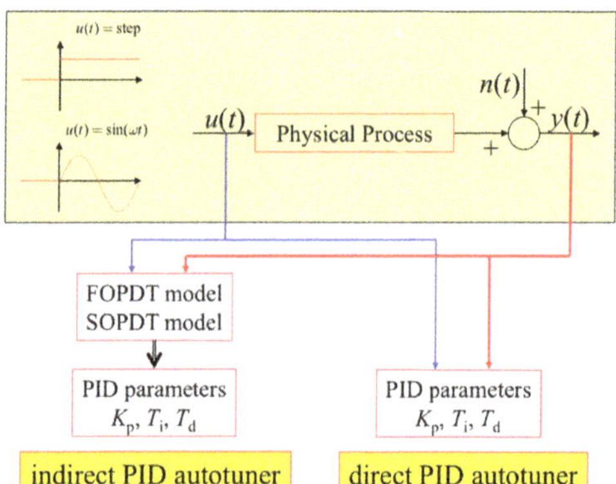

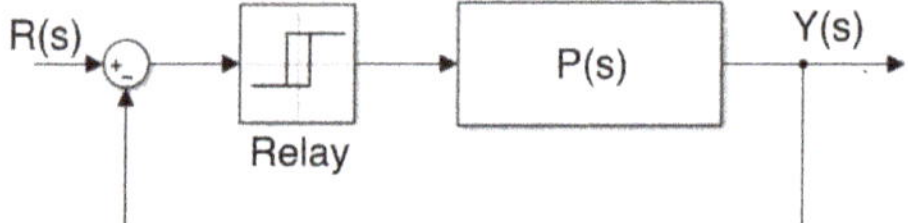

Figure 1. Autotuning approaches.

Figure 2. Relay experiment.

Several extensions of this approach and alternative solutions have been developed over the years. One of these uses the describing function analysis and a simple relay feedback test to estimate the process critical gain and corresponding frequency [4]. A solution for noisy signals was proposed based on a relay with hysteresis [5]. An artificial time delay is added within the relay closed-loop system in order to determine the process gain and phase at a random oscillation frequency. Then, a PI (proportional-integrative) controller is tuned according to this process data. A modified Ziegler-Nichols method [6]—where the ratio between the integral and derivative time constants is $r = 4$—was also developed. Other research papers discuss the impact the ratio value has upon the control performance [7]. Solutions to improve the robustness of the control system have been addressed [2]. Åström and Hägglund [1] use the relay test to design controllers based on robust loop shaping, with a clear tradeoff between robustness and performance.

The second method developed by Ziegler and Nichols [3] consists in applying a step signal on the process input and collecting the output data. The method is suitable for processes that have FOPDT dynamics or exhibit an S-shaped response, as indicated in Figure 3. The approach goes through an indirect step, where the parameters of the FOPDT model are estimated. Finally, the PID controller parameters are computed using a set of tuning rules that depend on the FOPDT model parameters.

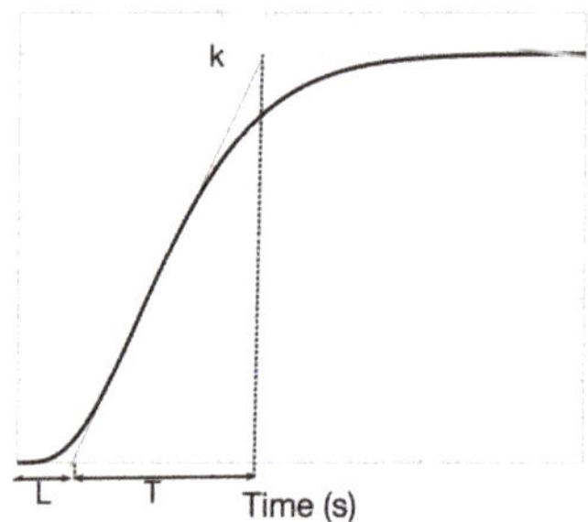

Figure 3. S-shaped response.

The demand for better control performance and increased robustness has led to several modifications of the standard PID controller, including a generalization to fractional order [8]. Research on fractional order PID (FO-PID) controllers has demonstrated that this generalization allows for more flexibility in the design, due to the two supplementary tuning parameters involved, the fractional orders of integration and differentiation [9–13]. This flexibility comes with important advantages, such as better closed-loop performance, disturbance rejection capabilities, improved control of time-delay systems and increased robustness [9–14]. The fractional order PID transfer function is given as:

$$C_{FO-PID}(s) = k_p\left(1 + \frac{1}{T_i s^\lambda} + T_d s^\mu\right) \tag{1}$$

where $0 < \lambda < 2$ and $0 < \mu < 1$ are the fractional orders of integration and differentiation, respectively, and k_p is the proportional gain, and T_i and T_d are the integral and derivative time constants. The "classical" tuning rules used to determine the five controller parameters are derived from the following performance specifications [9,12,15–17]:

1. A gain crossover frequency ω_c. This leads to the magnitude condition:

$$|H_{ol}(j\omega_c)| = 1 \tag{2}$$

   with $H_{ol}(s)$ the open-loop transfer function is defined as: $H_{ol}(s) = P(s) \cdot C_{FO-PID}(s)$, where $P(s)$ is the process transfer function;

2. A phase margin PM. This leads to the phase condition:

$$\angle H_{ol}(j\omega_c) = -\pi + PM \tag{3}$$

3. Iso-damping property (or robustness to gain variations). This is specified through:

$$\left.\frac{d(\angle H_{ol}(j\omega))}{d\omega}\right|_{\omega=\omega_c} = 0 \tag{4}$$

   where ω denotes the frequency. This last condition ensures that the overshoot of the closed-loop system remains approximately constant in the case of gain variations;

4. Good output disturbance rejection. This leads to a constraint on the sensitivity function S:

$$\left|S(j\omega) = \frac{1}{1 + P(j\omega)H_{FO-PID}(j\omega)}\right| \leq B \; dB \tag{5}$$

   for frequencies $\omega \leq \omega_s$, with B a scalar;

5. High frequency noise rejection. This leads to a constraint on the complementary sensitivity function T as:

$$\left|T(j\omega) = \frac{P(j\omega)H_{FO-PID}(j\omega)}{1 + P(j\omega)H_{FO-PID}(j\omega)}\right| \leq A \; dB \tag{6}$$

   for frequencies $\omega \geq \omega_T$, with A a scalar.

Further information regarding the tuning, implementation and related topics to fractional order PIDs can be found in some excellent review papers [15,16,18–24]. The phase shaper [25] is among the first automatic controller designs that uses fractional calculus tools. The autotuning method is based on the iso-damping property, but the final controller is an integer order PID. Throughout the past two decades, a couple of FO-PID autotuning methods have emerged. Some of these provide direct and indirect tuning rules for FO-PIDs in general or for fractional order PI (FO-PI) controllers. The purpose of this manuscript is to offer a comprehensive review of these autotuning methods, to compare them and to discuss which method is ranked best for controlling a specific type of process.

The paper is structured as follows. Sections 2 and 3 provide for a review of the most widely known indirect and direct autotuning methods for FO-PIDs, while Section 4 provides for some numerical examples. Possible applications of autotuning methods are reviewed in Section 5, along with a survey on self-tuning FO-PIDs. The last section concludes the paper.

2. Indirect Fractional Order Autotuning Methods

A popular indirect autotuning method, suitable only for FO-PIs, was developed by extending the M_s constrained integral (MIGO)-based controller design approach [26]. F-MIGO tuning determines the optimum parameters of the FO-PI controller such that the load disturbance rejection is optimized, with a constraint on the maximum or peak sensitivity. The F-MIGO method provides the tuning rules for the FO-PI controller provided that the process step response has an S-shaped form, as indicated in Figure 3, that could be approximated by the following transfer function:

$$P(s) = \frac{k}{Ts + 1}e^{-Ls} \tag{7}$$

where k is the process gain, L is the delay and T is the time constant of the process. The relative dead time can be computed as:

$$\tau = \frac{L}{T + L} \tag{8}$$

Systems where $L \gg T$ are delay dominant, whereas systems in which $T \gg L$ are lag dominant. Research studies performed in [26] revealed that the FO-PI fractional order is almost independent of L, but depends on the relative dead time. For some particular situations, where $0.4 \leq \tau < 0.6$, an integer order PI controller was determined to be more suitable for controlling the process. A summary of the results is indicated below:

$$\lambda = \begin{cases} 1.1, & \tau \geq 0.6 \\ 1, & 0.4 \leq \tau < 0.6 \\ 0.9, & 0.1 \leq \tau < 0.4 \\ 0.7, & \tau < 0.1 \end{cases} \tag{9}$$

The proportional and integrative gains of the FO-PI controller were also determined as a function of the relative dead time:

$$k_p = \frac{1}{k}\frac{0.2978}{\tau + 0.00037} \text{ and } T_i = T\frac{0.8578}{\tau^2 - 3.402\tau + 2.405} \tag{10}$$

An indirect autotuning method that applies to the S-shaped step response process was developed in [27]. The tuning is unnecessarily complicated as the parameters of (7) are firstly estimated and then used to determine the process critical frequency ω_{cr} and critical gain k_{cr}, according to:

$$\omega_{cr}T = -\tan(\omega_{cr}L) \text{ and } k_{cr} = -\frac{1 + \omega_{cr}T^2}{k(\cos(\omega_{cr}L) - \omega_{cr}T\sin(\omega_{cr}L))} \tag{11}$$

Then, the parameters of an integer order PID are determined using the previously computed process critical frequency and gain, as well as three additional design parameters referring to the ratio of the integral and derivative time constants, loop phase and gain:

$$k_p = k_{cr}r_b\cos\varnothing_b, \; T_i = -\frac{T_{cr}}{\pi}\frac{\pi\cos\varnothing_b}{\sin\varnothing_b + 1} \text{ and } T_d = \alpha T_i \tag{12}$$

where α, r_b and $\varnothing_b$ are design parameters [28]. Once the PID controller parameters are computed, a possible range for the fractional orders in the FO-PID is selected and an optimization routine is performed. The algorithm attempts to minimize the integral time absolute error with the open loop gain and phase margin imposed as design specifications.

Another indirect tuning method is proposed in [29] for processes that produce an S-shaped step response. The method is based on determining first the process dead time L and time constant T, as well as the value at which the system reaches steady state k. The standard Ziegler-Nichols equations are used then to estimate the k_p, T_i and T_d parameters of an integer order PID. Then, the fractional orders of differentiation and integration are determined by the Nelder-Mead optimization algorithm in order to meet certain phase and gain margins. A second approach based on the standard Cohen-Coon method is also used in [29], for processes that exhibit first order plus dead time dynamics. Based on the process parameters, the integer order PID parameters are first computed. The Nelder-Mead optimization algorithm is used afterwards to estimate the fractional orders of differentiation and integration based on certain phase and gain margin requirements. The Cohen-Coon tuning method is proposed as an alternative to the Ziegler-Nichols approach in order to improve the slow, steady state response of the latter.

An indirect autotuning method for designing only FO-PI controllers using the Ziegler-Nichols open-loop approach is described in [30]. The parameters of the integer order PI controller are firstly determined using the standard Ziegler-Nichols approach. In order to improve the overall closed-loop response, the research suggests that the PI performance can be improved a lot with a fractional order of integration. An error filter as proposed in [31] is used for steady state error compensation:

$$G_e(s) = \frac{s+n}{s} \tag{13}$$

where n is chosen to be small enough so that high frequency specifications are maintained and the system gain will not be altered drastically. The research in [30] proposed a modification of (13) such that the value of n is adjustable with respect to the fractional order of integration. The tuning of the fractional order and of the filter is performed by trial and error for a specific type of process. The method is evaluated experimentally on a steam temperature process and compared to the F-MIGO method [26] in terms of robustness for set point changes and disturbance rejection. The proposed controller shows better performance compared to the F-MIGO autotuning method, but it also requires higher control effort.

In [32], two existing analytical methods for tuning the parameters of fractional PIDs are reviewed. Then, for two specific sets of performance criteria similar to (2)–(6), the corresponding sets of tuning rules are developed based on an optimization method applied to the FO-PID control of an S-shaped process dynamics similar to (7). The newly developed tuning rules for fractional order PIDs use the time delay L value and the estimated process time constant, T, much like the standard S-shaped Ziegler-Nichols approach, to produce the controller parameters. The method works provided the step response of the process is S-shaped. These two methods were initially presented in [33]. The first set of rules developed works if $0.1 \leq T \leq 50$ and $L \leq 2$, while the second set of rules can be applied for processes with $0.1 \leq T \leq 50$ and $L \leq 0.5$. Both sets of rules are determined in a similar manner. For a batch of process described as FOPDT systems, a set of performance specifications is imposed. The set included values for the gain crossover frequency, for the phase margin, a high-frequency value for the improved high-frequency noise cancellation and the corresponding maximum magnitude limit, as well as a low frequency value for improved output disturbance and the corresponding maximum magnitude. Tuning by minimization is then applied using the fmincon Matlab® (Natick, Massachusetts, USA) function, where the magnitude equation in (2) is used as the main function to minimize, whereas the remaining conditions in (3)–(6) are used as constraints. Using least squares fit, polynomials are determined to compute the controller parameters based on the process time constant T and time delay L: $P = -0.0048 + 0.2664L + 0.4982T + 0.0232L^2 - 0.0720T^2 - 0.0348LT$. Using Figures 4 and 5, for the first set of rules and for the second one, the FO-PID controller parameters, as indicated in (1), can be finally computed:

$$k_p = P, \; T_i = \frac{k_p}{I} \text{ and } T_d = \frac{D}{k_p} \tag{14}$$

	Parameters to use when $0.1 \leq T \leq 5$					Parameters to use when $5 \leq T \leq 50$				
	P	I	λ	D	μ	P	I	λ	D	μ
1	−0.0048	0.3254	1.5766	0.0662	0.8736	2.1187	−0.5201	1.0645	1.1421	1.2902
L	0.2664	0.2478	−0.2098	−0.2528	0.2746	−3.5207	2.6643	−0.3268	−1.3707	−0.5371
T	0.4982	0.1429	−0.1313	0.1081	0.1489	−0.1563	0.3453	−0.0229	0.0357	−0.0381
L^2	0.0232	−0.1330	0.0713	0.0702	−0.1557	1.5827	−1.0944	0.2018	0.5552	0.2208
T^2	−0.0720	0.0258	0.0016	0.0328	−0.0250	0.0025	0.0002	0.0003	−0.0002	0.0007
LT	−0.0348	−0.0171	0.0114	0.2202	−0.0323	0.1824	−0.1054	0.0028	0.2630	−0.0014

Figure 4. Parameters for the first set of tuning rules for S-shaped response processes ($P = k_p$, $I = k_p/T_i$, $D = k_pT_d$) [32].

	P	I	λ	D	μ
1	-1.0574	0.6014	1.1851	0.8793	0.2778
L	24.5420	0.4025	-0.3464	-15.0846	-2.1522
T	0.3544	0.7921	-0.0492	-0.0771	0.0675
L^2	-46.7325	-0.4508	1.7317	28.0388	2.4387
T^2	-0.0021	0.0018	0.0006	-0.0000	-0.0013
LT	-0.3106	-1.2050	0.0380	1.6711	0.0021

Figure 5. Parameters for the second set of tuning rules for S-shaped response processes ($P = k_p$, $I = k_p/T_i$, $D = k_p T_d$) [32].

3. Direct Fractional Order Autotuning Methods

Most of the direct autotuning methods are based on using the relay test to determine the process critical gain and critical period of oscillations, but other methods have been developed [34].

Several generalizations of the Ziegler-Nichols ultimate gain method have been proposed over the years for the tuning of FO-PIDs. A new tuning method for such a controller that combines both the Ziegler-Nichols as well as the Astrom and Hagglund methods has been proposed in [35]. The idea is based on obtaining the process critical frequency and critical gain and then computing the k_p and T_i parameters using the classical Ziegler-Nichols method. For a specified phase margin, the T_d parameter is computed using the Astrom and Hagglund method. Two equations referring to the controller's real and imaginary parts are obtained. Fine tuning of T_d is employed to achieve the best numerical solution of the equations, for each specified phase margin. Matlab®'s built in functions, such as fsolve, are used to solve the two equations to obtain numerical values of λ and μ by considering the new value of T_d for each specified phase margin. An optimization Simulink model is used to obtain a better step response. The least squares method is used in the optimization model and the optimized FO-PID parameters are obtained. The approach is tedious and involves three controller designs before the final optimized FO-PID is obtained. However, the design allows for a direct specification of the loop phase margin.

In [36], an extension of the modified Ziegler-Nichols tuning rules for fractional-order controllers is presented. The proposed design approach is only suitable for tuning fractional order PI controllers. The tuning rules are derived without any knowledge of the process model, but they require the critical frequency ω_{cr}, as well as the corresponding critical gain k_{cr}. Based on this process information, the FO-PI autotuning objective is to determine the controller parameters such that the loop frequency response is moved to a point in the Nyquist plane where a performance criterion is minimized, according to a constraint. The performance criterion is mathematically expressed as a measure of the system ability to handle low-frequency load disturbances, subjected to a robustness constraint referring to the maximum sensitivity function of the closed-loop system. The tuning rules are given by:

$$\lambda = \frac{1.11k_{180} + 0.084}{k_{180} + 0.07}, \quad k_p = k_{cr}r_b \cos \beta + k_{cr}r_b \cot \lambda \sin \beta \text{ and } T_i = \frac{k_p}{k_i} \tag{15}$$

where $k_{180} = \frac{1}{k_{cr}k}$, $r_b = \frac{0.34k_{180}+0.03}{k_{180}+0.52}$, $\beta = \frac{-0.92k_{180}-0.012}{k_{180}+0.6}$, $k_i = \frac{-k_{cr}r_b\omega_{cr}^{\lambda}}{\sin \gamma} \sin \beta$ and $\gamma = \frac{\pi}{2}\lambda$, with k the process gain as indicated in (7).

The method is compared with several other direct and indirect autotuning methods for integer order PIDs and it provides good performance results. The method is also compared to some similar autotuning approaches developed in [37,38] and the results demonstrate the superiority of the current approach.

A similar idea as the one used in [31] is employed in [39], where an error filter is cascaded with a FO-PI controller. Unlike the autotuning approach taken in [31], the research in [39] is focused on estimating the parameters of an integer order PI controller using the relay method. An estimation of the process critical gain and period of oscillation is firstly determined, which in turn leads to the computation of an integer order PI controller parameters according to the standard Ziegler-Nichols approach. The same error filter is used in [39] as in [31] with the same advantage. Various values for the fractional order integration are used and the results evaluated on a steam temperature process.

Experimental results show that the FO-PI controller leads to better performance during the set-point change and load disturbance test in terms of output and control effort. However, poor closed-loop performance is obtained if λ is set too low. Even though both the direct [39] and the indirect [31] autotuning methods are simple enough for designing the FO-PI controllers, there is no clear advice on the selection of the fractional order of integration.

A modification of the Ziegler-Nichols closed-loop method is proposed in [40]. The method provides for an improvement of the standard Ziegler-Nichols results. The idea is based on the fact that a fractional order can help shape the "direction" of the loop frequency response in a fixed point in the Nyquist plot and thus keep the loop frequency response further away from the -1 point. The process critical frequency of oscillation, as well as the critical gain are obtained based on the relay test. To simplify the tuning method, the same fractional order of integration and differentiation is used in the FO-PID, similarly to [41]:

$$C_{\text{FO-PID}}(s) = k_p \left(1 + \frac{1}{(\tau_i s)^\lambda} + (\tau_d s)^\lambda \right) \tag{16}$$

where $T_i = \tau_i^\lambda$ and $T_d = \tau_d^\lambda$. The ratio $r = \frac{\tau_i}{\tau_d}$ between the integral and derivative time constants is considered to be a design parameter. The final tuning rules are exemplified for a ratio $r = 4$, similarly to [6,41]. Unlike the standard Ziegler-Nichols approach, the tuning rules depend not only on the process critical gain K_{cr} and critical period of oscillation P_{cr}, but also on the fractional order. The parameters of the FO-PID controller can thus be easily computed, without any complex optimization procedure [40], as indicated in Table 1.

Table 1. FO-PID parameters according to the modified Ziegler-Nichols method and for different values of the fractional order λ [40].

λ	k_p	T_i	T_d
0.4	0.16 kcr	$2.27 T_{cr}^{0.4}$	0.57 Ti
0.5	0.23 kcr	$1.55 T_{cr}^{0.5}$	0.50 Ti
0.6	0.29 kcr	$1.12 T_{cr}^{0.6}$	0.44 Ti
0.7	0.36 kcr	$0.87 T_{cr}^{0.7}$	0.38 Ti
0.8	0.42 kcr	$0.71 T_{cr}^{0.8}$	0.33 Ti
0.9	0.50 kcr	$0.59 T_{cr}^{0.9}$	0.29 Ti
1	0.6 kcr	$0.5 T_{cr}^{1.0}$	0.25 Ti

The critical process gain and period of oscillation are used in [42] to determine the parameters of a FO-PID controller. Three sets of tuning rules are developed. Processes described as FOPDT systems are used for two of the sets, whereas for the third one, integrative processes are considered. The first set of tuning rules applies when the critical period of oscillations $P_{cr} \leq 8$ and $P_{cr} K_{cr} \leq 640$. For the case when $P_{cr} \leq 2$, a second set of tuning rules is developed. Both of these are quite restrictive and do not often work properly for plants with a pole at the origin [42]. The third set of rules is designed specifically for integrative processes (without time delay), but can be used only when $0.2 \leq P_{cr} \leq 5$ and $1 \leq K_{cr} \leq 200$. The research in [42] concludes that the closed-loop performance can be poor near the borders of the mentioned range. All of these rules were developed in order to meet certain performance specifications regarding the loop gain crossover frequency, phase margin, iso-damping, rejection of high-frequency noise and output disturbance. All tuning rules are developed similarly to those in [32], by minimizing the magnitude equation in (2) and using the remaining conditions in (3)–(6) as design constraints. The controller parameters are obtained by polynomial fitting using least squares. The coefficients of the polynomials for the three sets of tuning rules are indicated in Figures 6–8.

	Parameters to use when $K_{cr}P_{cr} \le 64$					Parameters to use when $64 \le K_{cr}P_{cr} \le 640$				
	P	I	λ	D	μ	P	I	λ	D	μ
1	0.4139	0.7067	1.3240	0.2293	0.8804	-1.4405	5.7800	0.4712	1.3190	0.5425
K_{cr}	0.0145	0.0101	-0.0081	0.0153	-0.0048	0.0000	0.0238	-0.0003	-0.0024	-0.0023
P_{cr}	0.1584	-0.0049	-0.0163	0.0936	0.0061	0.4795	0.2783	-0.0029	2.6251	-0.0281
$1/K_{cr}$	-0.4384	-0.2951	0.1393	-0.5293	0.0749	32.2516	-56.2373	7.0519	-138.9333	5.0073
$1/P_{cr}$	-0.0855	-0.1001	0.0791	-0.0440	0.0810	0.6893	-2.5917	0.1355	0.1941	0.2873

Figure 6. Parameters for the first set of tuning rules for processes with critical gain and period ($P = k_p$, $I = k_p/T_i$, $D = k_p T_d$) [42].

	P	I	λ	D	μ
1	1.0101	10.5528	0.6213	15.7620	1.0101
K_{cr}	0.0024	0.2352	-0.0034	-0.1771	0.0024
P_{cr}	-0.8606	-17.0426	0.2257	-23.0396	-0.8606
P_{cr}^2	0.1991	6.3144	0.1069	8.2724	0.1991
$K_{cr}P_{cr}$	-0.0005	-0.0617	0.0008	0.1987	-0.0005
$1/K_{cr}$	-0.9300	-0.9399	1.1809	-0.8892	-0.9300
$1/P_{cr}$	-0.1609	-1.5547	0.0904	-2.9981	-0.1609
K_{cr}/P_{cr}	-0.0009	-0.0687	0.0010	0.0389	-0.0009
P_{cr}/K_{cr}	0.5846	3.4357	-0.8139	2.8619	0.5846

Figure 7. Parameters for the second set of tuning rules for processes with critical gain and period ($P = k_p$, $I = k_p/T_i$, $D = k_p T_d$) [42].

	P	I	λ	D	μ
1	-1.6403	-92.5612	0.7381	-8.6771	0.6688
K_{cr}	0.0046	0.0071	-0.0004	-0.0636	0.0000
P_{cr}	-1.6769	-33.0655	-0.1907	-1.0487	0.4765
$K_{cr}P_{cr}$	0.0002	-0.0020	0.0000	0.0529	-0.0002
$1/K_{cr}$	0.8615	-1.0680	-0.0167	-2.1166	0.3695
$1/P_{cr}$	2.9089	133.7959	0.0360	8.4563	-0.4083
K_{cr}/P_{cr}	-0.0012	-0.0011	0.0000	0.0113	-0.0001
P_{cr}/K_{cr}	-0.7635	-5.6721	0.0792	2.3350	0.0639
$\log_{10}(K_{cr})$	0.4049	-0.9487	0.0164	-0.0002	0.1714
$\log_{10}(P_{cr})$	12.6948	336.1220	0.4636	16.6034	-3.6738

Figure 8. Parameters for the third set of tuning rules for processes with critical gain and period ($P = k_p$, $I = k_p/T_i$, $D = k_p T_d$) [42].

The relay test is also used in [43], but with a variation that includes also a time delay, as indicated in Figure 9. The process frequency response at any frequency can be identified using this scheme. The main issue is to determine the correct value of the time delay that corresponds to a specific frequency. An iterative method is used [44] and two initial values for the time delay and their corresponding frequencies are needed to start the iteration.

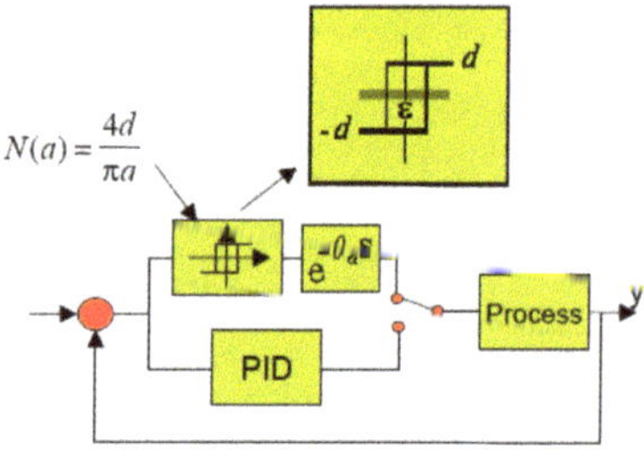

Figure 9. Relay autotuning scheme with delay [43].

The autotuning method is based on specifying an iso-damping property, a gain crossover frequency and a phase margin. A fractional order PI controller is designed first, followed by a fractional order PD controller with a filter. The fractional order PI controller will be used to ensure the iso-damping property around the gain crossover frequency w_{cg}. The slope of the phase of the plant is computed using the gain crossover frequency and the corresponding phase and a supplementary frequency and its corresponding process phase as resulting from the relay experiment. Once the slope is cancelled using the FO-PI controller, the FO-PD controller is designed to fulfill the design specifications of gain crossover frequency and phase margin. To ensure a maximum robustness to plant gain variations, a robustness criterion based on the flatness of the phase curve of the FO-PD controller is used such that the resulting phase of the open-loop system will be the flattest possible. The procedure is rather lengthy. A mechanical unit consisting mainly of a servo motor is used to experimentally validate the proposed method. The experimental results illustrate the effectiveness of this method.

The same method is described in [45], where experimental results with the FO-PID on a similar servo motor are used to validate the efficiency of the approach. A refinement of the relay feedback test in [43,45] is introduced in [46]. The improvement is based on adding a moving average filter. Simulation results for the control of a position servo with time delay are presented and validate the autotuning algorithm. The same autotuning method for determining a FO-PID controller for the servo system in [46] is presented in [47]. A similar approach is detailed in [48] for the design of FO-PID controllers. Two numerical case studies are provided for a double-integrator process and a fractional order integrative process. The simulation results validate the autotuning method.

Instead of using the relay test to determine the process magnitude, phase and phase slope, a single sine test at the gain crossover frequency is used in [34]. Novel filtering techniques are used to determine the process phase slope, as indicated in Figure 10. To determine the parameters of either a FO-PI or a FO-PD controller, performance specifications regarding the phase margin, gain crossover frequency and iso-damping property are used. The process magnitude, phase and phase slope previously determined are used in the resulting nonlinear equations. Optimization techniques or graphical methods are then employed to determine the controller parameters. Numerical examples are used to validate the approach. A different approach is presented in [49], where a forbidden region circle is defined based on the iso-damping property and phase margin specifications. The same sine test used in [34] is required here as well, in order to estimate the process phase, magnitude and phase slope. Instead of using optimization routines, the parameters of the optimal fractional order PID controller are determined by minimizing the slope difference between the circle border and the loop-frequency response. Numerical results are presented to validate the approach.

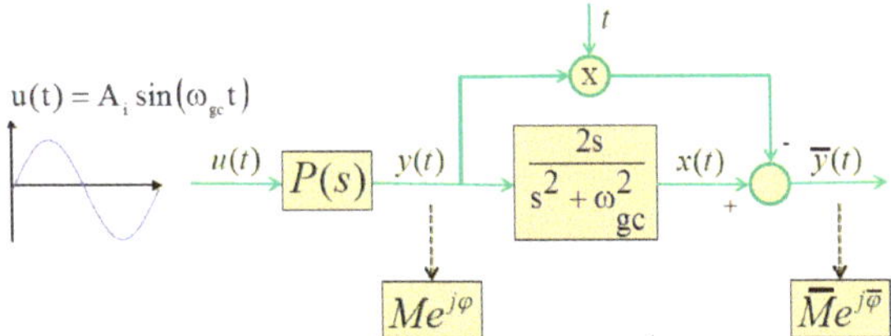

Figure 10. Experimental scheme used to compute the phase slope of the process at the gain crossover frequency (refer to [34] for details).

4. Numerical Examples

Some of the previously presented autotuning methods are used to determine the parameters of various types of fractional-order controllers for a series of processes that exhibit time delays, integrative effects, overdamped and poorly damped responses, higher orders. For simplicity, only the most recent and widely used autotuning methods are considered. All resulting fractional-order controllers are implemented using the same method and the same approximation parameters [50]. For the numerical examples considered in this manuscript, all fractional-order controllers are implemented with the proportional, fractional integration and differentiation actions on the error signal. The peaks in some

FO-PID controller output signals are not the result of the tuning; they are simply the result of using derivative action on a setpoint step. These can be removed by implementing the FO-PI action on the error signal and FO-D action on the output signal.

4.1. The FOPDT Lag-Dominant Process

The following FOPDT lag-dominant process taken from [26] is considered:

$$P(s) = \frac{2.4351}{12.5688s + 1}e^{-1.0787s} \tag{17}$$

In this case, $k = 2.4315$, $L = 1.0787$, $T = 12.5688$. Based on the relay test, the critical gain is $K_{cr} = 7.78$ and $P_{cr} = 4.175$. The parameters of the fractional-order controllers used for comparative purposes are indicated in Table 2. Indirect [26,27,32] and direct tuning methods [36,40,42,49] are used. The direct autotuning method in [34] produces the same result as in [49] and, therefore, was omitted from the comparison. Only the first set of tuning rules in [32] is used, as the second set of tuning rules cannot be applied. The same is valid for the direct autotuning method in [42], where only the first set of tuning rules is used, since the other two sets of tuning rules cannot be applied to this particular process.

Table 2. FO-PID parameters computed for the lag-dominant process.

Controller Type	k_p	T_i	λ	T_d	μ
FO-PID Tepljakov [27]	1.9255	0.8653	0.8	0.4010	0.6935
FO-PI F-MIGO [26]	1.5413	5.0326	0.7	-	-
FO-PI Gude [36]	0.6641	6.6277	1.1613	-	-
FO-PID ZN-FOC [40]	2.8008	2.3658	0.7	0.8990	0.7
FO-PID [42] critical first set	1.1112	1.5809	1.2298	0.5944	0.8976
FO-PID [32] S-shaped first set	−5.2747	0.8678	0.1903	1.7823	−1.5272
FO-PI FO KC [49]	1.0922	13.7410	1.15	-	-

The FO-PID [42] leads to a highly oscillating closed-loop response, while the FO-PID [32] is an unstable controller, which suggests that the proposed tuning rules work poorly for the lag-dominant system in (17). In fact, in both cases the expected phase margin is 38° [32,42], which explains the highly oscillating character. The FO-PI [49] was tuned to meet the iso-damping property, as well as a gain crossover frequency of 0.2 rad/s and a phase margin of 70°. These performance specifications were selected in order to obtain a small overshoot, as well as the fastest possible settling time. The closed-loop results considering step reference tracking and disturbance rejection are given in Figure 11, while the numerical values of the overshoot, settling time and disturbance rejection time are given in Table 3. The results show that the smallest overshoot is obtained using the FO-PI in [49], at the expense of a large settling time and the time required to reject the load disturbance. Small overshoot is obtained also using the FO-PID of Tepljakov in [27] or using the F-MIGO method [26], while the settling time is slightly larger in the latter case. However, the required control effort for Tepljakov's FO-PID [27] is extremely large compared to the other methods, as indicated in Figure 11b). A larger control effort is also observed in the case of the FO-PI controller in [40], which achieves the fastest settling time and the smallest disturbance rejection time. A decent control effort is necessary when using FO-PI controllers tuned according to [26,36,49]. Among these, the fastest settling time is obtained with the FO-PI controller [26], while the smallest overshoot is achieved by the FO-PI controller [49]. Improved settling time might be possible in this last case, if a FO-PD controller is designed and implemented in series with the FO-PI.

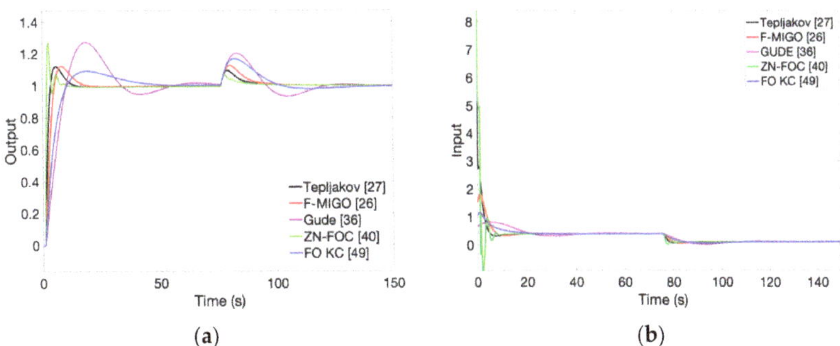

Figure 11. (a) Output signals for FO-PID control of lag-dominant process (b) Input signals for FO-PID control of lag-dominant process. Controllers tuned according to [26,27,36,40,49].

Table 3. Closed-loop results obtained with the FO-PID controller for the lag-dominant process.

Controller	Overshoot	Settling Time	Disturbance Rejection Time
FO-PID Tepljakov [27]	12%	11.5	13.5
FO-PI F-MIGO [26]	13%	15.7	17
FO-PI Gude [36]	28%	51	41.5
FO-PID ZN-FOC [40]	27%	7.2	10
FO-PI FO KC [49]	9%	41	23.5

4.2. The Higher Order Process

The following higher order process taken from [36] is considered:

$$P(s) = \frac{1}{(s+1)^4} \tag{18}$$

In this case, $k = 1$, $L = 1.42$, $T = 2.92$ and the critical gain is $K_{cr} = 4$ and $P_{cr} = 6.28$ [36]. The parameters of the fractional-order controllers used for comparative purposes are indicated in Table 4. Indirect [26,32] and direct tuning methods [36,40,42,49] are used. The direct autotuning method in [34] produces the same result as in [49] and therefore was omitted from the comparison. Only the first set of tuning rules in [32] is used, as the second set of tuning rules cannot be applied. The same is valid for the direct autotuning method in [42], where only the first set of tuning rules is used, since the other two sets of tuning rules cannot be applied to this particular processes.

Table 4. FO-PID parameters computed for the higher order process.

Controller Type	k_p	T_i	λ	T_d	μ
FO-PI F-MIGO [26]	0.9093	1.7905	0.9	-	-
FO-PI Gude [36]	0.6080	3.5486	1.13	-	-
FO-PID ZN-FOC [40]	2	3.0847	0.9	0.5861	0.9
FO-PID [42] critical first set	1.3439	2.1448	1.2366	0.5501	0.9311
FO-PID [32] S-shaped first set	1.1168	1.1449	1.1	1.2152	1.0373
FO-PI FO KC [49]	0.8162	4.5733	1.18	-	-

Figure 12 shows the closed-loop results obtained with the first three controllers in Table 4, while Figure 13 presents the closed-loop simulations obtained with the last three controllers. Note that the FO-PI [49] controller was first tuned for a gain crossover frequency of 0.5 rad/s and a phase margin of 38°. This is in agreement to the performance specifications used in [32,42] for the first set of tuning rules. The results in Figure 13 show that indeed similar overshoot and settling times are obtained with the fractional-order controllers tuned according to [32,42,49].

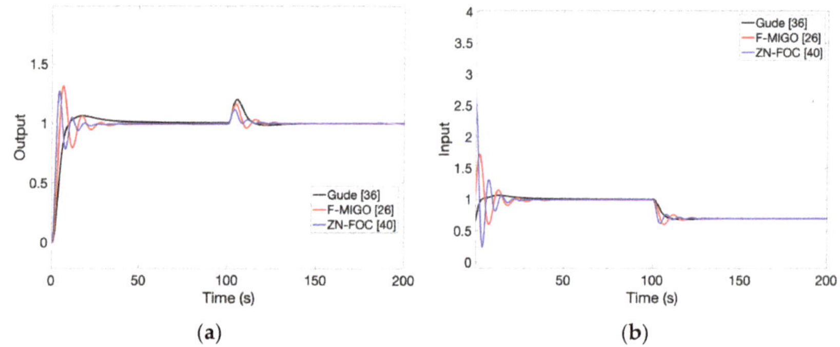

Figure 12. (**a**) Output signals for FO-PID control of higher order process (**b**) Input signals for FO-PID control of higher order process (controllers tuned according to [26,36,40]).

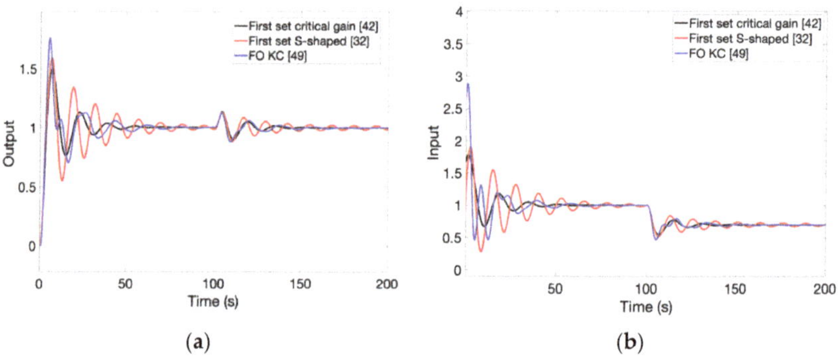

Figure 13. (**a**) Output signals for FO-PID control of higher order process (**b**) Input signals for FO-PID control of higher order process (controllers tuned according to [32,42,49]).

Table 5 contains the performance evaluation of the fractional-order controllers from Figure 12. The remaining three controllers in Figure 13 are not evaluated due to the large overshoot and settling time. The direct autotuning method from [49] can be used to tune a better FO-PI controller. Table 4 shows the resulting parameters of this improved FO-PI controller, which was tuned to meet a gain crossover frequency of 0.2 rad/s and a phase margin of 75°. The performance of this better FO-PI controller is compared to that of the FO-PI controller in [36], which achieves the best overshoot and settling time. The comparative simulation results are given in Figure 14 and in Table 5.

Table 5. Closed-loop results obtained with the FO-PID controller for the higher order process.

Controller	Overshoot	Settling Time	Disturbance Rejection Time
FO-PI F-MIGO [26]	31%	33.5	17.5
FO-PI Gude [36]	6.5%	34.3	13.1
FO-PID ZN-FOC [40]	27%	23.3	12.7
FO-PI FO KC [49]	6.5%	42.8	13.5

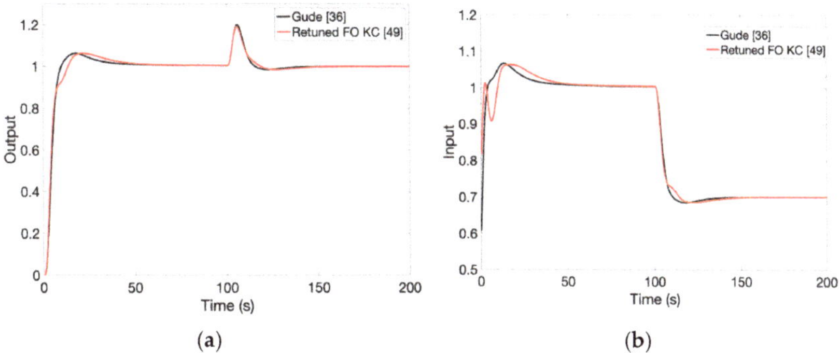

Figure 14. (**a**) Comparative results for the output signals for FO-PID control of higher order process. (**b**) Input signals for FO-PID control of higher order process (controllers tuned according to [36,49]).

To estimate the quantitative results given in Table 5 for the FO-PI and FO-PID controllers designed according to [26,40], the closed-loop simulation results from Figure 12 were considered. The quantitative results in Table 5 show that the smallest overshoot is possible using the FO-PI controller tuned using the methods in [36,49], which are also suitable to achieve a quick disturbance rejection. Similarly to the lag-dominant case study, in this case as well, the FO-PID controller [40] achieves the smallest settling time and the fastest disturbance rejection time, at the expense of an increased control effort similar to that of the FO-PID [42] and larger compared to the other controllers.

4.3. The Integrating Time-Delay Process

An integrating time-delay process [46] is considered here. The transfer function is:

$$P(s) = \frac{0.55}{s(0.6s + 1)}e^{-0.05s} \tag{19}$$

The classical Ziegler-Nichols autotuning method has a major disadvantage: poor results are obtained regarding setpoint tracking, especially when used with integrating systems [1]. Several extensions and improvements have been developed over the years to deal with such systems. The autotuning methods based on an S-shaped response of the process cannot be used in this particular situation.

In [47], an iterative experiment of a relay with delay is applied to the process in order to determine the process magnitude, phase and phase slope at a specific gain crossover frequency 2.3 rad/s. Then, a FO-PI in series with a FO-PD controller are designed to meet the iso-damping property, a gain crossover frequency of 2.3 rad/s for the open-loop system and a phase margin of 72°. The resulting fractional-order controller is given by [47]:

$$C_{MONJE}(s) = \left(\frac{0.4348s + 1}{s}\right)^{1.1803} \left(\frac{3.7282s + 1}{0.0037s + 1}\right)^{1.1580} \tag{20}$$

Four other direct autotuning methods are used for comparison purposes. First, based on the relay test, the critical gain is $K_{cr} = 36.88$ and $P_{cr} = 1.1043$. The parameters of the fractional-order controllers used for comparative purposes are indicated in Table 6, where the fractional-order controllers have been determined using [40,42,49]. The first and the third set of tuning rules in [42] are used to estimate the FO-PID controller parameters, as the second set cannot be applied for the process in (19). The tuning rules in [42] were developed for integrative processes without time delays (third set) and for FOPDT processes (first and second set). For the process in (19), the third set of tuning rules [42] leads to an unstable controller. Figure 15 shows the simulation results. The quantitative performance results are indicated in Table 7. Similar overshoot is obtained for the fractional-order controller in (20) designed using [47] and for the FO-PI controller [49], despite the latter having a large settling time. The fastest FO-PID controller is yet again the one designed according to [40]. The poorest overshoot along with a significant settling time is obtained with the FO-PID [42]. A comparison of the required control effort based on Figure 15b,c shows the increased amplitudes of the input signals are necessary for the FO-PIDs designed based on [40,42,47]. The smallest control effort is required by the FO-PI controller tuned according to [49], which also exhibits the largest settling time and a significant disturbance rejection time. However, this controller is also the simplest one, without any derivative effect.

Table 6. FO-PID parameters computed for the integrative time-delay process.

Controller Type	k_p	T_i	λ	T_d	μ
FO-PID ZN-FOC [40]	5.9002	2.6760	0.4	1.5242	0.4
FO-PID [42] critical first set	1.0342	1.0606	1.0827	0.8148	0.7855
FO-PID [42] critical third set	0.4616	0.0919	0.5929	−2.7379	0.9360
FO-PI FO KC [49]	0.3812	2.7988	0.71	-	-

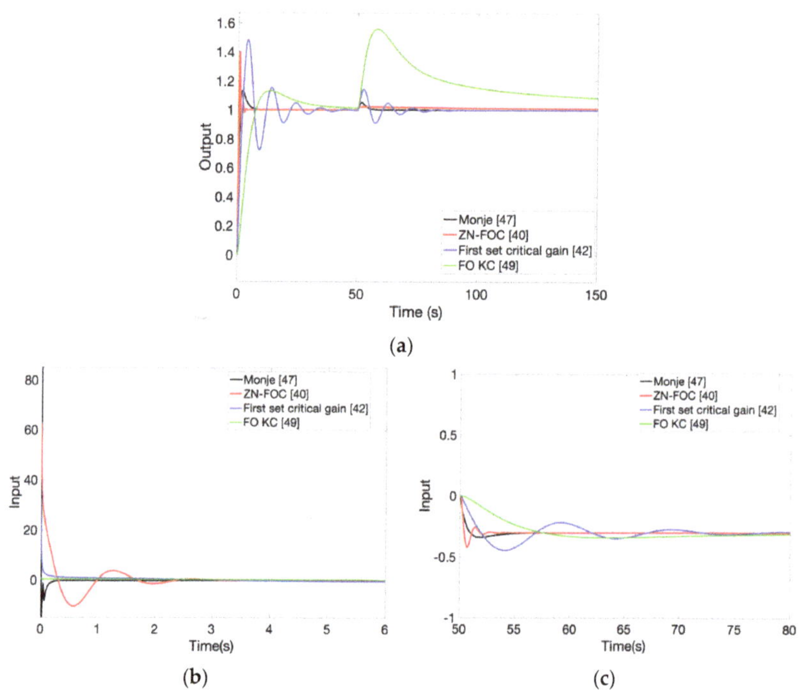

Figure 15. (**a**) Output signals for FO-PID control of integrative time-delay process. (**b**) Input signals for FO-PID control of integrative time-delay process required for setpoint tracking. (**c**) Input signals for FO-PID control of integrative time-delay process required for disturbance rejection. Controllers tuned according to [40,42,47,49].

Table 7. Closed-loop results obtained with the FO-PID controller for the integrative time-delay process.

Controller	Overshoot	Settling Time	Disturbance Rejection Time
FO-PID ZN-FOC [40]	40%	2.3	4.5
FO-PID [42] critical first set	48.5%	30.7	19.1
FO-PI FO KC [49]	13%	42	>70
FO-PID Monje [47]	13.5%	5.3	3.8

4.4. The FOPDT Delay-Dominant Process

A FOPDT delay-dominant process is considered in the comparison, with the transfer function given by:

$$P(s) = \frac{1}{0.2s + 1} e^{-0.4s} \tag{21}$$

In this case, $k = 1$, $L = 0.4$, $T = 0.2$ and the critical gain is $K_{cr} = 1.5202$ and $P_{cr} = 1.0985$. Three indirect autotuning methods are used in the comparisons with the FO-PID controllers computed according to [32] using the first and the second set of rules for S-shaped process response. The third method is the F-MIGO method described in [26]. The resulting controller parameters are indicated in Table 8. Five direct autotuning methods are also used for the comparison, namely: FO-PID tuned according to the first and second set of rules in [42], FO-PID computed based on the method in [40] and two FO-PI controllers determined using [36,49]. The controller parameters for these cases are also given in Table 8. The FO-PI [49] is tuned to meet the iso-damping property, a gain crossover frequency 1.2 rad/s and a 70° phase margin. The closed-loop results are indicated in Figures 16 and 17, while the performance is evaluated using quantitative measures as indicated in Table 9. The FO-PID controller obtained using the second set of tuning rules in [42] is not included in the comparison, due to its highly oscillating nature.

Table 8. FO-PID parameters computed for the integrative time-delay process.

Controller Type	k_p	T_i	λ	T_d	μ
FO-PI F-MIGO [26]	0.4465	0.2951	1.1	-	-
FO-PI Gude [36]	0.3179	0.2878	1.1187	-	-
FO-PID ZN-FOC [40]	0.7601	0.6420	0.9	0.1220	0.9
FO-PID [42] critical first set	0.2437	0.5649	1.4574	−0.1349	1.0028
FO-PID [42] critical second set	−0.0293	−0.4905	1.2668	23.7347	−0.0293
FO-PI FO KC [49]	0.6573	0.3471	1.186	-	-
FO-PID [32] S-shaped first set	0.1994	0.4622	1.4788	0.0845	0.9847
FO-PID [32] S-shaped second set	1.3281	1.7653	1.3168	−0.4142	−0.1793

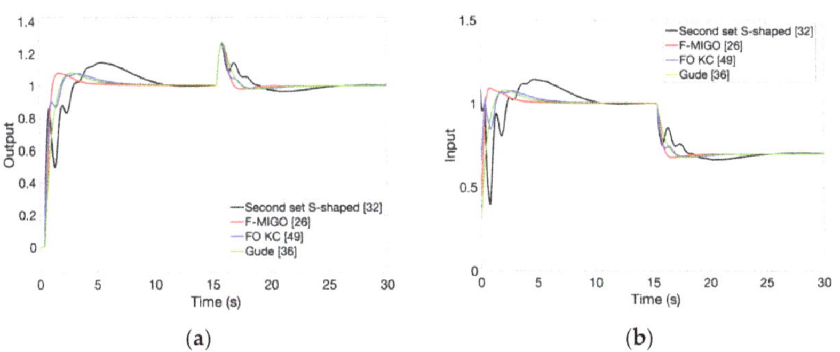

Figure 16. (a) Output signals for FO-PID control of delay-dominant process. (b) Input signals for FO-PID control of the delay-dominant process (controllers tuned according to [26,32,36] second set and [49]).

A small overshoot is obtained with the FO-PI controllers [26,36,49], combined with small settling times and fast disturbance rejection. The control effort in all these cases is similar, according to Figure 16b. FO-PID tuned using [40] manages to achieve a small settling time for this case study, as well. Good results are also obtained for disturbance rejection, at the expense of a larger control effort, compared to the other controllers (Figure 17b). FO-PIDs determined according to [32,42] lead to larger overshoots and increased settling times, as well as a poorer disturbance rejection, as indicated in Figure 17a. The required control effort for these controllers is small (Figure 17b), comparable to the input amplitudes given in Figure 16b.

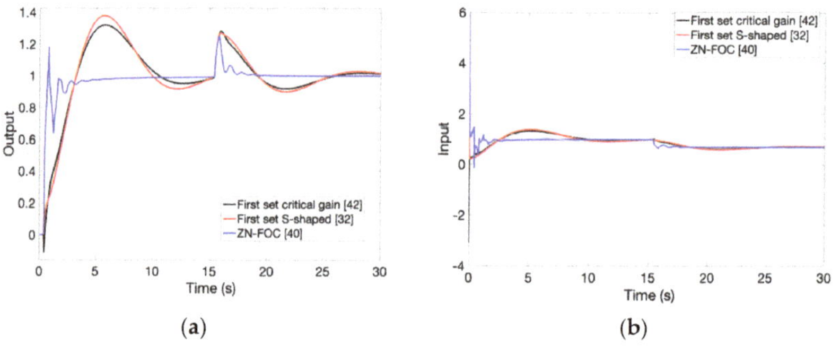

(a)   (b)

Figure 17. (a) Output signals for FO-PID control of delay-dominant process. (b) Input signals for FO-PID control of the delay-dominant process (controllers tuned according to [40,42] first set and [32] first set).

Table 9. Closed-loop results obtained with the FO-PID controller for the delay-dominant process.

Controller	Overshoot	Settling Time	Disturbance Rejection Time
FO-PI F-MIGO [26]	8%	4	1.5
FO-PI Gude [36]	8.2%	5.8	2.2
FO-PID ZN-FOC [40]	18%	5.5	2.9
FO-PID [42] critical first set	32%	17	9.9
FO-PI FO KC [49]	7.5%	6.7	1.8
FO-PID [32] S-shaped first set	37.8%	18.5	14.6
FO-PID [32] S-shaped second set	14.5%	10.1	8.4

4.5. The Poorly Damped Process

A final case study is considered in this section, with the process described by the following transfer function:

$$P(s) = \frac{22.24}{s^2 + 0.6934s + 5.244} e^{-0.8s} \tag{22}$$

The indirect autotuning methods based on an S-shaped response cannot be applied for (21). An FO-PI controller tuned according to [36] is compared with a FO-PID obtained using the method in [40] and a FO-PI controller determined using [49]. First, the relay method is used to estimate the critical gain as $K_{cr} = 0.0709$ and $P_{cr} = 2.8$. These critical gain and period of oscillations allow the design of a FO-PID controller according to the first set of tuning rules in [42]. However, the proportional gain obtained in this way is negative and destabilizes the closed-loop system. Thus, the design is not included in this comparison. To tune the FO-PI controller [49], a sine test is firstly applied to the process to determine its phase, magnitude and phase slope. Then, the parameters of the FO-PI controller are determined such that the open-loop system achieves a gain crossover frequency of 0.09 rad/s and a phase margin of 75°, along with the iso-damping property. The parameters of the fractional-order controllers are given in Table 10. Figure 18 shows the closed-loop results, as well as the required input signals. The performance measures are indicated in Table 11. The simulation results in Figure 18 and Table 11 show that the fastest settling time is achieved by the FO-PID controller [40], with a zero overshoot. However, in this case, the required control effort is the largest. The two FO-PI controllers determined using [36,49] have a similar overshoot, as well as control effort. For the latter, the settling and the disturbance rejection times are larger.

Table 10. FO-PID parameters computed for the poorly damped process.

Controller Type	k_p	T_i	λ	T_d	μ
FO-PI Gude [36]	0.0147	0.8348	1.1190	-	-
FO-PID ZN-FOC [40]	0.0355	1.4904	0.9	0.2832	0.9
FO-PI FO KC [49]	0.6573	0.3471	1.186	-	-

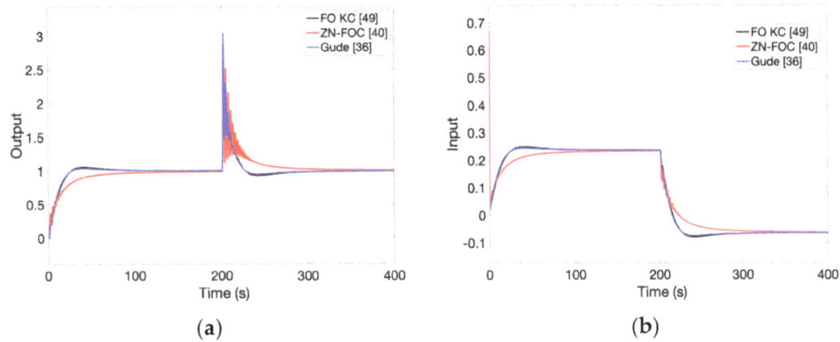

Figure 18. (**a**) Output signals for FO-PID control of the poorly damped process. (**b**) Input signals for FO-PID control of poorly damped process. Controllers tuned according to [36,40,49].

Table 11. Closed-loop results obtained with the FO-PID controller for the poorly damped process.

Controller	Overshoot	Settling Time	Disturbance Rejection Time
FO-PI Gude [36]	4%	68	74
FO-PID ZN-FOC [40]	0%	33.5	85
FO-PI FO KC [49]	4%	79.5	113.5

For second-order poorly damped processes, most fractional order autotuning methods cannot be applied, except for [47,49]. The direct autotuning method in [47] leads to a FO-PI in series with a FO-PD controller, of the form given in (20), whereas the method in [49] produces a simpler FO-PI controller. Similarly to the results in Table 7, a faster settling time and better disturbance rejection are achieved using the fractional-order controller in [47], due to the FO-PD component.

4.6. Remarks on Comparative Simulation Results

The simulations results and closed-loop performance analysis shows that some of these autotuning methods allow for greater flexibility in the design, such as [34,47,49]. A faster settling time is obtained in all case studies using the autotuning method in [40]. The drawback consists in a larger control effort. The simple tuning rules from [32,42] are generally outperformed by the other autotuning methods reviewed, except for delay dominant systems, where the performance is close to the best one. For higher order systems and poorly damped ones, the best closed-loop results are obtained using the autotuning methods in [36,49] for both reference tracking and disturbance rejection. For FOPDT delay-dominant processes, the results show that the parameters of the fractional-order controllers should be estimated using the autotuning methods in [26,36,49]. In this case, improved reference tracking and disturbance rejection are obtained. For integrating time-delay processes, the best results in terms of overshoot are obtained using either a FO-PID determined based on the autotuning method in [47] or in [49]. The best settling time is obtained using either the autotuning method in [40,47]. However, the FO-PID controller autotuned according to [47] requires a significant control effort, larger than those in [40,49]. For FOPDT lag-dominant processes, the autotuning methods from [26,27,49] ensure a small overshoot, whereas fast settling and disturbance rejection times are achieved using FO-PIDs determined according to [26,27,40]. However, the control effort required when using the FO-PIDs tuned based on [27,40] is larger compared to the FO-PID tuned using the approach in [26].

5. Applications and Self-Tuned FO-PIDs

Autotuning methods have been used to produce fractional-order controllers for different processes. The purpose of this section is to provide some applicative examples of autotuning methods for fractional-order controllers designed mostly according to the methods presented in Sections 2 and 3. The autotuning method in [34] is applied to a multivariable time-delay process to tune the FO-PI controllers for each loop [51]. The method in [49] is applied for designing fractional-order controller for a multivariable refrigeration system using vapor compression [52], a heterogeneous dynamic system [53] and to a highly coupled multivariable system [54]. A robust autotuning method is described and implemented for controlling an aerodynamic system in [55]. An experimental validation of the direct autotuning method in [49] is provided in [55] for controlling an UR10 robot. The autotuning method in [34] is applied to tune a FO-PD controller for vibration suppression in a smart beam [56]. An autotuning method designed for poorly damped systems that shapes the closed-loop system in order to achieve better damping is proposed in [57]. The design is performed in the frequency domain and requires information regarding the process magnitude and phase for five frequencies. Experimental results are given to validate the efficiency of the method.

A "plug and play" solution for a multivariable FO-PI controller is developed in [58] for controlling a multivariable twin-rotor aerodynamical system. A decentralized approach is considered and three performance specifications, as in (2)–(4), are used to compute the parameters of the two FO-PI controllers, one for azimuth and one for pitch angle control. The design is based on a novel, simplified algorithm using vector theory, where the proportional $z_1 = |k_p|$ and integral $z_2 = \left| \frac{1}{T_i}(j\omega)^{-\lambda} \right|$ terms are defined as vectors. The vectorial representation of the FO-PI controller as the sum of z_1 and z_2 is indicated in Figure 19.

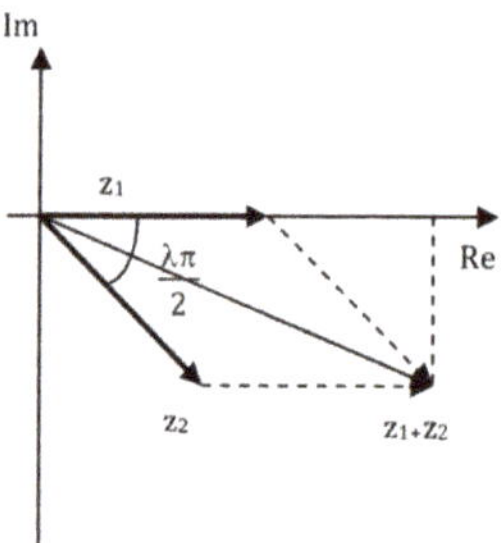

Figure 19. Vector form of a FO-PI controller.

Then, using classical trigonometric equations based on Figure 1, the proportional gain and integral time constant of the FO-PI controller are determined as a function of the fractional order λ, using the gain crossover equation (2) and the phase margin equation in (3). The procedure is iterative and computes the k_p and T_i parameters for small increments of $0 < \lambda < 1$. Then, the iso-damping property in (4) is evaluated and λ is selected to be the value that minimizes (4). Finally, k_p and T_i are computed using the selected value of λ. The fractional-order controller is implemented in a self-tuning structure as indicated in Figure 20, where the "Controller designer" block includes the iterative procedure. The "System identifier" block is used to estimate the process parameters online which are then used in the iterative procedure to determine the new values for the FO-PI controller. A recursive simple least squares algorithm is implemented in the "System identifier" block.

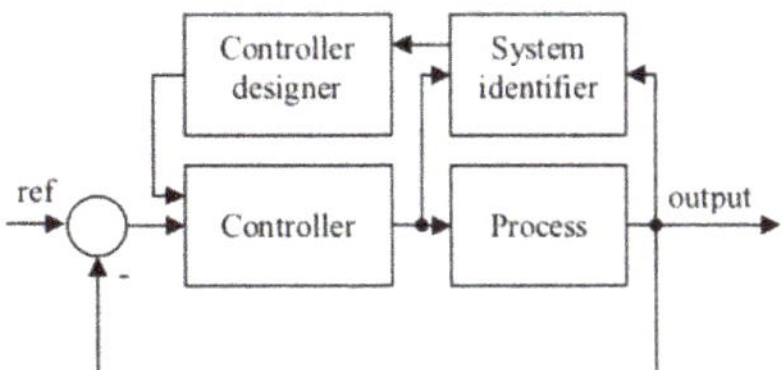

Figure 20. The self-tuning FO-PI controller.

Experimental results are provided to demonstrate the efficiency of the autotuning method. A step reference change of −1 rad for the azimuth angle and a step change of 0.2 rad for the pitch angle is considered, with the experimental results provided in Figure 21, demonstrating that reference tracking can be achieved successfully using the proposed multivariable self-tuned FO-PI control strategy.

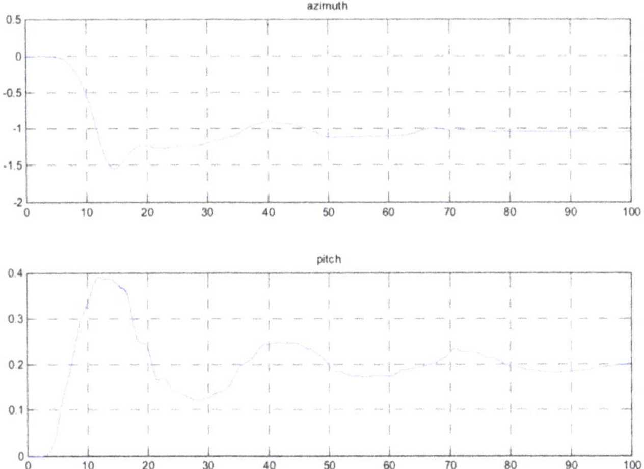

Figure 21. Experimental results of self-tuned FO-PI controllers for a twin-rotor system.

An autotuning approach for FO-PIDs is used to control the air-conditioning fan coil unit [59]. A basic differential evolution algorithm is modified by varying the mutation factor and crossover rate and used to tune the five parameters of indoor temperature FO- PID controller. Numerical simulations are presented that show that the approach is reliable and the related control performance indexes meet the requirements of comfortable air-conditioning design and control criteria.

Improvements in FO-PID controller design have been considered in order to determine algorithms that perform a better tuning in real time. One solution to this issue is the self-tuned FO-PID controller. The purpose of this last part of the manuscript is to present some ideas regarding additional solutions to autotuning methods that could facilitate the industrial acceptance of FO-PID controllers. In what follows, the manuscript covers an important part of adaptive control algorithms, namely, self-tuning methods, applied to fractional-order controllers. Only the most recent findings in this area of research are reviewed.

Fuzzy logic is usually used to achieve the self-tuning property, such a FO-PI self-tuned controller is presented in [60], in a differential mobile robot. Three different types of controllers are evaluated and compared to a classical controller, with its parameters being acquired through traditional methods. A similar self-tuned fuzzy FO-PI controller for a steam distillation process is evaluated in [61]. The numerical results show that this controller leads to better closed-loop performance in comparison to the integer order PI the FO-PI and self-tuning fuzzy PI. The control of the horizontal motion of a dual-axis photo voltaic sun-tracker is presented in [62]. A new technique for online self-tuning of a FO-PID controller based on both a type-1 fuzzy and a Takaji-Sugeno Fuzzy is developed Satisfactory results were obtained in numerical simulations. Takagi-Sugeno (TS) fuzzy technique combined with interval type-2 fuzzy sets is used in [63] to design a new adaptive self-tuning FO-PID controller. A modified FO-PID controller is obtained using TS, while the interval type-2 fuzzy sets are used as a tuner to update the gains of the FO-PID. Three types of interval type-2 fuzzy sets tuning methods are used and applied to load-frequency control as a case study of a power system comprising a single area. Comparative studies with type-1 fuzzy sets are carried. The simulation results show that the proposed approach works well considering disturbance changes and parameter uncertainties. A fuzzy FO-PID is used in [64] to control the position of a robotic manipulator. A fuzzy system combined with the particle swarm optimization method is used to determine the parameters of a FO-PID controller. Numerical simulations and comparisons with a fuzzy PID are performed. The simulation results show that the FO-PID is able to reduce the overshoot and the oscillatory dynamics, compared to the fuzzy PID. Three self-tuned fuzzy controllers are implemented in [65], namely, a FO-PD, a FO-PI and a FO-PID. The controllers are then evaluated in a servo-regulatory mechanism. The simulation results show that the self-tuned fuzzy FO-PID leads to the best closed-loop performance. The control of a mover position of a direct drive linear voice coil motor (VCM) is performed in [66] using a self-tuning FO-PID. The five FO-PID control parameters are optimized dynamically and concurrently using an adaptive differential evolution algorithm. Experimental results are provided and demonstrate that the proposed self-tuning FO-PID achieves better performance compared to PID and FOPID controllers, under both nominal and payload conditions.

The control of an inverted pendulum system is described in [67], where two self-tuned FO-PD controllers are designed to vertically balance the pendulum and for accurate positioning. The proportional and derivative gains of the two controllers are dynamically adjusted using particle swarm optimization after each sampling interval using piecewise nonlinear functions of their respective state-variations. Hardware-in-the-loop experiments are performed and the proposed approach is compared to fixed gain dual-PD and dual-FO-PD control schemes.

A direct autotuning method for a FO-PI controller is used in [68] to control the speed of a permanent magnet synchronous motor. Only the measured input-output data of the closed-loop servo system is required to tune the FO-PI controller. The FO-PI parameters are determined using a virtual reference feedback tuning with an incorporated Bode ideal transfer function, which allows the properties of the resulting system to be approximated to the desired fractional-order reference model. Optimal performance constraints, such as sensitivity criteria, frequency-domain and time-domain characteristics are considered in the autotuning. Experimental results are provided to illustrate the efficiency of the proposed model-free FO-PI control method for the servo system. The extremum seeking approach is used as a non-model-based method that searches online for the FO-PID parameters that minimizes a cost function related to the performance of the controller [69]. Simulation examples are provided to demonstrate the effectiveness of the proposed algorithm.

A novel self-tuning FO-PID controller using the optimal model reference adaptive control (MRAC) is applied to power system load-frequency control [70]. The requirements for the control systems are embedded in the model reference, mathematically described as a first- or second-order system. A harmony search optimization method is used to determine the parameters of MRAC. Three methods for self-tuning FO-PID control are presented. The first two methods assume some of the FO-PID parameters to be fixed and adjust the remaining ones, while the third method was developed to adjust all five parameters of the FO-PID controller, at the same time. The simulation results show that the latter method achieves better disturbance rejection, as well as improved handling of system uncertainty.

The control of coupled and non-linear 2-link rigid robot is tackled in [71] using a novel non-linear FO-PID that includes a non-linear hyperbolic function cascaded with a FO-PID. The fractional orders allow for greater flexibility in the controller design, while the adaptive feature is incorporated in the non-linear function. The parameters of the FO-PID are determined according to the multi-objective non-dominated sorting genetic algorithm II (NSGA-II) for small variations in control and error signal. Comparisons with a non-linear PID, FO-PID, non-linear hyperbolic function cascaded with an integer order PID or traditional PID are performed. The simulation results demonstrate that the proposed method provides robust and efficient control of the robotic arm.

A fractional fuzzy controller is designed in [72], without using an actual model of the robot and only well-known structural properties of mechanical systems. The entire implementation is model-free and tackles the control of robotic manipulators. To ensure improved disturbance rejection, a fuzzy logic formulation is used with an online adaptation of the outputs to achieve a better closed-loop response. To demonstrate the efficiency of the approach, simulations and experimental results are presented. An innovative design method, suitable for many industrial applications is presented in [73]. A self-tuning fractional-order controller is designed using fractional order pole placement and indirect adaptation profiles. Simulation results are provided for an air-lubricated capstan drive for precision positioning. The results show that, indeed, better closed-loop performance is possible using the proposed method instead of a similar one based on integer order pole placement.

A fractional-order self-tuned fuzzy PID controller is designed for a three-link rigid robotic manipulator system in [74]. The controller is tuned using a cuckoo search algorithm to minimize the weighted sum of the integral of absolute error and the integral of absolute change in controller output. The same tuning procedure is used to tune a fractional-order fuzzy PID and an integer-order self-tuning fuzzy PID. Comparative simulation results are provided and demonstrate better trajectory tracking, disturbance rejection, noise suppression and robustness to model uncertainty in the case of the proposed fractional-order self-tuned fuzzy PID controller.

In [75], an online identification of the parameters of a fractional order process is performed based on a particle swarm optimization algorithm. Then, a fractional order self-tuning regulator is designed using differential evolution algorithms. Simulation results show that the proposed method is robust and leads to good closed-loop results.

A self-tuning controller is designed in [76] using fuzzy logic for the control of micro-grid systems. A fractional-order controller is developed in combination with a fuzzy logic algorithm for load-frequency control of the off-grid microgrid. An optimal way to estimate the input and output scale coefficients of the fuzzy controller and fractional orders of the fractional-order controller is developed based on a novel meta-heuristic whale algorithm. The case study consists in a microgrid containing a diesel generator, wind turbine, photovoltaic systems and energy storage devices. Simulation results show that the proposed optimized fractional-order self-tuning fuzzy controller manages to outperform the classical PID controller in terms of operation characteristics, settling time and load-disturbance attenuation.

The active suspension system of a quarter car is considered as the case study in [77], where a self-tuned robust fractional-order fuzzy proportional-derivative controller is developed. The design of the controller attempts to minimize the root mean square of vertical vibration acceleration of car body. Tracking force, ratio between tire dynamics and static loads and suspension travel are considered as design constraints. Genetic algorithms are used to optimize the parameters online for a sinusoidal road surface. However, simulations were performed for random road surfaces and bumps. The proposed self-tuned fractional-order fuzzy proportional-derivative controller achieved better results compared to passive solutions, as well as to its integer order counterpart.

The cuckoo search algorithm is also proposed in [78] in the design of a self-tuned fractional-order fuzzy PID controller. The optimization algorithm is based on the minimization of an objective function defined as the sum of integral of squared error and integral of the squared deviation of controller output. The final controller consists in a Takagi-Sugeno model-based fuzzy adaptive controller containing non-integer-order differ-integral operators. For comparative purposes, the integer order counterpart of this controller is also designed. Simulation results indicate the increased robustness of the self-tuned fractional-order fuzzy PID controller when applied to the control of an integrated power system.

6. Conclusions

Fractional-order controllers have emerged as a generalization of the standard PID, allowing for greater flexibility and improved performance and robustness. The tuning of these FO-PIDs is not an easy task, since the complexity of the design increases along with the number of tuning parameters. Several tuning methods have been developed, but the majority of them require a process transfer function. In some cases, obtaining an accurate mathematical model of the process is time consuming and tedious, especially in the industrial sector. To cope with this issue, autotuning methods for FO-PIDs have emerged. In this paper, a survey of the existing autotuning methods for FO-PIDs is presented. Several autotuning approaches are compared for lag-dominant and delay-dominant FOPDT processes, for higher order systems, for integrative time-delay processes or poorly damped ones.

For each type of process, the autotuning methods are compared in terms of closed-loop performance regarding reference tracking and disturbance rejection. Robustness was not considered as a means for comparison, since some of the reviewed methods do not address directly this issue, while others do. This aspect would have led to unfair comparisons and possibly different remarks on the opportunity of using one autotuning method, instead of another.

Some of these autotuning methods have also been validated experimentally. Research in this area is still under way and the current autotuning methods stand as the premises for further innovation in this area. Further research regarding the robustness of the autotuning methods will be considered.

Author Contributions: Conceptualization, C.I.M. and I.B.; methodology, C.I.M.; software, C.I.M. and I.B.; validation, C.I. and R.D.K.; formal analysis, E.H.D.; writing—original draft preparation, C.I.M. and I.B.; writing—review and editing, E.H.D.; supervision, C.I. and R.D.K.; funding acquisition, C.I.M. All authors have read and agreed to the published version of the manuscript.

Funding: This research was funded by a grant of the Romanian Ministry of Education and Research, CNCS-UEFISCDI, project number PN-III-P1-1.1-TE-2019- 0745, within PNCDI III. This research was also supported by Research Foundation Flanders (FWO) under grant 1S04719N.

Institutional Review Board Statement: Not applicable.

Informed Consent Statement: Not applicable.

Data Availability Statement: Not applicable.

Conflicts of Interest: The authors declare no conflict of interest.

References

1. Åström, K.J.; Hägglund, T. Revisiting the Ziegler-Nichols step response method for PID control. *J. Process Contr.* **2004**, *14*, 635–650. [CrossRef]
2. Åström, K.J.; Hägglund, T. The future of PID control. *Control Eng. Pract.* **2001**, *9*, 1163–1175. [CrossRef]
3. Ziegler, J.G.; Nichols, N.B. Optimum settings for automatic controllers. *Trans. ASME* **1942**, *64*, 759–768. [CrossRef]
4. Åström, K.J.; Hägglund, T. Automatic tuning of simple regulators with specifications on phase and amplitude margins. *Automatica* **1984**, *20*, 645–651. [CrossRef]
5. Tan, K.K.; Lee, T.H.; Wang, Q.G. Enhanced automatic tuning procedure for process control of PI/PID controllers. *AIChE J.* **1996**, *42*, 2555–2562. [CrossRef]
6. Hang, C.C.; Åström, K.J.; Ho, W.K. Refinements of the Ziegler-Nichols tuning formula. *IEE Proc. D* **1991**, *138*, 111–118. [CrossRef]
7. Wallen, A.; Åström, K.J.; Hägglund, T. Loopshaping design of PID controllers with constant ti/td ratio. *Asian J. Control* **2002**, *4*, 403–409. [CrossRef]
8. Podlubny, I. Fractional-order systems and PIλDμ-controllers. *IEEE Trans. Autom. Control* **1999**, *44*, 208–214. [CrossRef]

9. Monje, C.A.; Chen, Y.Q.; Vinagre, B.; Xue, D.; Feliu, V. *Fractional Order Systems and Controls: Fundamentals and Applications*; Springer: Berlin, Germany, 2010.
10. Zheng, W.; Luo, Y.; Chen, Y.; Wang, X. A Simplified Fractional Order PID Controller's Optimal Tuning: A Case Study on a PMSM Speed Servo. *Entropy* **2021**, *23*, 130. [CrossRef] [PubMed]
11. Li, X.; Gao, L. Robust Fractional-order PID Tuning Method for a Plant with an Uncertain Parameter. *Int. J. Control Autom. Syst.* **2021**, *19*, 1302–1310. [CrossRef]
12. Garrido, S.; Monje, C.A.; Martín, F.; Moreno, L. Design of Fractional Order Controllers Using the PM Diagram. *Mathematics* **2020**, *8*, 2022. [CrossRef]
13. Flores, C.; Muñoz, J.; Monje, C.A.; Milanés, V.; Lu, X.Y. Iso-damping fractional-order control for robust automated car-following. *J. Adv. Res.* **2020**, *25*, 181–189. [CrossRef] [PubMed]
14. Tepljakov, A.; Alagoz, B.B.; Yeroglu, C.; Gonzalez, E.A.; Hosseinnia, S.H.; Petlenkov, E.; Ates, A.; Cech, M. Towards Industrialization of FOPID Controllers: A Survey on Milestones of Fractional-Order Control and Pathways for Future Developments. *IEEE Access* **2021**, *9*, 21016–21042. [CrossRef]
15. Tepljakov, A.; Alagoz, B.B.; Yeroglu, C.; Gonzalez, E.; Hassan HosseinNia, H.; Petlenkov, E. FOPID Controllers and Their Industrial Applications: A Survey of Recent. *IFAC-PapersOnLine* **2018**, *51*, 25–30. [CrossRef]
16. Birs, I.; Muresan, C.I.; Nascu, I.; Ionescu, C. A Survey of Recent Advances in Fractional Order Control for Time Delay Systems. *IEEE Access* **2019**, *7*, 30951–30965. [CrossRef]
17. Ionescu, C.; Dulf, E.H.; Ghita, M.; Muresan, C.I. Robust Controller Design: Recent Emerging Concepts for Control of Mechatronic Systems. *J. Frankl. Inst.* **2020**, *357*, 7818–7844. [CrossRef]
18. Shah, P.; Agashe, S. Review of fractional PID controller. *Mechatronics* **2016**, *38*, 29–41. [CrossRef]
19. Sondhi, S.; Hote, Y.V. Fractional order controller and its applications: A review. In Proceedings of the 2nd IASTED Asian Conference on Modelling, Identification, and Control, Phuket, Thailand, 2–4 April 2012; pp. 118–123.
20. Soukkou, A.; Belhour, M.C.; Leulmi, S. Review, Design, Optimization and Stability Analysis of Fractional-Order PID Controller. *Int. J. Intell. Syst. Appl.* **2016**, *8*, 73. [CrossRef]
21. Valerio, D.; da Costa, J.S. A review of tuning methods for fractional PIDs. In Proceedings of the 4th IFAC Workshop on Fractional Differentiation and Its Applications, Badajoz, Spain, 18–20 October 2010.
22. Dastjerdi, A.A.; Vinagre, B.M.; Chen, Y.Q.; HosseinNia, S.H. Linear fractional order controllers; A survey in the frequency domain. *Annu. Rev. Control* **2019**, *47*, 51–70. [CrossRef]
23. Almeida, A.M.d.; Lenzi, M.K.; Lenzi, E.K. A Survey of Fractional Order Calculus Applications of Multiple-Input, Multiple-Output (MIMO) Process Control. *Fractal Fract* **2020**, *4*, 22. [CrossRef]
24. Muresan, C.I.; Birs, I.R.; Dulf, E.H.; Copot, D.; Miclea, L. A Review of Recent Advances in Fractional-Order Sensing and Filtering Techniques. *Sensors* **2021**, *21*, 5920. [CrossRef]
25. Chen, Y.Q.; Moore, K.L.; Vinagre, B.M.; Podlubny, I. Robust PID controller autotuning with a phase shaper. In Proceedings of the First IFAC Workshop on Fractional Differentiation and Its Applications, Bordeaux, France, 19–21 July 2004; pp. 162–167.
26. Chen, Y.Q.; Bhaskaran, T.; Xue, D. Practical Tuning Rule Development for Fractional Order Proportional and Integral Controllers. *ASME J. Comput. Nonlinear Dynam.* **2008**, *3*, 021403. [CrossRef]
27. Tepljakov, A.; Petlenkov, E.; Belikov, J. Development of analytical tuning methods for fractional-order controllers. In Proceedings of the Sixth IKTDK Information and Communication Technology Doctoral School Conference, Tallin, Estonia, 2012.
28. Xue, D.; Chen, Y.Q.; Atherton, D.P. *Linear Feedback Control: Analysis and Design with MATLAB (Advances in Design and Control)*, 1st ed.; Society for Industrial and Applied Mathematics: Philadelphia, PA, USA, 2008.
29. Basu, A.; Mohanty, S. Tuning of FOPID (PIλDμ) Controller for Heating Furnace. *Int. J. Electron. Eng. Res.* **2017**, *9*, 1415–1437.
30. Tajjudin, M.; Rahiman, M.H.F.; Arshad, N.M.; Adnan, R. Robust fractional-order PI controller with Ziegler-Nichols rules. *Int. J. Electr. Comput. Eng.* **2013**, *7*, 1034–1041.
31. Feliu-Batlle, V.; Pérez, R.R.; Rodríguez, L.S. Fractional robust control of main irrigation canals with variable dynamic parameters. *Control Eng. Pract.* **2007**, *15*, 673–686. [CrossRef]
32. Valério, D.; Sá da Costa, J. Tuning of fractional PID controllers with Ziegler-Nichols-type rules. *Signal Process.* **2006**, *86*, 2771–2784. [CrossRef]
33. Valério, D.; Sá da Costa, J. Ziegler-Nichols type tuning rules for fractional PID controllers. In Proceedings of the International Design Engineering Technical Conferences and Computers and Information in Engineering Conference, Long Beach, CA, USA, 24–28 September 2005; Volume 47438, pp. 1431–1440.
34. De Keyser, R.; Muresan, C.I.; Ionescu, C. A Novel Auto-tuning Method for Fractional Order PI/PD Controllers. *ISA Trans.* **2016**, *62*, 268–275. [CrossRef] [PubMed]
35. Yeroglu, C.; Onat, C.; Tan, N. A new tuning method for PIλDμ controller. In Proceedings of the International Conference on Electrical and Electronics Engineering, Bursa, Turkey, 5–8 November 2009; pp. 312–316.
36. Gude, J.J.; Kahoraho, E. Modified Ziegler-Nichols method for fractional PI controllers. In Proceedings of the IEEE 15th Conference on Emerging Technologies & Factory Automation, Bilbao, Spain, 13–16 September 2010; pp. 1–5.
37. Gude, J.J.; Kahoraho, E. Comparison between Ziegler-Nichols type tuning rules for PI and fractional PI controllers. In Proceedings of the 3rd Seminar for Advanced Industrial Control Applications, Madrid, Spain, 16–17 November 2009; pp. 165–180.

38. Gude, J.J.; Kahoraho, E. Comparison between Ziegler-Nichols type tuning rules for PI and fractional PI controllers—Part 2. In Proceedings of the 3rd Seminar for Advanced Industrial Control Applications, Madrid, Spain, 16–17 November 2009; pp. 181–194.

39. Tajjudin, M.; Tahir, S.F.; Rahiman, M.H.F.; Arshad, N.M.; Adnan, R. Fractional-order PI controller with relay auto-tuning method. In Proceedings of the IEEE 4th Control and System Graduate Research Colloquium, Shah Alam, Malaysia, 19–20 August 2013. [CrossRef]

40. Muresan, C.I.; De Keyser, R. Revisiting Ziegler-Nichols. A fractional order approach. *ISA Trans.* 2021, *under review*.

41. Chevalier, A.; Francis, C.; Copot, C.; Ionescu, C.M.; De Keyser, R. Fractional-order PID design: Towards transition from state-of-art to state-of-use. *ISA Trans.* **2019**, *84*, 178–186. [CrossRef]

42. Valério, D.; Sá da Costa, J. Tuning rules for fractional PID controllers. *IFAC Proc. Vol.* **2006**, *39*, 28–33. [CrossRef]

43. Monje, C.A.; Vinagre, B.M.; Feliu, V.; Chen, Y.Q. Tuning and auto-tuning of fractional order controllers for industry applications. *Control Eng. Pract.* **2008**, *16*, 798–812. [CrossRef]

44. Chen, Y.Q.; Moore, K.L. Relay feedback tuning of robust PID controllers with iso-damping property. *IEEE Trans. Syst. Man Cybern. Part B* **2005**, *35*, 23–31. [CrossRef] [PubMed]

45. Monje, C.A.; Vinagre, B.M.; Santamaría, G.E.; Tejado, I. Auto-tuning of fractional order PI·D controllers using a PLC. In Proceedings of the IEEE Conference on Emerging Technologies & Factory Automation, Palma de Mallorca, Spain, 22–25 September 2009. [CrossRef]

46. Santamaria, G.E.; Tejado, I.; Vinagre, B.M.; Monje, C.A. Fully Automated Tuning and Implementation of Fractional PID Controllers. In Proceedings of the ASME 2009 International Design Engineering Technical Conferences and Computers and Information in Engineering Conference. Volume 4: 7th International Conference on Multibody Systems, Nonlinear Dynamics, and Control, Parts A, B and C, San Diego, CA, USA, 30 August–2 September 2009; pp. 1275–1283.

47. Monje, C.A.; Vinagre, B.M.; Feliu, V.; Chen, Y.Q. On Auto-Tuning Of Fractional Order PIλDμ Controllers. In Proceedings of the 2nd IFAC Workshop on Fractional Differentiation and Its Applications, Porto, Portugal, 19–21 July 2006.

48. Caponetto, R.; Dongola, G.; Pappalardo, F.L.; Tomasello, V. Autotuning method for PIλDμ controllers design. *Int. J. Innov. Comput. Inf. Control* **2013**, *9*, 4043–4055.

49. De Keyser, R.; Muresan, C.I.; Ionescu, C.M. Autotuning of a Robust Fractional Order PID Controller. *IFAC-PapersOnLine* **2018**, *51*, 466–471. [CrossRef]

50. De Keyser, R.; Muresan, C.I.; Ionescu, C.M. An efficient algorithm for low-order discrete-time implementation of fractional order transfer function. *ISA Trans.* **2018**, *74*, 229–238. [CrossRef]

51. Muresan, C.I.; De Keyser, R.; Ionescu, C. Autotuning Method for a Fractional Order Controller for a Multivariable 13C Isotope Separation Column. In Proceedings of the 15th European Control Conference (ECC16), Aalborg, Denmark, 29 June–1 July 2016; pp. 358–363. [CrossRef]

52. Muresan, C.I.; De Keyser, R.; Birs, I.R.; Copot, D.; Ionescu, C. Benchmark Challenge: A robust fractional order control autotuner for the Refrigeration Systems based on Vapor Compression. *IFAC-PapersOnLine* **2018**, *51*, 31–36. [CrossRef]

53. Cajo, R.; Muresan, C.I.; Ionescu, C.; De Keyser, R.; Plaza, D. Multivariable Fractional Order PI Autotuning Method for Heterogeneous Dynamic Systems. *IFAC-PapersOnLine* **2018**, *51*, 865–870. [CrossRef]

54. Juchem, J.; Muresan, C.I.; De Keyser, R.; Ionescu, C.M. Robust fractional-order auto-tuning for highly-coupled MIMO systems. *Heliyon* **2019**, *5*, e02154. [CrossRef] [PubMed]

55. Muresan, C.I.; Copot, C.; Birs, I.R.; De Keyser, R.; Vanlanduit, S.; Ionescu, C. Experimental validation of a novel auto-tuning method for a fractional order PI controller on an UR10 robot. *Algorithms* **2018**, *11*, 95. [CrossRef]

56. Muresan, C.I.; De Keyser, R.; Birs, I.R.; Folea, S.; Prodan, O. An Autotuning Method for a Fractional Order PD Controller for Vibration Suppression. In *Mathematical Methods in Engineering. Nonlinear Systems and Complexity*; Taş, K., Baleanu, D., Tenreiro Machado, J.A., Eds.; Springer: Berlin, Germany, 2018; Volume 24, pp. 245–256.

57. Birs, I.; Folea, S.; Prodan, O.; Dulf, E.; Muresan, C. An experimental tuning approach of fractional order controllers in the frequency domain. *Appl. Sci.* **2021**, *10*, 2379. [CrossRef]

58. Dulf, E.H.; Muresan, C.I.; Both-Rusu, R.; Dulf, F.V. Robust Auto-tuning Fractional Order Control of an Aerodynamical System. In Proceedings of the 2016 International Conference on Mechatronics, Control and Automation Engineering, Bangkok, Thailand, 24–25 July 2016; Volume 58, pp. 42–45. [CrossRef]

59. Li, S.; Wang, D.; Han, X.; Cheng, K.; Zhao, C. Auto-Tuning Parameters of Fractional PID Controller Design for Air-Conditioning Fan Coil Unit. *J. Shanghai Jiao Tong Univ. (Sci.)* **2021**, *26*, 186–192. [CrossRef]

60. Bernardes, N.D.; Castro, F.A.; Cuadros, M.A.; Salarolli, P.F.; Almeida, G.M.; Munaro, C.J. Fuzzy Logic in Auto-tuning of Fractional PID and Backstepping Tracking Control of a Differential Mobile Robot. *J. Intell. Fuzzy Syst.* **2019**, *37*, 4951–4964. [CrossRef]

61. Tajjudin, M.; Ishak, N.; Fazalul Rahiman, M.H.; Mohd Arshad, N.; Adnan, R. Self-tuning fuzzy fractional-order PI controller: Design and application in steam distillation process. In Proceedings of the 2014 IEEE International Conference on Control System, Computing and Engineering (ICCSCE 2014), Penang, Malaysia, 28–30 November 2014; pp. 316–321. [CrossRef]

62. Gaballa, M.S.; Bahgat, M.; Abdel-Ghany, A.-G.M. A novel technique for online self-tuning of fractional order PID, based on takaji-sugeno fuzzy. In Proceedings of the Nineteenth International Middle East Power Systems Conference, Cairo, Egypt, 19–21 December 2017; pp. 1362–1368. [CrossRef]

63. Ghany, M.A.A.; Bahgat, M.E.; Refaey, W.M.; Sharaf, S. Type-2 fuzzy self-tuning of modified fractional-order PID based on Takagi-Sugeno method. *J. Electr. Syst. Inf. Technol.* **2020**, *7*, 2. [CrossRef]

64. Ardeshiri, R.R.; Kashani, H.N.; Reza-Ahrabi, A. Design and simulation of self-tuning fractional order fuzzy PID controller for robotic manipulator. *Int. J. Autom. Control* **2019**, *13*, 595–618. [CrossRef]

65. Agrawal, A. Analytical Study of the Robustness of the Different Variants of Fractional-Order Self-Tuned Fuzzy Logic Controllers. In Proceedings of the 1st International Conference on Computational Research and Data Analytics, Rajpura, India, 24 October 2020; IOP Conference Series: Materials Science and Engineering. IOP: Bristol, UK, 2020; Volume 1022.

66. Chen, S.Y.; Chia, C.S. Precision Position Control of a Voice Coil Motor Using Self-Tuning Fractional Order Proportional-Integral-Derivative Control. *Micromachines* **2016**, *7*, 207. [CrossRef] [PubMed]

67. Saleem, O.; Mahmood-ul-Hasan, K. Robust stabilisation of rotary inverted pendulum using intelligently optimised nonlinear self-adaptive dual fractional-order PD controllers. *Int. J. Syst. Sci.* **2019**, *50*, 1399–1414. [CrossRef]

68. Xie, Y.; Tang, X.; Song, B.; Zhou, X.; Guo, Y. Model-free tuning strategy of fractional-order PI controller for speed regulation of permanent magnet synchronous motor. *Trans. Inst. Meas. Control* **2019**, *41*, 23–35. [CrossRef]

69. Neçaibia, A.; Ladaci, S. Self-tuning fractional order PIλDμ controller based on extremum seeking approach. *Int. J. Autom. Control* **2014**, *8*, 99–121. [CrossRef]

70. Shamseldin, M.A.; Sallam, M.; Abdel Halim, B.; Abdel Ghany, A.M. A novel self-tuning fractional order PID control based on optimal model reference adaptive system. *Int. J. Power Electron. Drive Syst.* **2019**, *10*, 230–241. [CrossRef]

71. Mohan, V.; Chhabra, H.; Rani, A.; Singh, V. Robust Self-tuning Fractional Order PID Controller Dedicated to Non-linear Dynamic System. *J. Intell. Fuzzy Syst.* **2018**, *34*, 1467–1478. [CrossRef]

72. Muñoz-Vázquez, A.J.; Treesatayapun, C. Model-free discrete-time fractional fuzzy control of robotic manipulators. *J. Frankl. Inst.* 2021. Available online: https://www.sciencedirect.com/science/article/abs/pii/S0016003221007365 (accessed on 1 December 2021).

73. Ladaci, S.; Bensafia, Y. Fractional order self-tuning control. In Proceedings of the IEEE 13th International Conference on Industrial Informatics, Cambridge, UK, 22–24 July 2015; pp. 544–549. [CrossRef]

74. Kumar, J.; Kumar, V.; Rana, K.P.S. Fractional-order self-tuned fuzzy PID controller for three-link robotic manipulator system. *Neural Comput. Applic* **2020**, *32*, 7235–7257. [CrossRef]

75. Maiti, D.; Chakraborty, M.; Acharya, A.; Konar, A. Design of a fractional-order self-tuning regulator using optimization algorithms. In Proceedings of the 2008 11th International Conference on Computer and Information Technology, Khulna, Bangladesh, 24–27 December 2008; pp. 470–475. [CrossRef]

76. Naderipour, A.; Abdul-Malek, Z.; Davoodkhani, I.F.; Kamyab, H.; Ali, R.R. Load-frequency control in an islanded microgrid PV/WT/FC/ESS using an optimal self-tuning fractional-order fuzzy controller. *Environ. Sci. Pollut. Res. Int.* **2021**, *28*, 1–12. [CrossRef] [PubMed]

77. Kumar, V.; Rana, K.P.S.; Kumar, J.; Mishra, P. Self-tuned robust fractional order fuzzy PD controller for uncertain and nonlinear active suspension system. *Neural Comput. Appl.* **2016**, *30*, 1827–1843. [CrossRef]

78. Nithilasaravanan, K.; Thakwani, N.; Mishra, P.; Kumar, V.; Rana, K.P.S. Efficient control of integrated power system using self tuned fractional order fuzzy PID controller. *Neural Comput. Appl.* **2018**, *31*, 4137–4155. [CrossRef]

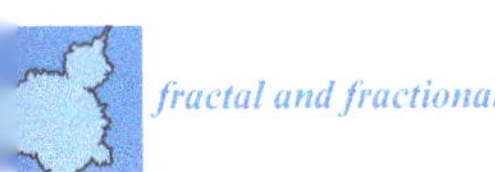 *fractal and fractional*

Article

Exponential Enclosures for the Verified Simulation of Fractional-Order Differential Equations

Andreas Rauh

Group: Distributed Control in Interconnected Systems, School II—Department of Computing Science, Carl von Ossietzky Universität Oldenburg, D-26111 Oldenburg, Germany; andreas.rauh@uni-oldenburg.de

Abstract: Fractional-order differential equations are powerful tools for the representation of dynamic systems that exhibit long-term memory effects. The verified simulation of such system models with the help of interval tools allows for the computation of guaranteed enclosures of the domains of reachable pseudo states over a certain finite time horizon. In the previous work of the author, an iteration scheme—derived on the basis of the Picard iteration—was published that makes use of Mittag-Leffler functions to determine guaranteed pseudo-state enclosures. In this paper, the corresponding iteration is generalized toward the use of exponential functions during the evaluation of the iteration scheme. Such exponential functions are well-known from a verified solution of integer-order sets of differential equations. The aim of this work is to demonstrate that the use of exponential functions instead of pure box-type interval enclosures for Mittag-Leffler functions does not only improve the tightness of the computed pseudo-state enclosures but also reduces the required computational effort. These statements are demonstrated with the help of a close-to-life simulation model for the charging/discharging dynamics of Lithium-ion batteries.

Keywords: fractional-order differential equations; interval analysis; verification of pseudo-state enclosures; Mittag-Leffler-type enclosures; exponential enclosures

Citation: Rauh, A. Exponential Enclosures for the Verified Simulation of Fractional-Order Differential Equations. *Fractal Fract.* **2022**, *6*, 567. https://doi.org/10.3390/fractalfract 6100567

Academic Editor: Gani Stamov

Received: 5 September 2022
Accepted: 30 September 2022
Published: 5 October 2022

Publisher's Note: MDPI stays neutral with regard to jurisdictional claims in published maps and institutional affiliations.

1. Introduction

Fractional differential equations have the property of an infinite-horizon memory of the previous evolution of the system dynamics [1–4]. As opposed to integer-order models, where only initial conditions for the state variables at a certain point of time and the external system inputs after that time instant are necessary for determining a unique solution for the evaluation of the system states, fractional-order models need to be initialized with the complete past behavior as the initialization function [5].

Due to this reason, the system state at a specific point in time is typically referred to as a pseudo state for fractional system models as the complete history of its evolution is required for a unique solution [5]. To solve this difficulty when initializing a simulation at a specific point in time, additive (interval-valued) correction terms of the right-hand sides of explicit fractional-order models have been employed in [6] to account for the past pseudo-state evolution. Simulation methods, which make use of further extensions on the basis of a (pseudo) state observer concept for tightening these additive corrections after the reset of fractional integrators, were developed in [7]. These extensions exploit a formula derived by Podlubny in [1] for shifting the reference point in time, associated with the definition of a fractional differentiation operator.

On the one hand, the infinite-horizon memory property of fractional differential equations allows to efficiently model dynamic systems with long-term memory effects. On the other hand, however, interval-based simulations, allowing for enclosing the domains of reachable (pseudo) states in a guaranteed way, are significantly complicated because classical Taylor series-based simulation approaches, such as those employed in tools such as AWA [8], VNODE-LP [9–11], or VSPODE [12], for systems of integer-order ordinary differential equations, can no longer be employed without modifications.

Therefore, the author has developed an iterative pseudo-state enclosure approach in [6,13] which is based on the Picard iteration [14]. Under the assumption that the pseudo-state at the point of time $t = 0$ corresponds also to the complete previous, temporally constant state evolution for $t < 0$, corresponding to Caputo's definition of fractional derivatives, the iterative solution makes use of Mittag-Leffler functions [15–19] to represent pseudo-state enclosures in a guaranteed manner. This kind of function represents the true pseudo-state trajectories of linear fractional differential equations with the aforementioned Caputo-type initialization [20,21]. For nonlinear models, the iteration procedure developed in [6,13] yields outer bounds for the actually reachable pseudo states.

It has to be noted that this iteration scheme is a natural generalization of a counterpart making use of classical exponential functions in the integer-order case, cf. [6]. However, the drawback is an increased complexity of the numerical evaluation because the typically arising quotients of Mittag-Leffler functions with different arguments cannot be simplified further in an analytic manner to reduce the overestimation that is well-known in the domain of interval methods [22,23]. This issue is further discussed in the current paper and resolved by outer exponential enclosures of Mittag-Leffler functions.

In this paper, Section 2 summarizes the Mittag-Leffler function representation of guaranteed solution enclosures for fractional-order system models as presented in [6,13]. It is extended in Section 3 toward exponential functions for the computation of guaranteed pseudo-state enclosures. The representative simulation results, focusing on the tightness of the resulting pseudo-state enclosures and the required computational effort, are presented in Section 4 for a close-to-life quasi-linear fractional model of the charging/discharging behavior of Lithium-ion batteries before conclusions and an outlook on future work are given in Section 5.

2. Fundamentals of Verified Mittag-Leffler-Type Pseudo-State Enclosures for Fractional Differential Equations

2.1. System Models under Consideration

Throughout this article, we analyze and derive set-valued simulation approaches for commensurate fractional-order differential equations of the form

$$\mathbf{x}^{(\nu)}(t) = \mathbf{f}(\mathbf{x}(t)) , \quad \mathbf{f} : \mathbb{R}^n \mapsto \mathbb{R}^n , \tag{1}$$

with the order $0 < \nu \leq 1$, where the right-hand side $\mathbf{f}(\mathbf{x}(t))$ is assumed to be given by a continuous function. Moreover, we assume that the system model (1) is initialized by the pseudo state $\mathbf{x}(0)$ at the time instant $t = 0$, where $\mathbf{x}(t) = \mathbf{x}(0)$ holds for all $t < 0$. This case corresponds to the Caputo definition of fractional derivatives as described, for example, in [1,2].

To account for uncertainty in the pseudo-state initialization, we use the interval representation

$$\mathbf{x}(0) \in [\mathbf{x}](0) := \left[\underline{\mathbf{x}}(0) ; \overline{\mathbf{x}}(0)\right] , \tag{2}$$

for which $\underline{x}_i(0) \leq \overline{x}_i(0)$ holds for each vector component $i \in \{1, \ldots, n\}$. Note that the property of temporally constant initializations for $t < 0$ is still assumed to hold.

2.2. Linear Scalar System Models

For the special case of scalar fractional differential equations

$$x^{(\nu)}(t) = \lambda \cdot x(t) \quad \text{with} \quad x(0) = x_0 \tag{3}$$

which are linear in the pseudo state $x(t)$ with the parameter $\lambda \in \mathbb{R}$ and which are initialized according to the previous subsection in the Caputo-type sense, it is well-known according to [20,21] that the exact solution is given in the form

$$x(t) = E_{\nu,1}\left(\lambda t^{\nu}\right) \cdot x(0) . \tag{4}$$

In (4), $E_{v,\beta}(\zeta)$ is the two-parameter Mittag-Leffler function. It is defined by the infinite series

$$E_{v,\beta}(\zeta) = \sum_{i=0}^{\infty} \frac{\zeta^i}{\Gamma(vi+\beta)} \tag{5}$$

for the general argument $\zeta \in \mathbb{C}$. In (4), $\Gamma(vi+\beta)$ denotes the gamma function of the respective argument and v is the derivative order as introduced in (1). To obtain the solution in (4), the parameter β is set to the value $\beta = 1$.

Remark 1. *The classical exponential function e^{ζ} is obtained as a special case of the Mittag-Leffler function* (5) *when setting $v = 1$ and $\beta = 1$.*

Remark 2. *For an interval extension of the Mittag-Leffler function on the basis of the accurate floating-point* MATLAB *implementation by R. Garrappa [18,24], see [13].*

2.3. Mittag-Leffler Functions as Pseudo-State Enclosures for Fractional-Order Differential Equations

For nonlinear scalar and vector-valued system models (1), the Mittag-Leffler functions introduced in the previous subsection can be used to define guaranteed pseudo-state enclosures according to Definition 1.

Definition 1 (Mittag-Leffler-type pseudo-state enclosure). *A verified Mittag-Leffler-type pseudo-state enclosure for the system model* (1) *with* (2) *is defined by the time-dependent enclosure function*

$$\mathbf{x}^*(t) \in [\mathbf{x}_e](t) = \mathbf{E}_{v,1}([\mathbf{\Lambda}] \cdot t^v) \cdot [\mathbf{x}_e](0), \quad [\mathbf{x}_e](0) = [\mathbf{x}_0] \tag{6}$$

with the diagonal parameter matrix $[\mathbf{\Lambda}] := \mathrm{diag}\{[\lambda_i]\}, i \in \{1, \ldots, n\}$, if it is determined according to Theorem 1*. In* (6)*, the generalization of the scalar Mittag-Leffler function $E_{v,1}$ to the matrix case $\mathbf{E}_{v,1}$ is given by the following diagonal matrix*

$$\mathbf{E}_{v,1}([\mathbf{\Lambda}] \cdot t^v) = \mathrm{diag}\left\{ \left[E_{v,1}([\lambda_1] \cdot t^v) \quad \ldots \quad E_{v,1}([\lambda_n] \cdot t^v) \right] \right\}. \tag{7}$$

Theorem 1 ([6,13,25] Iteration for Mittag-Leffler-type enclosures). *All reachable pseudo states $\mathbf{x}^*(T)$ are enclosed in accordance with Theorem* 1 *by the Mittag-Leffler-type pseudo-state enclosure*

$$\mathbf{x}^*(T) \in [\mathbf{x}_e](T) = \mathbf{E}_{v,1}([\mathbf{\Lambda}] \cdot T^v) \cdot [\mathbf{x}_e](0) \tag{8}$$

at the point of time $t = T > 0$ if $[\mathbf{\Lambda}]$ is set to the outcome of the converging iteration

$$[\lambda_i]^{\langle\kappa+1\rangle} := \frac{f_i\left(\mathbf{E}_{v,1}\left([\mathbf{\Lambda}]^{\langle\kappa\rangle} \cdot [t]^v\right) \cdot [\mathbf{x}_e](0) \right)}{E_{v,1}\left([\lambda_i]^{\langle\kappa\rangle} \cdot [t]^v\right) \cdot [x_{e,i}](0)}, \tag{9}$$

$i \in \{1, \ldots, n\}$, with the prediction horizon $[t] = [0 \, ; \, T]$. To ensure convergence, the value $x_i^* = 0$ must not belong to the solution for any vector component $i \in \{1, \ldots, n\}$.

Remark 3. *Typically, the iteration according to Theorem* 1 *is initialized with intervals centered around the eigenvalues of the Jacobian of the right-hand side of* (1)*, evaluated for the midpoint of the interval vector* (2)*.*

As a preparation for the derivation of the exponential enclosure approach presented for the first time in this paper, the following proof of Theorem 1, according to [6,13,25], is summarized.

Proof. Formulate a Picard iteration (iteration index κ) for computing pseudo-state enclosures in the differential form

$$[\mathbf{x}^{(\nu)}]^{\langle\kappa\rangle}\left([0\,;\,T]\right) \supset [\mathbf{x}^{(\nu)}]^{\langle\kappa+1\rangle}\left([0\,;\,T]\right) = \mathbf{f}\left([\mathbf{x}^{(\nu)}]^{\langle\kappa\rangle}\left([0\,;\,T]\right)\right)\,, \tag{10}$$

where $[\mathbf{x}^{(\nu)}]^{\langle\kappa\rangle}\left([0\,;\,T]\right)$ is an interval extension of the temporal derivative of order ν of the inclusion function $[\mathbf{x}]^{\langle\kappa\rangle}(t)$ over the time interval $t \in [t] = [0\,;\,T]$. Substituting the pseudo-state enclosure given in Definition 1 into (10) yields the expression

$$\begin{aligned}
\mathbf{x}^{(\nu)}(t) &\in \left([\mathbf{\Lambda}]^{\langle\kappa+1\rangle}\right) \cdot \mathbf{E}_{\nu,1}\left([\mathbf{\Lambda}]^{\langle\kappa+1\rangle}\cdot t^{\nu}\right) \cdot [\mathbf{x}_e](0) \\
&= \mathbf{f}\left(\mathbf{E}_{\nu,1}\left([\mathbf{\Lambda}]^{\langle\kappa\rangle}\cdot t^{\nu}\right)\cdot[\mathbf{x}_e](0)\right) \quad \text{for all}\quad t \in [t]\,.
\end{aligned} \tag{11}$$

A converging iteration implies the set-valued relation

$$[\mathbf{x}_e]^{\langle\kappa+1\rangle}\left([t]\right) \subset [\mathbf{x}_e]^{\langle\kappa\rangle}\left([t]\right)\,, \tag{12}$$

which corresponds to the relation

$$[\lambda_i]^{\langle\kappa+1\rangle} \subset [\lambda_i]^{\langle\kappa\rangle} \tag{13}$$

for the unknown intervals of the solution parameters λ_i.

Overapproximating the interval evaluation of the Mittag-Leffler-type enclosure

$$[\mathbf{x}]^{\langle\kappa+1\rangle}(t) = \mathbf{E}_{\nu,1}\left([\mathbf{\Lambda}]^{\langle\kappa+1\rangle}\cdot t^{\nu}\right)\cdot[\mathbf{x}_e](0) \tag{14}$$

in the iteration step $\kappa+1$ on the first line of (11) by the enclosure $[\mathbf{x}_e]^{\langle\kappa\rangle}\left([t]\right)$ in the case of convergence, i.e., using the relation

$$\left([\mathbf{\Lambda}]^{\langle\kappa+1\rangle}\right)\cdot\mathbf{E}_{\nu,1}\left([\mathbf{\Lambda}]^{\langle\kappa+1\rangle}\cdot t^{\nu}\right)\cdot[\mathbf{x}_e](0) \subset \left([\mathbf{\Lambda}]^{\langle\kappa+1\rangle}\right)\cdot\mathbf{E}_{\nu,1}\left([\mathbf{\Lambda}]^{\langle\kappa\rangle}\cdot t^{\nu}\right)\cdot[\mathbf{x}_e](0)\,, \tag{15}$$

leads to the new iteration formula

$$\mathrm{diag}\left\{[\tilde{\lambda}_i]^{\langle\kappa+1\rangle}\right\}\cdot[\mathbf{x}_e]^{\langle\kappa\rangle}\left([t]\right) = \mathbf{f}\left([\mathbf{x}_e]^{\langle\kappa\rangle}\left([t]\right)\right)\,, \tag{16}$$

where

$$\mathrm{diag}\left\{[\tilde{\lambda}_i]^{\langle\kappa+1\rangle}\right\} \supseteq \mathrm{diag}\left\{[\lambda_i]^{\langle\kappa+1\rangle}\right\}\,. \tag{17}$$

Solving expression (16) for $[\tilde{\lambda}_i]^{\langle\kappa+1\rangle}$ with subsequent renaming of this parameter into $[\lambda_i]^{\langle\kappa+1\rangle}$ completes the proof of Theorem 1. For further details, the reader is referred to the references [6,13,25]. $\square$

Corollary 1. *In the case that the fractional-order differential equations given in Equation (1) with the initial conditions (2) can be rewritten into the quasi-linear form*

$$\mathbf{x}^{(\nu)}(t) = \mathscr{A}\left(\mathbf{x}(t)\right)\cdot\mathbf{x}(t) \quad \text{with}\quad 0 < \nu \leq 1\,, \tag{18}$$

with an equilibrium at $\mathbf{x} = \mathbf{0}$ and the state-dependent matrix $\mathscr{A}\left(\mathbf{x}(t)\right)$, Theorem 1 simplifies to the iteration scheme

$$[\lambda_i]^{\langle\kappa+1\rangle} := a_{ii}\left([\mathbf{x}_e]^{\langle\kappa\rangle}\left([t]\right)\right) + \sum_{\substack{j=1\\j\neq i}}^{n}\left\{a_{ij}\left([\mathbf{x}_e]^{\langle\kappa\rangle}\left([t]\right)\right)\cdot\frac{E_{\nu,1}\left([\lambda_j]^{\langle\kappa\rangle}\cdot[t]^{\nu}\right)}{E_{\nu,1}\left([\lambda_i]^{\langle\kappa\rangle}\cdot[t]^{\nu}\right)}\cdot\frac{[x_{e,j}](0)}{[x_{e,i}](0)}\right\}\,. \tag{19}$$

Remark 4. *In (19), the quotient of two Mittag-Leffler functions can usually only be simplified further for $\nu = 1$. In all other cases, where $\nu \neq 1$, overestimation due to the so-called dependency effect [23] (which arises due to multiple dependencies on common interval parameters) can be reduced by exploiting the monotonicity properties of the Mittag-Leffler function that were analyzed in detail in [13].*

3. Exponential Enclosures for Fractional-Order System Models

Exponential functions can be introduced in general at two different stages of the solution procedure presented in the previous section. These are:

1. The replacement of the solution representation $[\mathbf{x}_e](t)$ given so far by Mittag-Leffler functions by exponential functions; or
2. The introduction of exponential enclosures for the interval evaluation of the Mittag-Leffler function instead of the currently employed box-type representations.

These two alternative options are further discussed in the current section.

3.1. Exponential Pseudo-State Enclosures

To directly replace the Mittag-Leffler functions by exponential ones in the enclosure technique according to Theorem 1 and Corollary 1, it is necessary to determine the Caputo fractional derivative of order ν (initialized at $t = 0$) of a classical exponential function.

According to [26,27], the derivative of $e^{\lambda t}$ is given by the closed-form representation

$$\frac{\mathrm{d}^{\nu} e^{\lambda t}}{\mathrm{d} t^{\nu}} = \lambda \cdot \left(t^{1-\nu} \cdot E_{1,2-\nu}(\lambda t) \right) \tag{20}$$

with $t \geq 0$ and $\lambda \in \mathbb{R}$. In (20), the right-hand side again depends on the two-parameter Mittag-Leffler function (5). Note that this two-parameter Mittag-Leffler function, if substituted into the first line of Equation (11), leads to the same difficulty already observed in Equation (19) of Corollary 1 that the arising quotients of functions cannot be simplified, even in the case of (quasi-)linear system models in which $a_{ij}\left([\mathbf{x}_e]^{\langle \kappa \rangle}([t]) \right) \neq 0$ holds for at least one $i \neq j$.

Moreover, it has to be pointed out that an interval evaluation of this fractional derivative of the exponential function on the time interval $t \in [0 \; ; \; T]$ always contains the value 0 at the left end point of this time interval, so that the division of the differential formulation of the Picard iteration according to (11) by the term in parentheses in (20) is undefined. Therefore, solution representations in the form $e^{\lambda t}$ are not useful to generalize the iteration scheme according to Theorem 1.

At least theoretically, one could try to resolve this second problem by changing the solution template from $e^{\lambda t}$ to $e^{\lambda t^{\nu}}$ with a non-integer power of the time variable t. Unfortunately, however, its Caputo derivative of order ν does not have a closed-form, exponential-type solution in the general case so the problem persists that the solution presented in (19) can still not be simplified further. For this reason, we are switching to the idea of the following subsection in which the interval box enclosures of Mittag-Leffler functions employed in (9) as well as in (19) are replaced by tubes parameterized in terms of exponential functions.

3.2. Exponential Enclosures of the Mittag-Leffler Function

For the computation of exponential tube enclosures of time-dependent Mittag-Leffler functions, the following monotonicity theorem is employed. It is a simplified version of Theorem 4 published in [13], where also the fractional derivative order ν was considered as a (temporally constant) interval parameter.

Theorem 2 (Box-type interval bounds for the Mittag-Leffler function). *The range of function values for the Mittag-Leffler function in (4) with the uncertain real-valued parameter $\lambda \in \left[\underline{\lambda} \; ; \; \overline{\lambda} \right]$,*

$\overline{\lambda} < 0$ *and the non-negative time argument* $t \in \left[\underline{t} \,;\, \overline{t}\right]$, $\underline{t} \geq 0$, *with* $0 < v \leq 1$, *is bounded by the box-type interval enclosure*

$$E_{v,1}\left(\lambda t^v\right) \in \left[E^*_{v,1}\right]([\boldsymbol{\mathcal{X}}]) = \left[E^*_{v,1}\left(\inf([\boldsymbol{\mathcal{X}}])\right) \,;\, E^*_{v,1}\left(\sup([\boldsymbol{\mathcal{X}}])\right)\right] \tag{21}$$

with $[\boldsymbol{\mathcal{X}}] := [\lambda] \cdot [t]^{[v]}$, *where* $\sup([\boldsymbol{\mathcal{X}}]) \leq 0$ *holds and* $\left[E^*_{v,1}\right]$ *is obtained by the outward rounded interval extension of a floating-point evaluation of the two-parameter Mittag-Leffler function according to Equation (31) of [13]. Due to* $\overline{\lambda} < 0$, *Equation (21) simplifies to*

$$E_{v,1}\left(\lambda t^v\right) \in \left[E^*_{v,1}\left(\underline{\lambda} \cdot \overline{t}^{\,v}\right) \,;\, E^*_{v,1}\left(\overline{\lambda} \cdot \underline{t}^{\,v}\right)\right] . \tag{22}$$

Proof. For a detailed proof of this theorem, the reader is referred to the proof of Theorem 4 published in [13]. It is a direct consequence of the fact that the two-parameter Mittag-Leffler function is strictly monotonically decreasing for a growing time argument t with $\lambda < 0$. This property is also reflected by the so-called complete monotonicity of the Mittag-Leffler function that is reported, for example, in [16,17].

Furthermore, monotonicity with respect to the parameter λ is verified by differentiating $E_{v,1}(\lambda t^v)$ with respect to λ together with the change of variables $\tau := \lambda^{\frac{1}{v}} \cdot t < 0$. This leads to

$$\frac{\partial E_{v,1}(\lambda t^v)}{\partial \lambda} = \frac{\partial E_{v,1}(\tau^v)}{\partial \tau} \cdot \frac{\partial \tau}{\partial \lambda} = \frac{\partial E_{v,1}(\tau^v)}{\partial \tau} \cdot \frac{t \cdot \lambda^{\left(\frac{1}{v}-1\right)}}{v} . \tag{23}$$

In (23), the first factor is non-positive due to the complete monotonicity of the Mittag-Leffler function as shown in [16,17]; for any $t \geq 0$ and $\lambda \leq 0$, the second factor is also non-positive, leading to $\frac{\partial E_{v,1}(\lambda t^v)}{\partial \lambda} \geq 0$, which completes the proof. $\square$

Theorem 3 (Exponential enclosures for the Mittag-Leffler function). *The range of function values for the Mittag-Leffler function in (4) with the uncertain real-valued parameter* $\lambda \in \left[\underline{\lambda} \,;\, \overline{\lambda}\right]$, $\overline{\lambda} < 0$ *and the non-negative time argument* $t \in \left[\underline{t} \,;\, \overline{t}\right]$, $\underline{t} \geq 0$, $\overline{t} > \underline{t}$, *with* $0 < v \leq 1$, *is bounded by the exponential enclosure*

$$E_{v,1}\left(\lambda t^v\right) \in e^{\left[\underline{\eta} \,;\, \overline{\eta}\right] \cdot \left[\underline{t} \,;\, \overline{t}\right]} , \tag{24}$$

where

$$\underline{\eta} = \begin{cases} \inf\left(\dfrac{1}{\underline{t}^v} \cdot \ln\left(\left[E^*_{v,1}\right]\left(\underline{\lambda} \cdot \underline{t}^{\,v}\right)\right)\right) & \text{for } \underline{t} > 0 \\[4mm] \dfrac{\underline{\lambda}}{\Gamma(v+1)} & \text{for } \underline{t} = 0 \end{cases} \tag{25}$$

and

$$\overline{\eta} = \sup\left(\frac{1}{\overline{t}^v} \cdot \ln\left(\left[E^*_{v,1}\right]\left(\overline{\lambda} \cdot \overline{t}^{\,v}\right)\right)\right) \tag{26}$$

with $\left[E^*_{v,1}\right]$ *being the outward rounded interval extension of a floating-point evaluation of the two-parameter Mittag-Leffler function according to Equation (31) of [13].*

Proof. Consider the Mittag-Leffler function

$$f(\tilde{t}) = E_{v,1}(-\tilde{t}) . \tag{27}$$

Its derivative with respect to $\tilde{t}$ satisfies the following properties:

1. $\dfrac{\mathrm{d}f}{\mathrm{d}\tilde{t}}(0) = -\dfrac{1}{\Gamma(v+1)};$

2. $\lim\limits_{\tilde{t}\to\infty}\dfrac{\mathrm{d}f}{\mathrm{d}\tilde{t}}(\tilde{t}) = 0;$

3. $\dfrac{\mathrm{d}f}{\mathrm{d}\tilde{t}}(\tilde{t}) < 0$ for $\tilde{t} > 0;$ and

4. $\dfrac{\mathrm{d}^2 f}{\mathrm{d}\tilde{t}^2}(\tilde{t}) > 0$ for $\tilde{t} > 0.$

Property 1 is a consequence of the series definition (5) of the Mittag-Leffler function, while the properties 2–4 result from its complete monotonicity according to [16,17].

Case 1: For a fixed positive point $\tilde{t} = \tilde{t}^* > 0$, determine the intersection of an exponential function $e^{\eta\tilde{t}}$ and the Mittag-Leffler function $f(\tilde{t})$ according to

$$E_{v,1}(-\tilde{t}^*) = e^{\eta^*\cdot\tilde{t}^*} > 0 \quad\Longleftrightarrow\quad \eta^* = \frac{1}{\tilde{t}^*}\cdot\ln\big(E_{v,1}(-\tilde{t}^*)\big) < 0 \;, \tag{28}$$

where $E_{v,1}(0) = e^0 = 1$ obviously holds according to (5) and $0 < E_{v,1}(-\tilde{t}^*) < 1$ due to the property of complete monotonicity.

Case 2: In the case $\tilde{t} \to 0$, $\tilde{t} \in \mathbb{R}_0^+$, the limit value

$$\begin{aligned}
\eta^* &= \lim_{\tilde{t}^*\to 0}\left(\frac{1}{\tilde{t}^*}\cdot\ln\big(E_{v,1}(-\tilde{t}^*)\big)\right) \\
&= \lim_{\tilde{t}^*\to 0}\left(\frac{1}{E_{v,1}(-\tilde{t}^*)}\cdot\frac{\mathrm{d}\big(E_{v,1}(-\tilde{t}^*)\big)}{\mathrm{d}\tilde{t}^*}\right) = -\frac{1}{\Gamma(v+1)}
\end{aligned} \tag{29}$$

is obtained, where the second line results from the application of L'Hôpital's rule.

Moreover, the property $\eta < \bar{\eta}$, necessary for the interval definition in (24), is obvious due to the fact that η^* in (28) is defined as the quotient of a strictly monotonically decreasing numerator and a strictly monotonically increasing denominator.

Thus, the substitution $\tilde{t} := -\lambda t^v$ together with the monotonicity of the Mittag-Leffler function with respect to the parameter λ, as already also exploited in Theorem 2, cf. (23), concludes the proof. $\square$

The Figure 1a,b give a comparison of the box-type interval enclosures of Mittag-Leffler functions according to Theorem 2 with the exponential enclosures according to Theorem 3. According to the Figure 1c,d, it becomes obvious that for identical subdivisions of the time interval $t \in [0\,;\,1]$, the box-type enclosure is much more pessimistic at the beginning of the time horizon than at its end. Therefore, to obtain an identical degree of overestimation for both types of enclosures, a significantly larger number of subintervals would be required in the box-type case at the beginning of the considered time span. Moreover, the lower bound for the range of the Mittag-Leffler function is exactly represented at the beginning of each temporal subslice by the exponential enclosure, while the lower bound at the endpoint is represented exactly by the box-type enclosure, cf. Figure 1c. For the upper bound of the range, this property is reversed between both representations for the enclosure of the Mittag-Leffler function, cf. Figure 1d.

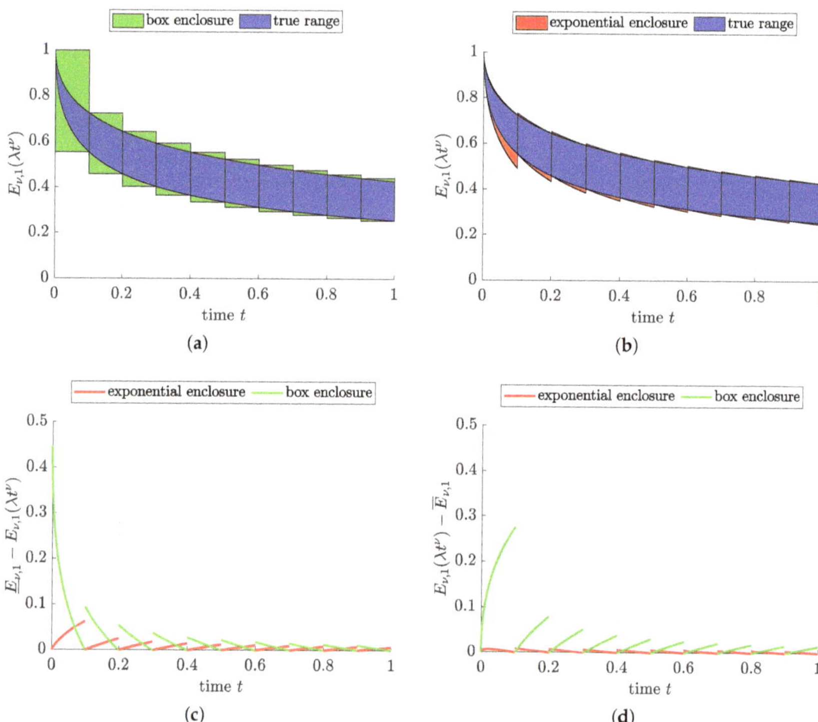

Figure 1. Comparison between box-type and exponential enclosures of the Mittag-Leffler function $E_{\nu,1}(\lambda t^\nu)$ with $\nu = 0.5$ for $t \in [0\,;\,1]$, $\underline{t} = k \cdot 0.1$, $\overline{t} = (k+1) \cdot 0.1$, $k \in \{0, 1, \ldots, 9\}$, and $\lambda \in [-2\,;\,-1]$: (**a**) Box-type enclosure of the Mittag-Leffler function; (**b**) Exponential enclosure of the Mittag-Leffler function; (**c**) Overestimation of the lower enclosure bound; (**d**) Overestimation of the upper enclosure bound.

Furthermore, Figure 2 illustrates the property stated in the proof above that η^* is a strictly monotonically increasing function for increasing values of the time argument $\tilde{t}^*$. To show that this property holds for all $0 < \nu < 1$, several values of this fractional differentiation order are depicted. Moreover, the limit case for $\tilde{t} \to 0$ according to (29) is also depicted in this graph by using a logarithmic time scale.

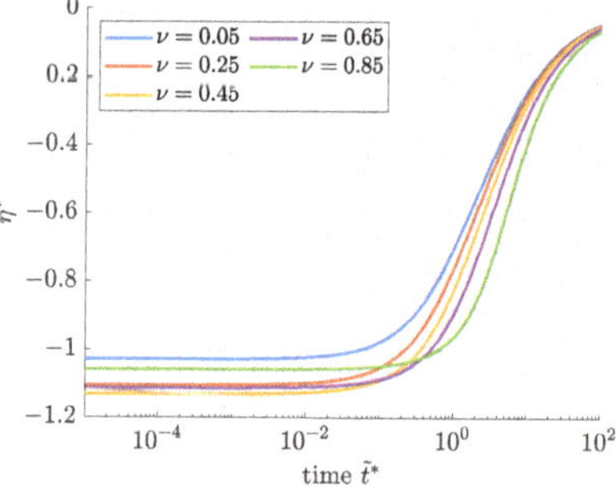

Figure 2. Evolution of the parameter η^* as a function of $\tilde{t}^*$ for different values of the fractional differentiation order ν.

3.3. Iterative Pseudo-State Enclosures for Box-Type and Exponential Representations of Mittag-Leffler Functions

Both box-type and exponential enclosures are used in this subsection to evaluate the iteration Formula (19) of Corollary 1. In this subsection, an interval subdivision scheme with respect to the time interval $[t]$ is employed to reduce the effect of overestimation.

For that purpose, we assume that $[t] = [0\,;\,T]$ is subdivided into Ξ not necessarily equally wide subintervals

$$[t] = \bigcup_{\zeta=1}^{\Xi} \left[t_{\zeta-1}\,;\,t_\zeta \right] = \bigcup_{\zeta=1}^{\Xi} [t]_\zeta\ , \tag{30}$$

where $t_0 := 0$, $t_\Xi := T$, and $t_0 < t_1 < \ldots < t_\Xi$.

Then, a subinterval-based evaluation of the iteration Formula (19) of Corollary 1 is given by

$$[\lambda_i]^{\langle \kappa+1 \rangle} := \bigsqcup_{\zeta=1}^{\Xi} \left(a_{ii}\left([\mathbf{x}_e]^{\langle \kappa \rangle}\left([t]_\zeta \right) \right) + \sum_{\substack{j=1\\j\neq i}}^{n} \left\{ a_{ij}\left([\mathbf{x}_e]^{\langle \kappa \rangle}\left([t]_\zeta \right) \right) \cdot \frac{E_{\nu,1}\left(\left[\lambda_j\right]^{\langle \kappa \rangle} \cdot [t]_\zeta^\nu \right)}{E_{\nu,1}\left([\lambda_i]^{\langle \kappa \rangle} \cdot [t]_\zeta^\nu \right)} \cdot \frac{\left[x_{e,j}\right](0)}{\left[x_{e,i}\right](0)} \right\} \right) , \tag{31}$$

where the symbol $\sqcup$ denotes the convex interval hull around all arguments of this operator.

Moreover, the expression

$$[\mathbf{x}_e]^{\langle \kappa \rangle}\left([t]_\zeta \right) = \mathbf{E}_{\nu,1}\left([\mathbf{\Lambda}]^{\langle \kappa \rangle} \cdot [t]_\zeta^\nu \right) \cdot [\mathbf{x}_e](0),\ [\mathbf{x}_e](0) = [\mathbf{x}_0] \tag{32}$$

in (31) represents the evaluation of the Mittag-Leffler-type pseudo-state enclosure for each temporal subinterval $[t]_\zeta$.

To replace the evaluation of the iteration Formula (31) by a counterpart that exploits the novel exponential enclosures of the Mittag-Leffler function for each temporal subinterval, define the enclosure

$$E_{\nu,1}\left(\lambda \cdot t^\nu \right) \in e^{\left[\underline{\eta}_{i,\zeta}^{\langle \kappa \rangle}\,;\,\overline{\eta}_{i,\zeta}^{\langle \kappa \rangle} \right] \cdot [t_{\zeta-1}\,;\,t_\zeta]^\nu} \tag{33}$$

for $\lambda \in [\lambda]_i^{\langle \kappa \rangle}$ and $t \in [t]_\zeta$. The interval bounds $\underline{\eta}_{i,\zeta}^{\langle \kappa \rangle}$ and $\overline{\eta}_{i,\zeta}^{\langle \kappa \rangle}$ on the right-hand side of (33) are obtained by replacing $\underline{\lambda}$ with $\underline{\lambda}_i^{\langle \kappa \rangle}$, $\overline{\lambda}$ with $\overline{\lambda}_i^{\langle \kappa \rangle}$, $\underline{t}$ with $t_{\zeta-1}$, and $\overline{t}$ with t_ζ in the Equations (25) and (26) that are defined in Theorem 3.

Then, the iteration Formula (19) of Corollary 1 is replaced with the expression

$$[\lambda_i]^{\langle\kappa+1\rangle} := \bigsqcup_{\zeta=1}^{\Xi} \left(a_{ii}\left([\mathbf{r}_e]^{\langle\kappa\rangle}\left([t]_\zeta\right)\right) + \right.$$

$$\left. \sum_{\substack{j=1 \\ j\neq i}}^{n} \left\{ a_{ij}\left([\mathbf{r}_e]^{\langle\kappa\rangle}\left([t]_\zeta\right)\right) \cdot e^{\left(\left[\underline{\eta}_{j,\zeta}^{\langle\kappa\rangle} \,;\, \overline{\eta}_{j,\zeta}^{\langle\kappa\rangle}\right] - \left[\underline{\eta}_{i,\zeta}^{\langle\kappa\rangle} \,;\, \overline{\eta}_{i,\zeta}^{\langle\kappa\rangle}\right]\right)\cdot[t]_\zeta^{\nu}} \cdot \frac{[x_{e,j}]\,(0)}{[x_{e,i}]\,(0)} \right\} \right) , \tag{34}$$

where

$$[\mathbf{r}_e]^{\langle\kappa\rangle}\left([t]_\zeta\right) = \exp\left(\mathrm{diag}\left\{ \left[\underline{\eta}_{1,\zeta}^{\langle\kappa\rangle} \,;\, \overline{\eta}_{1,\zeta}^{\langle\kappa\rangle}\right] \quad \cdots \quad \left[\underline{\eta}_{1,\zeta}^{\langle\kappa\rangle} \,;\, \overline{\eta}_{1,\zeta}^{\langle\kappa\rangle}\right]\right\} \cdot [t]_\zeta^{\nu} \right) \tag{35}$$

is a direct substitute for (32) that was included before in (31).

As a preparation for the evaluation of Formulas (31) and (34) in the following section, the true range as well as both the box-type and exponential enclosures of the quotient

$$\frac{E_{\nu,1}(\lambda_1 t^{\nu})}{E_{\nu,1}(\lambda_2 t^{\nu})} \tag{36}$$

are illustrated in Figure 3 for the mutually independent interval parameters $\lambda_1 \in [-2 \,;\, -1]$ and $\lambda_2 \in [-1.9 \,;\, -1]$. It becomes obvious that the box-type enclosure (as already discussed in Figure 1) is much more pessimistic for small points in time than for larger ones. Therefore, it can be expected that an intersection of both enclosure approaches leads to less pessimism during the evaluation of the iteration Formula (34). Note that this comes with practically no additional computational effort because the box-type range bounds form the basis for the application of Theorem 3.

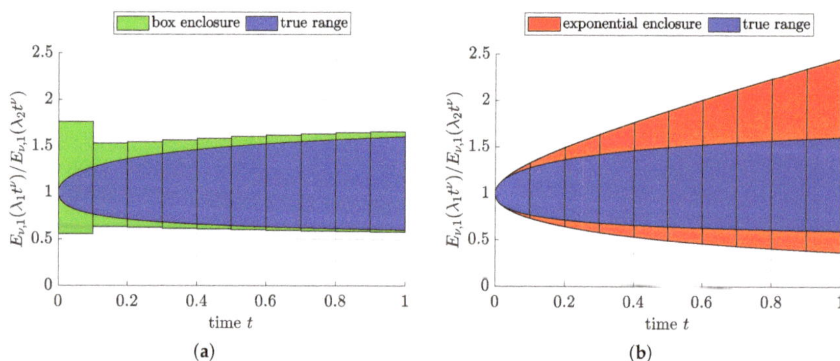

Figure 3. Illustration of the two considered guaranteed enclosure methods for the quotient of two Mittag-Leffler functions with uncertain parameters $\lambda_1 \in [\lambda_1]$ and $\lambda_2 \in [\lambda_2]$ for $\nu = 0.5$: (**a**) Box-type enclosures vs. exact range for the quotient (36); (**b**) Exponential enclosures vs. exact range for the quotient (36).

In the following section, both subdivision-based Formulas (31) and (34) are compared for the computation of guaranteed pseudo-state enclosures for a quasi-linear model of the charging/discharging dynamics of a Lithium-ion battery. The comparison is based on a quantification of the tightness of the obtained solution enclosures and a count of the numbers of iterations κ for identical numbers Ξ of temporal subintervals.

4. Simulation Results

The benchmark application considered in this section for the comparison of both iteration Formulas (31) and (34) is the simplified model of a Lithium-ion battery originally investigated in [25]. For the sake of completeness, this model is summarized in the following subsection before the respective simulation results are presented.

4.1. Simplified Fractional-Order Battery Model

Figure 4 illustrates a fractional-order equivalent circuit model of the charging/discharging behavior of a Lithium-ion battery, where the state of charge $\sigma(t)$, its fractional derivative $_0\mathscr{D}_t^{0.5}\sigma(t)$ of order $\nu = 0.5$, and the voltage drop $v_1(t)$ across the fractional-order constant phase element Q (as a generalization of an ideal capacitor [28–30]) are employed as pseudo-state variables.

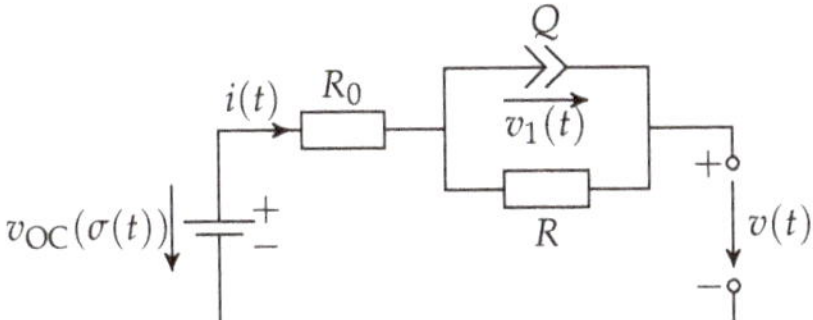

Figure 4. Basic fractional-order equivalent circuit model of batteries according to [25].

These pseudo-state variables are summarized in the vector

$$\mathbf{x}(t) = \begin{bmatrix} \sigma(t) & _0\mathscr{D}_t^{0.5}\sigma(t) & v_1(t) \end{bmatrix}^T \in \mathbb{R}^3 \ . \tag{37}$$

Using the modeling steps described in [28] and generalizing the charging/discharging dynamics to

$$_0\mathscr{D}_t^1\sigma(t) = -\frac{\eta_0 \cdot i(t) + \eta_1 \cdot \sigma(t) \cdot \text{sign}(i(t))}{3600 C_\text{N}} \tag{38}$$

as described in [25] with the terminal current $i(t)$ as the system input, the commensurate fractional-order quasi-linear state equations

$$_0\mathscr{D}_t^{0.5}\mathbf{x}(t) = \mathscr{A} \cdot \mathbf{x}(t) + \mathscr{b} \cdot i(t) \tag{39}$$

with the system and input matrices

$$\mathscr{A} = \begin{bmatrix} 0 & 1 & 0 \\ \frac{\eta_1 \cdot \text{sign}(i(t))}{3600 C_\text{N}} & 0 & 0 \\ 0 & 0 & -\frac{1}{RQ} \end{bmatrix} \quad \text{and} \quad \mathscr{b} = \begin{bmatrix} 0 \\ -\frac{\eta_0}{3600 C_\text{N}} \\ \frac{1}{Q} \end{bmatrix} \tag{40}$$

are obtained.

The simulations discussed in the following two subsections consider the parameters listed in Table 1 which are a subset of those used in [25]. To make the evaluation based on the novel exponential enclosures for the Mittag-Leffler function according to Theorem 3 comparable with the previous work reported in [25], we employ the same linear state feedback controller for the discharging phase ($i(t) > 0$) that was designed in terms of an assignment of the asymptotically stable eigenvalues $\lambda \in \{-0.0001; -0.0002; -0.4832\}$ to the closed-loop dynamics. This leads to the closed-loop dynamic model

$$\mathscr{D}_t^{0.5}\mathbf{x}(t) = \left(\mathscr{A} - \mathscr{b}\mathscr{k}^T\right) \cdot \mathbf{x}(t) =: \mathscr{A}_\text{C} \cdot \mathbf{x}(t) \ , \tag{41}$$

where the matrix entries $\mathscr{A}_{C,2,1}$, $\mathscr{A}_{C,2,2}$, $\mathscr{A}_{C,2,3}$ are inflated to interval parameters by symmetric bounds of a 1% radius of the respective nominal quantity and $\mathscr{A}_{C,3,3}$ to 10%, respectively.

Table 1. Parameters of the Lithium-ion battery model.

$R_0\ [\Omega]$	$Q\ [\mathrm{F/s}^{0.5}]$	$R\ [\Omega]$	$\eta_0\ [-]$	$\eta_1\ [-]$	$C_N\ [\mathrm{Ah}]$
$+1.7{\cdot}10^{-5}$	$+20.591$	$+0.1005$	$+1.0000$	$+0.1000$	$+3.1000$

To investigate the effect of the novel exponential enclosure approach for Mittag-Leffler functions, the following simulations also make use of the intersection of the solution to the system model (41) with an alternative representation resulting from a time-invariant similarity transformation of the system matrix $\mathscr{A}_C$ into an interval-valued diagonally dominant representation. This transformation, as detailed in [25], employs the matrix of eigenvectors for a matrix containing the elementwise-defined interval midpoints of $\mathscr{A}_C$. As discussed in [25], this transformation reduces the overestimation due to the wrapping effect of interval analysis [23,31] when evaluating both iteration Formulas (31) and (34).

For the rest of this paper, consider the two sets of initial pseudo-state vectors

$$\mathbf{x}(0) \in [\mathbf{x}]_1(0) = \begin{bmatrix} [\ \ 0.9000\ ;\ \ \ 1.1000] \\ [-0.0011\ ;\ -0.0009] \\ [\ \ 0.0900\ ;\ \ \ 0.1100] \end{bmatrix} \tag{42}$$

and

$$\mathbf{x}(0) \in [\mathbf{x}]_2(0) = \begin{bmatrix} [\ \ 0.99000\ ;\ \ \ 1.01000] \\ [-0.00101\ ;\ -0.00099] \\ [\ \ 0.09900\ ;\ \ \ 0.10100] \end{bmatrix} \tag{43}$$

in the sense of a Caputo-type initialization of the controlled battery model (41), i.e., with $\mathbf{x}(0)$ corresponding to $\mathbf{x}(t)$ for all $t < 0$.

In all simulations summarized in the following two subsections, the final points in time T for the evaluation of the iteration Formulas (31) and (34) are chosen as either of the ten values

$$T \in \left\{ 0.5, 0.5 + \frac{10 - 0.5}{9}, 0.5 + 2 \cdot \frac{10 - 0.5}{9}, \ldots, 10 \right\} \tag{44}$$

with $t_\xi - t_{\xi-1} = 0.005$. Note that this latter choice for the interval subdivision is not identical to the one used in [25], where for all the different values T the considered time ranges were always subdivided into 100 equally spaced slices.

4.2. Simulation with the Help of Box-Type Enclosures

In Figure 5, the iteration Formula (31) has been used to compute guaranteed enclosures for the pseudo states of the controlled Lithium-ion battery for the two differently wide enclosures of initial values.

Due to the fact that the solution parameters λ_i are determined in such a way that they are valid for the complete time interval $[0\ ;\ T]$, longer simulation horizons lead to inflating interval bounds as long as the integrator reset approach derived in [25] and its extended version published in [7] are not employed. The simulations show clearly that the overestimation of the range enclosures for the pseudo states is larger for those variables that change faster. In this scenario, the faster changing variable is the voltage $v_1(t)$ as compared to the state of charge $\sigma(t)$. However, it can also be seen that the blow-up of the solution bounds can be reduced by tighter bounds for the pseudo-state initialization which gives rise to the option to repeat the simulations after paving the possible initial pseudo-state domain.

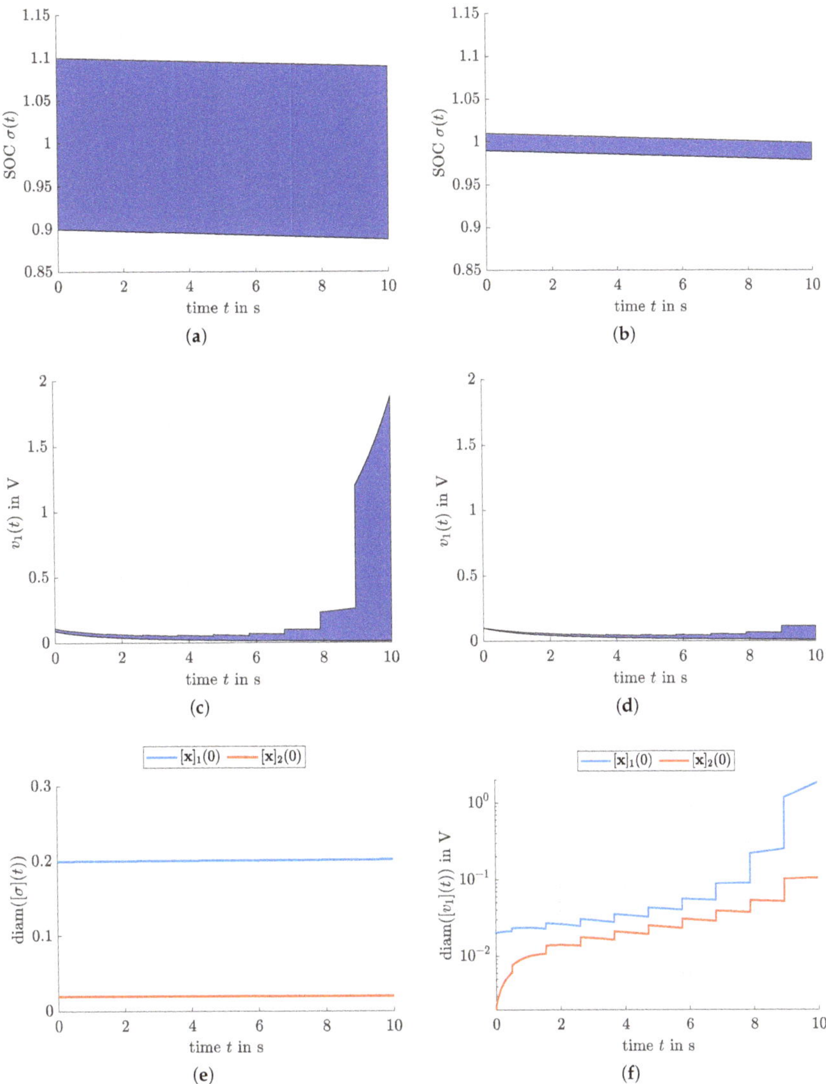

Figure 5. Use of box-type enclosures for the evaluation of the iteration Formula (31) for the computation of guaranteed pseudo-state enclosures: (**a**) State of charge $\sigma(t)$ for $\mathbf{x}(0) \in [\mathbf{x}]_1(0)$; (**b**) State of charge $\sigma(t)$ for $\mathbf{x}(0) \in [\mathbf{x}]_2(0)$; (**c**) Voltage $v_1(t)$ for $\mathbf{x}(0) \in [\mathbf{x}]_1(0)$; (**d**) Voltage $v_1(t)$ for $\mathbf{x}(0) \in [\mathbf{x}]_2(0)$; (**e**) Interval diameter for the enclosure of $\sigma(t)$; (**f**) Interval diameter for the enclosure of $v_1(t)$.

4.3. Simulation with the Help of Exponential Enclosures

Figure 6 shows that the pseudo-state enclosures are significantly tightened by using the intersection of the novel exponential enclosures of Mittag-Leffler functions with the box-type ones that were solely used in Figure 5. This advantage cannot only be observed for the case of the wide initialization $\mathbf{x}(0) \in [\mathbf{x}]_1(0)$ but also for the tighter one $\mathbf{x}(0) \in [\mathbf{x}]_2(0)$.

The second advantage of this new enclosure approach is illustrated in Figure 7, where the required maximum numbers of evaluations of both (31) and (34) are illustrated for the ten different choices of the simulation horizon T. There, it can be seen clearly that the novel exponential enclosure approach allows to reduce the number of required iterations significantly for longer simulation horizons T. To make this comparison fair, the iterations have been stopped in all cases if the diameters of all $[\lambda_i]^{\langle\kappa+1\rangle}$ and $[\lambda_i]^{\langle\kappa\rangle}$ deviate from each other by less than the value 10^{-4}.

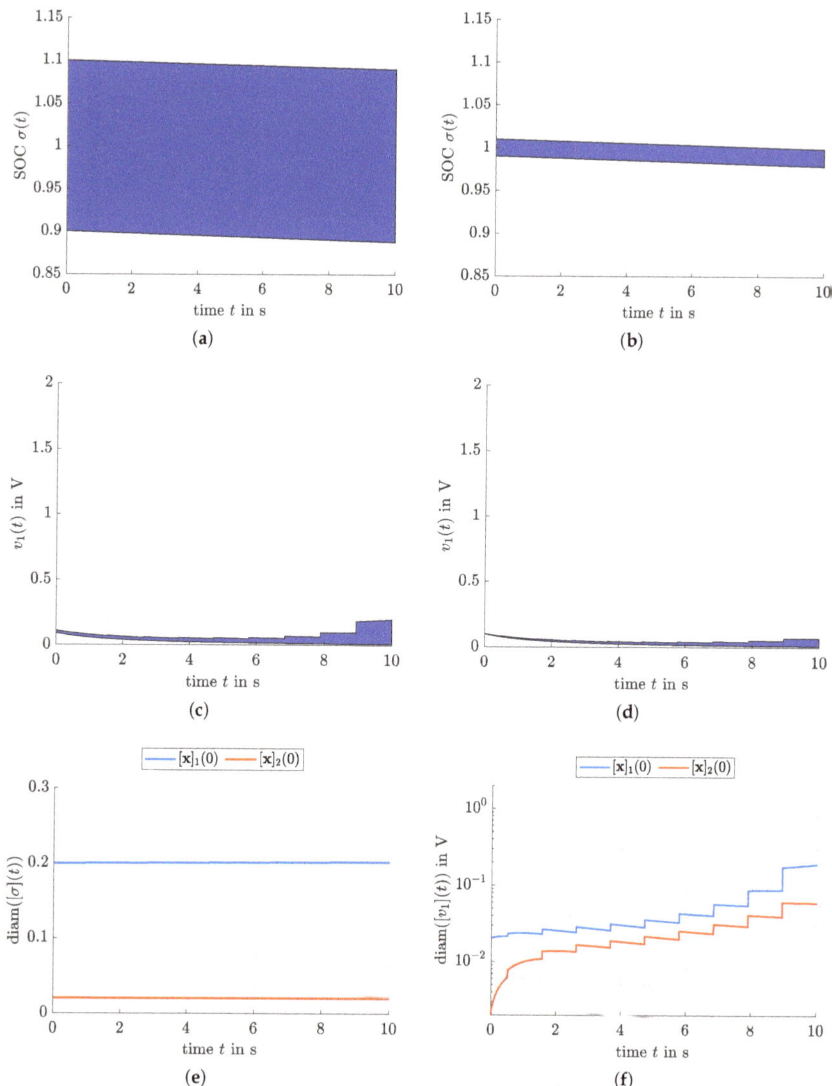

Figure 6. Use of exponential enclosures, intersected with the box-type ones, for the evaluation of the iteration Formula (34) for the computation of guaranteed pseudo-state enclosures: (**a**) State of charge $\sigma(t)$ for $\mathbf{x}(0) \in [\mathbf{x}]_1(0)$; (**b**) State of charge $\sigma(t)$ for $\mathbf{x}(0) \in [\mathbf{x}]_2(0)$; (**c**) Voltage $v_1(t)$ for $\mathbf{x}(0) \in [\mathbf{x}]_1(0)$; (**d**) Voltage $v_1(t)$ for $\mathbf{x}(0) \in [\mathbf{x}]_2(0)$; (**e**) Interval diameter for the enclosure of $\sigma(t)$; (**f**) Interval diameter for the enclosure of $v_1(t)$.

As a summary, the new enclosure approach does not only allow to significantly reduce the overestimation in the computation of guaranteed bounds for the pseudo states but—at least for the application scenario at hand—also reduces the computational effort by up to 80% due to significantly less required iterations.

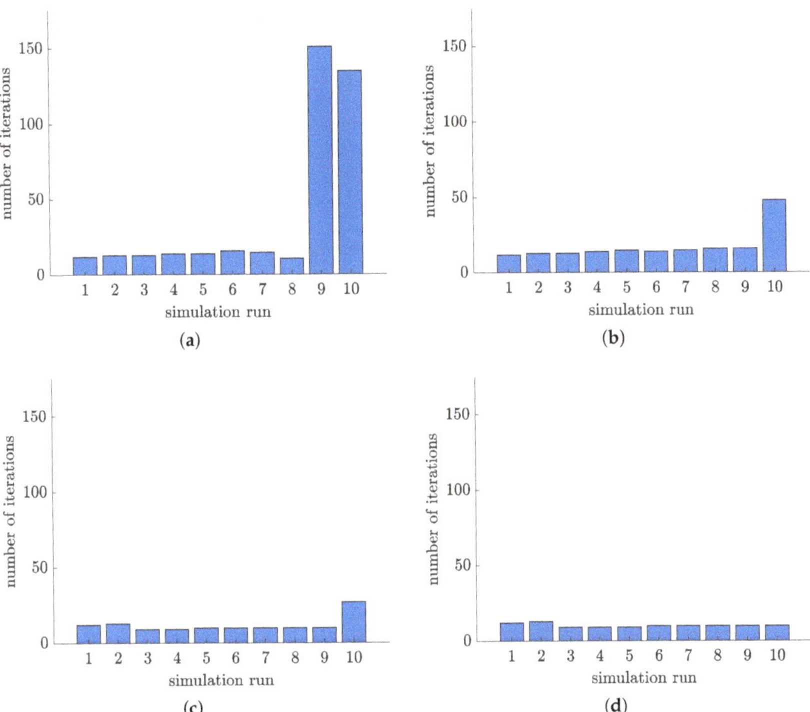

Figure 7. Comparison of the required number of iterations for both box-type and exponential enclosures: (**a**) Required iterations for the initialization $\mathbf{x}(0) \in [\mathbf{x}]_1(0)$ (box-type enclosure); (**b**) Required iterations for the initialization $\mathbf{x}(0) \in [\mathbf{x}]_2(0)$ (box-type enclosure); (**c**) Required iterations for the initialization $\mathbf{x}(0) \in [\mathbf{x}]_1(0)$ (exponential enclosure); (**d**) Required iterations for the initialization $\mathbf{x}(0) \in [\mathbf{x}]_2(0)$ (exponential enclosure).

5. Conclusions and Outlook on Future Work

In this paper, a novel enclosure approach for the pseudo states of fractional-order differential equations has been presented. It is based on enclosing Mittag-Leffler functions by exponential enclosures instead of box-type enclosures employed so far in previous work. By using a close-to-life model for the charging/discharging dynamics of a Lithium-ion battery, it has been shown that this new enclosure technique leads not only to significantly tighter enclosures, yet preserving the guaranteed enclosure property, but also leads to a noticeable reduction in the computational effort by significantly less iterations required to obtain an identical enclosure quality.

Future work will aim at extending the presented approach to system models with non-commensurate orders. In addition, the approach will be included into the observer-based technique presented in [7] for the quantification of truncation errors which allows for resetting fractional integrators in a guaranteed way. In such a way, it is planned to make the proposed simulation approach applicable to tasks such as the parameter identification of fractional-order differential equations and to the identification of their initialization functions for $t < 0$. Moreover, the application to more complex dynamic models from the domains of electrochemical energy storage and energy conversion will be investigated.

Funding: This research received no external funding.

Data Availability Statement: Data are contained within the article.

Conflicts of Interest: The author declares no conflict of interest.

References

1. Podlubny, I. *Fractional Differential Equations: An Introduction to Fractional Derivatives, Fractional Differential Equations, to Methods of Their Solution and Some of Their Applications*; Mathematics in Science and Engineering; Academic Press: London, UK, 1999.
2. Oustaloup, A. *La Dérivation Non Entière: Théorie, Synthèse et Applications*; Hermès: Paris, France, 1995. (In French)
3. Malti, R.; Victor, S. CRONE Toolbox for System Identification Using Fractional Differentiation Models. In Proceedings of the 17th IFAC Symposium on System Identification SYSID 2015, Beijing, China, 19–21 October 2015; Volume 48, pp. 769–774.
4. Sabatier, J.; Moze, M.; Farges, C. LMI Stability Conditions for Fractional Order Systems. *Comput. Math. Appl.* **2010**, *59*, 1594–1609. [CrossRef]
5. Bel Haj Frej, G.; Malti, R.; Aoun, M.; Raïssi, T. Fractional Interval Observers And Initialization Of Fractional Systems. *Commun. Nonlinear Sci. Numer. Simul.* **2020**, *82*, 105030. [CrossRef]
6. Rauh, A.; Kersten, J. Toward the Development of Iteration Procedures for the Interval-Based Simulation of Fractional-Order Systems. *Acta Cybern.* **2020**, *25*, 21–48. [CrossRef]
7. Rauh, A.; Malti, R. Quantification of Time-Domain Truncation Errors for the Reinitialization of Fractional Integrators. *Acta Cybern.* 2022, *accepted for publication*.
8. Lohner, R. Enclosing the Solutions of Ordinary Initial and Boundary Value Problems. In *Proceedings of the Computer Arithmetic: Scientific Computation and Programming Languages*; Kaucher, E.W., Kulisch, U.W., Ullrich, C., Eds.; Wiley-Teubner Series in Computer Science: Stuttgart, Germany, 1987; pp. 255–286.
9. Nedialkov, N.S. Implementing a Rigorous ODE Solver through Literate Programming. In *Modeling, Design, and Simulation of Systems with Uncertainties*; Rauh, A., Auer, E., Eds.; Mathematical Engineering; Springer: Berlin/Heidenberg, Germany, 2011; pp. 3–19.
10. Nedialkov, N.S. Interval Tools for ODEs and DAEs. In Proceedings of the 12th GAMM-IMACS Intl. Symposium on Scientific Computing, Computer Arithmetic, and Validated Numerics SCAN, Duisburg, Germany, 26–29 September 2006; IEEE Computer Society, 2007.
11. Nedialkov, N.S. Computing Rigorous Bounds on the Solution of an Initial Value Problem for an Ordinary Differential Equation. Ph.D. Thesis, Graduate Department of Computer Science, University of Toronto, Toronto, ON, Canada, 1999.
12. Lin, Y.; Stadtherr, M.A. Validated Solutions of Initial Value Problems for Parametric ODEs. *Appl. Numer. Math.* **2007**, *57*, 1145–1162. [CrossRef]
13. Rauh, A.; Kersten, J.; Aschemann, H. Interval-Based Verification Techniques for the Analysis of Uncertain Fractional-Order System Models. In Proceedings of the 18th European Control Conference ECC2020, Virtual Event, St. Petersburg, Russia, 12–15 May 2020.
14. Lyons, R.; Vatsala, A.; Chiquet, R. Picard's Iterative Method for Caputo Fractional Differential Equations with Numerical Results. *Mathematics* **2017**, *5*, 65. [CrossRef]
15. Haubold, H.; Mathai, A.; Saxena, R. Mittag-Leffler Functions and Their Applications. *J. Appl. Math.* **2011**, *2011*, 51. [CrossRef]
16. Gorenflo, R.; Kilbas, A.; Mainardi, F.; Rogosin, S. *Mittag-Leffler Functions, Related Topics and Applications*; Springer: Berlin/Heidelberg, Germany, 2014. [CrossRef]
17. Miller, K.; Samko, S. A Note on the Complete Monotonicity of the Generalized Mittag-Leffler Function. *Real Anal. Exch.* **1997**, *23*, 753–756. [CrossRef]
18. Garrappa, R. Numerical Evaluation of Two and Three Parameter Mittag-Leffler Functions. *SIAM J. Numer. Anal.* **2015**, *53*, 1350–1369. [CrossRef]
19. Gorenflo, R.; Loutchko, J.; Luchko, Y. Computation of the Mittag-Leffler Function and its Derivatives. *Fract. Calc. Appl. Anal. (FCAA)* 2002, *5*, 491–518.
20. Dorjgotov, K.; Ochiai, H.; Zunderiya, U. On Solutions of Linear Fractional Differential Equations and Systems Thereof. *arXiv* **2018**, arXiv:1803.09063.
21. Ghosh, U.; Sarkar, S.; Das, S. Solution of System of Linear Fractional Differential Equations with Modified Derivative of Jumarie Type. *Am. J. Math. Anal.* **2015**, *3*, 72–84.
22. Mayer, G. *Interval Analysis and Automatic Result Verification*; De Gruyter Studies in Mathematics; De Gruyter: Berlin, Germany; Boston, MA, USA, 2017.
23. Jaulin, L.; Kieffer, M.; Didrit, O.; Walter, É. *Applied Interval Analysis*; Springer: London, UK, 2001.
24. Garrappa, R. The Mittag-Leffler Function. MATLAB Central File Exchange. Available online: www.mathworks.com/matlabcentral/fileexchange/48154-the-mittag-leffler-function (accessed on 14 January 2021).
25. Rauh, A.; Kersten, J. Verification and Reachability Analysis of Fractional-Order Differential Equations Using Interval Analysis. In *Electronic Proceedings in Theoretical Computer Science, Proceedings of the 6th International Workshop on Symbolic-Numeric methods for Reasoning about CPS and IoT, online, 31 August 2020*; Dang, T., Ratschan, S., Eds.; Open Publishing Association: Den Haag, The Netherlands, 2021; Volume 331, pp. 18–32. [CrossRef]

26. Ishteva, M.; Boyadjiev, L.; Scherer, R. On the Caputo Operator of Fractional Calculus and C-Laguerre Functions. *Math. Sci. Res. J.* **2005**, *9*, 161–170.
27. Chen, M.; Sho, S.; Shi, P. *Robust Adaptive Control for Fractional-Order Systems with Disturbance and Saturation*; John Wiley & Sons, Ltd.: Chichester, UK, 2018.
28. Hildebrandt, E.; Kersten, J.; Rauh, A.; Aschemann, H. Robust Interval Observer Design for Fractional-Order Models with Applications to State Estimation of Batteries. In Proceedings of the 21st IFAC World Congress, Virtual Event, Berlin, Germany, 11–17 July 2020.
29. Andre, D.; Meiler, M.; Steiner, K.; Wimmer, C.; Soczka-Guth, T.; Sauer, D. Characterization of High-Power Lithium-Ion Batteries by Electrochemical Impedance Spectroscopy. I. Experimental Investigation. *J. Power Sources* **2011**, *196*, 5334–5341. [CrossRef]
30. Zou, C.; Zhang, L.; Hu, X.; Wang, Z.; Wik, T.; Pecht, M. A Review of Fractional-Order Techniques Applied to Lithium-Ion Batteries, Lead-Acid Batteries, and Supercapacitors. *J. Power Sources* **2018**, *390*, 286–296. [CrossRef]
31. Lohner, R. On the Ubiquity of the Wrapping Effect in the Computation of the Error Bounds. In *Proceedings of the Perspectives on Enclosure Methods*; Kulisch, U., Lohner, R., Facius, A., Eds.; Springer: Wien, Austria; New York, NY, USA, 2001; pp. 201–217.

fractal and fractional

Article

Fractional-Order Interval Observer for Multiagent Nonlinear Systems

Haoran Zhang, Jun Huang * and Siyuan He

The School of Mechanical and Electrical Engineering, Soochow University, Suzhou 215131, China;
zhrsudaedu@163.com (H.Z.); hesiyuan021108@163.com (S.H.)
* Correspondence: cauchyhot@163.com

Abstract: A framework of distributed interval observers is introduced for fractional-order multiagent systems in the presence of nonlinearity. First, a frame was designed to construct the upper and lower bounds of the system state. By using monotone system theory, the positivity of the error dynamics could be ensured, which implies that the bounds could trap the original state. Second, a sufficient condition was applied to guarantee the boundedness of distributed interval observers. Then, an extension of Lyapunov function in the fractional calculus field was the basis of the sufficient condition. An algorithm associated with the procedure of the observer design is also provided. Lastly, a numerical simulation is used to demonstrate the effectiveness of the distributed interval observer.

Keywords: distributed interval observer; fractional-order multiagent systems; monotone system theory

Citation: Zhang, H.; Huang, J.; He, S. Fractional-Order Interval Observer for Multiagent Nonlinear Systems. *Fractal Fract.* **2022**, *6*, 355. https://doi.org/10.3390/fractalfract6070355

Academic Editors: Thach Ngoc Dinh, Shyam Kamal and Rajesh Kumar Pandey

Received: 31 May 2022
Accepted: 23 June 2022
Published: 25 June 2022

Publisher's Note: MDPI stays neutral with regard to jurisdictional claims in published maps and institutional affiliations.

1. Introduction

Fractional calculus is the definition of the differential and integral of arbitrary order systems. During the last 20 years, fractional calculus grew into a hot topic and attracted increasing interest [1,2]. It plays an important role when people are faced with natural dynamics problems. It works well when it is applied in describing the memory and hereditary properties of manifold materials. In addition, the applications of fractional calculus are rapidly expanding, such as non-Fickian dynamics [3], fractional boundary value problems [4], and variable-order thermostat models [5].

Due to the inaccessibility of the system state in many physical backgrounds and the availability of output information, research on observation problems is very significant. With the indepth study of observation problems, interval observers are widely accepted as an efficient tool in reconstructing the system state with nonlinearity or bounded uncertainty [6,7]. The concept of interval observers was first proposed by Gouzé, and it was successfully applied to uncertain biological systems [8]. After that, research on interval observers can be divided into two parts. The first is based on monotone system theory, and the second was developed from set-membership estimation. The monotone-system-theory-based method requires researchers to design suitable observer gains in order to ensure that the error dynamics is positive and bounded. In this case, the bounds from the interval observer converge to the original system. Inspired by this idea, there are many new techniques for various systems emerging. For linear time-invariant systems, the time-varying coordinate transformation technique was used in interval observer design [9]. Efimov et al. also extended the interval estimation technique to nonlinear time-varying systems [10]. An interval observer for switched systems was also introduced in [11]. On the other hand, a set-membership estimation-based method combined robust observer design with reachability analysis [12]. Interval observers in this framework obviously improved estimation accuracy. Additionally, scholars presented a new integrated version of interval observers that achieved an expected tradeoff between robust estimation conservatism and computational complexity [13]. In general, research on interval observers for integer-order systems is fruitful, which is instructive for us to design interval observers of fractional-order systems.

A tremendous expansion for multiagent systems (MASs) has been witnessed, which mainly regards output regulation [14], observer design [15] and consensus control [16]. It started from simple two-order systems and attracted more attention on various complex MASs, such as switched [17], nonlinear [18], and time-delaying [19] systems. The observers designed for MASs are called distributed observers if the communication topology is taken into account. A distributed observer was designed for leader-following MASs, and an actual vehicle model was used to verify the function of the distributed observer [20]. In [21], the existence condition of the distributed observer was established to recover the state of nonlinear MASs. There are some works for fractional-order MASs in [22] and [23]. The consensus control problem for fractional-order MASs under a fixed topology was studied by stability theory and the Lyapunov method [22]. For unknown nonlinear dynamics, an adaptive control protocol was introduced that was used in a trajectory tracking problem [23]. There were also some interesting results about the observation problem in [24–27]. Traditional observers and distributed controllers were both used in fractional-order MASs [24]. Compared with [24], the time-varying formation control was investigated, and the consensus protocol was designed to match a switching topology [25]. For the fractional-order heterogeneous nonlinear MASs, the consensus control problem was converted into the stabilisation problem using the distributed control technique [26]. In [27], the object was the fractional-order system with unknown orders, and a robust observer-based controller was formulated to solve the consensus problem.

For these works [24–27], the introduced observers were all Luenberger-like. When the original system suffers from uncertain disturbance or nonlinearity, Luenberger-like observers are unable to recover the system state. It is challenging to recover the system state when the general fractional-order system has a complex communication with its neighborhood. Therefore, a distributed interval observer is a better choice for fractional-order MASs. In many physical problems such as trajectory tracking or output regulation, distributed interval observers provide a more accurate state information, which is a great step forward for the consensus control protocol design. However, there is not any paper focusing on distributed interval observers for fractional-order MASs. In the field of fractional calculus, there are some interesting works about fractional-order systems, including state reconstruction and algorithm programming [28–30]. Motivated by the interval observer mentioned above, we developed a framework of distributed interval observers for the general fractional-order MASs. The specific contributions are outlined as follows:

1. By investigating the fractional differential problem, a novel fractional-order Lyapunov function is proposed for the boundedness problem in the observer deign, which is an approach to prove that a matrix is Hurwitz.
2. Different from linear MASs, nonlinear MASs are considered, and the solution for Lipschitz functions is combined with the distributed interval observer design.
3. For fractional-order MASs, the communication topology is applied to the observer design. Each observer of the corresponding agent could accept the information from its adjacent observers. A novel distributed interval observer was first designed for the fractional-order MASs.

The rest of the paper is structured as follows. Section 2 mainly presents some preparation work, including the fractional differential, graph theory and fractional-order systems. Section 3 proposes the distributed interval observer design method for fractional-order MASs. An illustrative example is given in Section 4 to verify the validity of the designed observer. Lastly, conclusions and future work are drawn in Section 5.

Notations: For two vectors $x \in R^n$ and $y \in R^n$, $x < (\leq)y$ are understood with $x_i < (\leq)y_i$, $i \in \{1, \ldots, n\}$. For $A \prec 0$ and $A \succ 0$, symbol $\prec (\succ)$ means that matrix A is positive(negative) definite. For a matrix B, there exist three properties: $B^+ = max(B, 0)$, $B^- = B^+ - B$ and $|B| = B^+ + B^-$. Matrix B^T denotes the transpose of B, and $He(B)$ is defined as $He(B) = B + B^T$. $\lambda_{min}(A)$ represents the minimal nonzero eigenvalue of A. $\otimes$ is the Kronecker product used in MASs. 1_N means a N-order column vector, and all elements are 1.

2. Preliminaries

2.1. Fractional Calculus

There are several definitions used to regard the fractional differentiation operator, such as the Grünwald–Letnikov, Riemann–Liouville and Caputo definitions introduced in [31]. Among them, the Caputo definition is the most frequently used and has many applications in the engineering field. In the Caputo definition, the initial value of the differentiation is considered.

The Caputo fractional derivative for a function $f(t)$ is

$$
{}^C_{t_0}D^\alpha_t f(t) = \frac{1}{\Gamma(n-\alpha)} \int_{t_0}^t \frac{f^{(n)}(\tau)}{(t-\tau)^{\alpha-n+1}} d\tau,
\tag{1}
$$

where ${}^C_{t_0}D^\alpha_t$ is the fractional integration differentiation operator, t_0 represents the initial time, and $\alpha \in [n-1,n]$ is the order of the system. $\Gamma(\cdot)$ is a gamma function that is introduced in Definition 1. $f(t)$ is a differentiable function with the nth derivative. In this paper, we mainly focus on the fractional-order system with $n = 1$ and $t_0 = 0$, then we have $0 < \alpha < 1$. When $n = 1$, $f(t)$ is only required to have the first derivative. For the sake of simplicity, the differentiation operator ${}^C_{t_0}D^\alpha_t$ is replaced by D^α_t.

The Grünwald-Letnikov fractional derivative for a function $f(t)$ is

$$
{}^{GL}_{t_0}D^\alpha_t f(t) = \lim_{N\to\infty} \left[\frac{t-t_0}{N}\right]^{-\alpha} \sum_{j=0}^{N-1} (-1)^j f\left(t - j\left[\frac{t-t_0}{N}\right]\right).
\tag{2}
$$

The Riemann-Liouville fractional derivative for a function $f(t)$ is

$$
{}^{RL}_{t_0}D^\alpha_t f(t) = \begin{cases} \frac{1}{\Gamma(-\alpha)} \int_{t_0}^t (t-\tau)^{-\alpha-1} f(\tau)d\tau, & \alpha < 0 \\ f(t), & \alpha = 0 \\ D^n_t \left[{}^{RL}_{t_0}D^{\alpha-n}_t f(t)\right], & \alpha > 0 \end{cases}
\tag{3}
$$

with $n - 1 \le \alpha < n$.

For a wide class of functions, the Grünwald–Letnikov and Riemann–Liouville definitions are equivalent [32]. However, it is difficult for the Grünwald-Letnikov definition to have a Laplace transform. The Laplace transform of the Riemann–Liouville definition is

$$
L[{}^{RL}_0 D^\alpha_t f(t)] = \begin{cases} s^\alpha F(s), & q \le 0 \\ s^\alpha F(s) - \sum_{k=0}^{n-1} s^k {}^{RL}_0 D^{\alpha-k-1}_t f(0), & n-1 \le q < n \end{cases}
\tag{4}
$$

where $F(s)$ is the Laplace transform of $f(t)$.

The Laplace transform of the Caputo definition is

$$
L[{}^C_0 D^\alpha_t] = s^\alpha F(s) - \sum_{k=0}^{n-1} s^{\alpha-1-k} f^{(k)}(0).
\tag{5}
$$

Comparing (4) and (5), it is obvious that the Riemann–Liouville fractional derivative is unsuitable for the Laplace transform technique because it requires the knowledge of noninteger-order derivatives of the function at $t = 0$, while the Caputo fractional derivative only requires the knowledge of integer-order derivatives of the function. This is why Caputo fractional derivative was chosen.

Definition 1 ([33]). *Gamma function is an important element for the Caputo fractional differentiation operator, which is defined by*

$$\Gamma(a) = \int_0^\infty e^{-t} t^{a-1} dt, \tag{6}$$

where value $\Gamma(a)$ is in the convergence of the right-hand side of the complex plane.

2.2. Graph Theory

A communication topology $\mathcal{G}(\mathcal{N}, \xi, \mathcal{A})$ contains three elements: $\mathcal{N} \in \{1, \ldots, N\}$ is the node set of MASs, $\xi \subset \mathcal{N} \times \mathcal{N}$ represents an edge set for nodes in MASs, and $\mathcal{A} \in R^{N \times N}$ is an adjacent matrix of the graph. For $\mathcal{A} = [a_{ij}] \in R^{N \times N}$, a_{ji} is the weight from node i to node j. In this paper, the graph has no self loops, i.e., $a_{ii} = 0, i \in \mathcal{N}$. $a_{ji} \neq 0$ if and only if $(i, j) \in \xi$. For node i, its neighborhood is denoted as $\mathcal{N}_i = \{j | j \in \mathcal{N}, a_{ji} \neq 0, j \neq i\}$. For $\mathcal{G}$, it is defined as a strongly connected graph if and only if any node in it is mutually reachable. For adjacent matrix $\mathcal{A}$, the corresponding Laplacian matrix is defined as $\mathcal{L} = [l_{ij}]_{N \times N}$, where $l_{ii} = \Sigma_{j=1}^{N} a_{ij}$ and $l_{ij} = -a_{ij}, i \neq j$. If a topology is call as a balanced topology, the edges of it must be balanced, which means that edge (i, j) belongs to set ξ if edge (j, i) belongs to set ξ.

Lemma 1 ([34]). *for a Laplacian matrix $\mathcal{L}$, zero is one of the eigenvalues and it has a fixed right eigenvector 1_N. The other nonzero eigenvalues are all positive. If there exists a directed spanning tree in $\mathcal{G}$, zero is a simple eigenvalue of $\mathcal{L}$.*

Lemma 2 ([35]). *considering $\mathcal{G}$ as a strongly connected graph, suppose that $r = [r_1, \ldots, r_N]$ is the left eigenvector connected with the eigenvalue zero. Then, we have $R\mathcal{L} + \mathcal{L}^T R \geq 0$, where $R = diag\{r_1, \ldots, r_N\}$. For a balanced graph, we have $r_1 = \cdots = r_N$.*

Lemma 3 ([35]). *If $\mathcal{G}$ is a strongly connected graph; its generalized algebraic connectivity is define as $a(\mathcal{L}) = \min\limits_{r^T x = 0, x \neq 0} \frac{x^T(R\mathcal{L} + \mathcal{L}^T R)x}{2x^T Rx}$. Due to the topology in this paper being defined as balanced, matrix R can be converted into $R = r_1 I_N$. The generalized algebraic connectivity is equal to $a(\mathcal{L}) = \lambda_{min}(\frac{He(\mathcal{L})}{2})$.*

2.3. Fractional-Order Systems

Consider the following fractional-order system for the i-th agent.

$$\begin{cases} D_t^\alpha x_i(t) = Ax_i(t) + f(x_i(t)), \\ y_i(t) = Cx_i(t), \end{cases} \tag{7}$$

where $x_i \in R^n$ is the system state, $y_i \in R^m$ is the output, and $f(x_i) \in R^n$ is the Lipschitz function. $A \subset R^{n \times n}$ and $C \in R^{m \times n}$ are matrices with suitable dimensions. Communication topology $\mathcal{G}$ for (7) is a strongly connected balanced graph.

Property 1. *Given a matrix $M \in R^{n \times n}$, M is a Metzler matrix if all its elements outside the main diagonal are non-negative. For example, the following matrix is Metzler:*

$$M = \begin{bmatrix} -1 & 1 \\ 2 & 0.5 \end{bmatrix}.$$

Property 2. *Given a matrix $N \in R^{n \times n}$, M is a Hurwitz matrix if its all real parts of the eigenvalues are negative. For example, the following matrix is Hurwitz:*

$$N = \begin{bmatrix} -3 & -2 \\ -2 & -3 \end{bmatrix}.$$

For nonlinearity $f(x_i)$, we began the analysis with the following properties.

Property 3 ([36]). *the differentiable global Lipschitz function $f(x_i)$ can be divided into two increasing functions $a(x_i)$ and $b(x_i)$; their relationship is*

$$f(x_i) = a(x_i) - b(x_i). \tag{8}$$

Property 4 ([36]). *for a Lipschitz function $f(x_i)$, there exists a differentiable global Lipschitz function $\tilde{f}(x_i^{k_1}, x_i^{k_2})$, such that*

- $\tilde{f}(x_i^{k_1}, x_i^{k_2}) = a(x_i^{k_1}) - b(x_i^{k_2})$,
- $\tilde{f}(x_i, x_i) = f(x_i)$,
- $\frac{\partial \tilde{f}}{\partial a(x_i)} \geq 0$ *and* $\frac{\partial \tilde{f}}{\partial b(x_i)} \leq 0$.

Remark 1. *In Property 3, a Lipschitz function is transformed into two increasing functions, which is just for us to introduce the function $\tilde{f}(\cdot, \cdot)$. Because $a(x_i)$ and $b(x_i)$ are both increasing functions, one can deduce that $\frac{\partial \tilde{f}}{\partial a(x_i)} \geq 0$ and $\frac{\partial \tilde{f}}{\partial b(x_i)} \leq 0$ easily. For $\underline{x} \leq x \leq \overline{x}$, we have $\tilde{f}(\underline{x}, \overline{x}) \leq \tilde{f}(x, x) \leq \tilde{f}(\overline{x}, \underline{x})$, which are the upper and lower bounds of the Lipschitz function $f(x)$ in the structure of distributed interval observers.*

Assuming that the bounds of the system state satisfy $\underline{x}_i \leq x_i \leq \overline{x}_i$, on the basis of Properties 3 and 4, one can deduce that

$$\tilde{f}(\underline{x}_i, \overline{x}_i) \leq \tilde{f}(x_i, x_i) = f(x_i) \leq \tilde{f}(\overline{x}_i, \underline{x}_i). \tag{9}$$

By using the generalized Taylor formula, $\tilde{f}(\overline{x}_i, \underline{x}_i) - \tilde{f}(x_i, x_i)$ is written as

$$\tilde{f}(\overline{x}_i, \underline{x}_i) - \tilde{f}(x_i, x_i) = \int_0^1 \frac{\partial \tilde{f}}{\partial \delta_1}(\tau \delta_1 + (1 - \tau)\delta_2)d\tau(\delta_1 - \delta_2), \tag{10}$$

where $\delta_1 = \begin{bmatrix} \overline{x} \\ \underline{x} \end{bmatrix}$ and $\delta_2 = \begin{bmatrix} x \\ x \end{bmatrix}$. It follows from Property 4 that

$$\begin{aligned} \tilde{f}(\overline{x}_i, \underline{x}_i) - \tilde{f}(x_i, x_i) = [&\int_0^1 \frac{\partial a}{\partial x}(\tau \overline{x} + (1 - \tau)x)d\tau \\ &- \int_0^1 \frac{\partial b}{\partial x}(\tau x + (1 - \tau)\underline{x})d\tau](\delta_1 - \delta_2) \\ = &[\mathcal{F}_1(\overline{x}, x) - \mathcal{F}_2(x, \underline{x})](\delta_1 - \delta_2). \end{aligned} \tag{11}$$

Similarly, we have

$$\tilde{f}(x_i, x_i) - \tilde{f}(\overline{x}_i, \underline{x}_i) = [\mathcal{F}_3(x, \underline{x}) - \mathcal{F}_4(\overline{x}, x)](\delta_1 - \delta_2), \tag{12}$$

where matrices $\mathcal{F}_i, i \in \{1, \ldots, 4\}$ are non-negative and can be derived from (11). On the basis of the above discussion, the following property is presented.

Property 5 ([36]). *for $f(x_i)$ and $\tilde{f}(\cdot,\cdot)$ defined in Property 4, if Jacobian matrix $\frac{\partial \tilde{f}}{\partial \delta_1}$ is bounded, there exist matrices $F_i, i \in \{1,\ldots,4\}$ such that*

$$\begin{cases} \tilde{f}(\overline{x}_i,\underline{x}_i) - \tilde{f}(x_i,x_i) \le F_1\overline{e}_i + F_2\underline{e}_i, \\ \tilde{f}(x_i,x_i) - \tilde{f}(\underline{x}_i,\overline{x}_i) \le F_3\overline{e}_i + F_4\underline{e}_i, \end{cases} \tag{13}$$

where $\overline{e}_i = \overline{x}_i - x_i$ and $\underline{e}_i = x_i - \underline{x}_i$.

Example 1. *For nonlinear function $f(x) = sinx$, corresponding functions $\tilde{f}(\overline{x},\underline{x})$ and $\tilde{f}(\underline{x},\overline{x})$ are defined as*

$$\begin{cases} \tilde{f}(\overline{x},\underline{x}) = sin(\overline{x}) + \overline{x} - \underline{x}, \\ \tilde{f}(\underline{x},\overline{x}) = sin(\underline{x}) + \underline{x} - \overline{x}, \end{cases}$$

where $f(x)$, $\tilde{f}(\underline{x},\overline{x})$ and $\tilde{f}(\overline{x},\underline{x})$ satisfy Properties 3 and 4.

The functions mentioned in Property 3 are defined as $a(x) = sin(x) + x$ and $b(x) = x$. $a(x)$ and $b(x)$ are both obviously increasing functions. Then, $\tilde{f}(\overline{x},\underline{x}) - \tilde{f}(x,x)$ can be transformed into

$$\begin{aligned} \tilde{f}(\overline{x},\underline{x}) &- \tilde{f}(x,x) \\ &= sin(\overline{x}) + \overline{x} - \underline{x} - sin(x) \\ &= (sin(\overline{x}) - sin(x)) + (\overline{x} - x) + (x - \underline{x}) \\ &\le 2(\overline{x} - x) + (x - \underline{x}). \end{aligned} \tag{14}$$

From Property 5, we have $\tilde{f}(\overline{x},\underline{x}) - \tilde{f}(x,x) \le F_1\overline{e} + F_2\underline{e}$. Combining (14) with Property 5, we chose $F_1 = 2I$ and $F_2 = I$. Similarly, we have $F_3 = I$ and $F_4 = 2I$.

From Example 1, Property 5 is feasible, and the result from Formulas (9)–(12) stands.

For System (7), nonlinear function $f(x_i(t))$ was assumed to satisfy Properties 3–5; then, the interval observer for (7) was designed with

$$\begin{cases} D_t^\alpha \overline{x}_i(t) = A\overline{x}_i(t) + L(y_i(t) - C\overline{x}_i(t)) + \gamma M\sum_{i=1}^N a_{ij}(\overline{x}_j(t) - \overline{x}_i(t)) + \tilde{f}(\overline{x}_i,\underline{x}_i), \\ D_t^\alpha \underline{x}_i(t) = A\underline{x}_i(t) + L(y_i(t) - C\underline{x}_i(t)) + \gamma M\sum_{i=1}^N a_{ij}(\underline{x}_j(t) - \underline{x}_i(t)) + \tilde{f}(\underline{x}_i,\overline{x}_i), \end{cases} \tag{15}$$

where M and L are interval observer gains.

The error dynamics of the interval observer is

$$\begin{cases} D_t^\alpha \overline{e}_i(t) = D_t^\alpha \overline{x}_i(t) - D_t^\alpha x_i(t) \\ \qquad = (A - LC - \gamma M\sum_{j=1}^N \mathcal{L}_{ij})\overline{e}_i(t) + \tilde{f}(\overline{x}_i,\underline{x}_i) - \tilde{f}(x_i,x_i), \\ D_t^\alpha \underline{e}_i(t) = D_t^\alpha x_i(t) - D_t^\alpha \underline{x}_i(t) \\ \qquad = (A - LC - \gamma M\sum_{j=1}^N \mathcal{L}_{ij})\underline{e}_i(t) + \tilde{f}(x_i,x_i) - \tilde{f}(\underline{x}_i,\overline{x}_i). \end{cases} \tag{16}$$

Then, $e(t)$, $\tilde{f}(\overline{x},\underline{x})$, $\tilde{f}(\underline{x},\overline{x})$ and $\tilde{f}(x,x)$ are defined as

$$e(t) = \begin{bmatrix} e_1 \\ \vdots \\ e_N \end{bmatrix}, \tilde{f}(\overline{x},\underline{x}) = \begin{bmatrix} \tilde{f}(\overline{x}_1,\underline{x}_1) \\ \vdots \\ \tilde{f}(\overline{x}_N,\underline{x}_N) \end{bmatrix}, \tilde{f}(\underline{x},\overline{x}) = \begin{bmatrix} \tilde{f}(\underline{x}_1,\overline{x}_1) \\ \vdots \\ \tilde{f}(\underline{x}_N,\overline{x}_N) \end{bmatrix}, \tilde{f}(x,x) = \begin{bmatrix} \tilde{f}(x_1,x_1) \\ \vdots \\ \tilde{f}(x_N,x_N) \end{bmatrix}.$$

System (16) can be written in the following form:

$$\begin{cases} D_t^\alpha \overline{e}(t) = (I_N \otimes (A - LC) - \gamma(\mathcal{L} \otimes M)\overline{e}(t) + \tilde{f}(\overline{x},\underline{x}) - \tilde{f}(x,x), \\ D_t^\alpha \underline{e}(t) = (I_N \otimes (A - LC) - \gamma(\mathcal{L} \otimes M))\underline{e}(t) + \tilde{f}(x,x) - \tilde{f}(\underline{x},\overline{x}). \end{cases} \tag{17}$$

3. Main Results

Proving the boundedness of the distributed interval observer is equal to proving the convergence of the error dynamics. The Lyapunov function is a great choice for it. For error dynamics $\dot{e}(t) = A_e e(t)$, the following Lyapunov function is constructed:

$$V(t) = e^T(t)Pe(t), \tag{18}$$

where P is a positive symmetric matrix with suitable dimensions.

In the integer-order system, $\dot{V}(t) < 0$ is the sufficient condition to prove the convergence of error dynamics. $\dot{V}(t) < 0$ is equivalent to

$$A_e^T P + PA_e \prec 0. \tag{19}$$

Equation (19) is the strict LMI that can be computed with MATLAB. Formula (19) can yield that A_e is a Hurwitz matrix. By configuring matrix A_e, the convergence of the system can be reached.

However, the above details are mainly about the integer-order system. There are no corresponding lemmas about the fractional-order system. Therefore, the fractional-order extension of a Lyapunov candidate function is introduced to demonstrate the convergence of the error dynamics by referring [37].

Lemma 4. *Consider error dynamics $e(t) \in R^n$ to be a continuous and derivable function. Then, for any time $t \geq 0$, the fractional derivative of the Lyapunov function is*

$$D_t^\alpha V(t) \leq (D_t^\alpha e^T(t))Pe(t) + e^T(t)P(D_t^\alpha e(t)), \tag{20}$$

where $V(t) = e^T(t)Pe(t)$ is the Lyapunov function connected with $e(t)$.

Proof. Denote that $J = (D_t^\alpha e^T)Pe + e^T P(D_t^\alpha e) - D_t^\alpha V$.

According to the definition of fractional calculus, $D_t^\alpha V(t)$ is equivalent to

$$
\begin{aligned}
D_t^\alpha V(t) &= \frac{1}{\Gamma(1-\alpha)} \int_0^t \frac{\dot{V}(\tau)}{(t-\tau)^\alpha} d\tau \\
&= \frac{1}{\Gamma(1-\alpha)} \int_0^t \frac{e^T(\tau)P\dot{e}(\tau) + \dot{e}^T(\tau)Pe(\tau)}{(t-\tau)^\alpha} d\tau.
\end{aligned}
\tag{21}
$$

Considering (21), J is rewritten as

$$
\begin{aligned}
J &= \frac{1}{\Gamma(1-\alpha)} \int_0^t \frac{e^T(t)P\dot{e}(\tau) + \dot{e}^T(\tau)Pe(t) - e^T(\tau)P\dot{e}(\tau) - \dot{e}^T(\tau)Pe(\tau)}{(t-\tau)^\alpha} d\tau \\
&= \frac{1}{\Gamma(1-\alpha)} \int_0^t \frac{(e^T(t) - e^T(\tau))P\dot{e}(\tau) + \dot{e}^T(\tau)P(e(t) - e(\tau))}{(t-\tau)^\alpha} d\tau \\
&= -\frac{1}{\Gamma(1-\alpha)} \int_0^t \frac{\dot{z}^T(\tau)Pz(\tau) + z^T(\tau)P\dot{z}(\tau)}{(t-\tau)^\alpha} d\tau,
\end{aligned}
\tag{22}
$$

where $z(\tau) = e(t) - e(\tau)$.

Then, the proof for Lemma 1 is equivalent to proving that

$$\frac{1}{\Gamma(1-\alpha)} \int_0^t \frac{(\dot{z}^T(\tau))Pz(\tau) + z^T(\tau)P\dot{z}(\tau)}{(t-\tau)^\alpha} d\tau \leq 0. \tag{23}$$

For $u = z^T(t)Pz(t)$, one can deduce that

$$\frac{du}{dt} = \dot{z}^T Pz(t) + z^T(t)P\dot{z}(t). \tag{24}$$

Then, $v(\tau) = \frac{(t-\tau)^{-\alpha}}{\Gamma(1-\alpha)}$ is denoted, which yields

$$dv(\tau) = \frac{\alpha(t-\tau)^{-\alpha-1}}{\Gamma(1-\alpha)} d\tau. \tag{25}$$

On the basis of (24) and (25), (23) is equal to

$$\frac{z^T(\tau)Pz(\tau)}{\Gamma(1-\alpha)(t-\tau)^\alpha}\Big|_0^t - \frac{\alpha}{\Gamma(1-\alpha)}\int_0^t \frac{z^T(\tau)Pz(\tau)}{(t-\tau)^\alpha} d\tau \leq 0, \tag{26}$$

with

$$\frac{z^T(\tau)Pz(\tau)}{\Gamma(1-\alpha)(t-\tau)^\alpha}\Big|_0^t = \frac{z^T(\tau)Pz(\tau)}{\Gamma(1-\alpha)(t-\tau)^\alpha}\Big|_{\tau=t} - \frac{z^T(0)Pz(0)}{\Gamma(1-\alpha)(t)^\alpha}. \tag{27}$$

When $\tau = t$, the first term of (27), has nondeterminacy, and the analysis of its limitation is necessary:

$$\lim_{\tau \to t} \frac{z^T(\tau)Pz(\tau)}{\Gamma(1-\alpha)(t-\tau)^\alpha} = \frac{-\dot{e}^T(\tau)Pe(t) + \dot{e}^T(\tau)Pe(\tau) - e^T(t)P\dot{e}(\tau) + e^T(\tau)P\dot{e}(\tau)}{-\Gamma(1-\alpha)\alpha(t-\tau)^{\alpha-1}}$$

$$= \frac{-\dot{e}^T(\tau)Pe(t) + \dot{e}^T(\tau)Pe(\tau) - e^T(t)P\dot{e}(\tau) + e^T(\tau)P\dot{e}(\tau)}{-\Gamma(1-\alpha)\alpha}(t-\tau)^{1-\alpha}. \tag{28}$$

Due to $\alpha \in (0,1)$, it is obvious that $\frac{z^T(\tau)Pz(\tau)}{\Gamma(1-\alpha)(t-\tau)^\alpha} \to 0$ when $\tau = t$. Expression (26) can be simplified to

$$-\frac{z^T(0)Pz(0)}{\Gamma(1-\alpha)t^\alpha} - \frac{\alpha}{\Gamma(1-\alpha)}\int_0^t \frac{z^T(\tau)Pz(\tau)}{(t-\tau)^\alpha} d\tau \leq 0. \tag{29}$$

For $P \succ 0$ and $\Gamma(\cdot) > 0$, it is certain to stand. $\square$

Theorem 1. *For System* (15), *there exist bounds* $\overline{x}$ *and* $\underline{x}$, *such that* $\underline{x}(t) \leq x(t) \leq \overline{x}(t)$ *if the following conditions are satisfied:*

1. $\Omega = I_N \otimes (A - LC) - \gamma(\mathcal{L} \otimes M)$ *is Metzler;*
2. *The initial condition of* (7) *satisfies* $\underline{x}(0) \leq x(0) \leq \overline{x}(0)$;
3. *Nonlinear function* $f(x_i(t))$ *possesses the features introduced in Properties 3–5.*

Proof. From Properties 3 and 4, the upper bound of the nonlinear function is equal to

$$\tilde{f}(\overline{x}_i, \underline{x}_i) = a(\overline{x}_i) - b(\underline{x}_i). \tag{30}$$

Then, it follows from (30) that

$$\tilde{f}(\overline{x}_i, \underline{x}_i) - \tilde{f}(x_i, x_i) = (a(\overline{x}_i) - a(x_i)) - (b(\underline{x}_i) - b(x_i)). \tag{31}$$

Functions $a(\cdot)$ and $b(\cdot)$ are all increasing, which yields

$$\tilde{f}(\overline{x}_i, \underline{x}_i) - \tilde{f}(x_i, x_i) \geq 0. \tag{32}$$

Similarly, one can deduce that

$$\tilde{f}(x_i, x_i) - \tilde{f}(\underline{x}_i, \overline{x}_i) \geq 0. \tag{33}$$

From $\underline{x}(0) \leq x(0) \leq \overline{x}(0)$, the initial value of the error dynamics satisfies $\overline{e}(0) \geq 0$ and $\underline{e}(0) \geq 0$. If matrix Ω is Metzler, and (32) and (33) are true, it follows that $\overline{e}(t) \geq 0$ and $\underline{e}(t) \geq 0$, i.e., $\underline{x}(t) \leq x(t) \leq \overline{x}(t)$, for any $t \geq 0$. $\square$

Remark 2. *For Theorem 1, we constructed a frame for (7). The positivity of the error dynamics is guaranteed, which implies $\underline{x}(t) \leq x(t) \leq \overline{x}(t)$. However, an interval observer not only requires the error dynamics be positive, but also that the upper and lower errors are in a convergence of zero. Then, we give the following theorem.*

Theorem 2. *On the basis of the result in Theorem 1, given constant $\tau > 0$ and positive matrix $P = P^T$, if there exists a solution such that*

$$\hat{\Pi} = \begin{bmatrix} He(PA - UC + PN_1) - 2\tau I & PN_2 + N_3^T P \\ PN_3 + N_2^T P & He(PA - UC + PN_4) - 2\tau I \end{bmatrix} \prec 0,$$

$$\gamma > \frac{\tau}{a(\mathcal{L})},$$

where γ is the coupling strength, $L = P^{-1}U$ and $M = P^{-1}$ are the observer gains, then System (15) is a distributed interval observer.

Proof. The Lyapunov function is chosen as follows:

$$V(t) = \sum_{i=1}^{N} r_i \overline{e}_i^T(t) P \overline{e}_i(t) + \sum_{i=1}^{N} r_i \underline{e}_i^T(t) P \underline{e}_i(t). \tag{34}$$

From Lemma 1, the fractional derivative of $V(t)$ is

$$
\begin{aligned}
D_t^\alpha V(t) \leq {}& \sum_{i=1}^{N} r_i (D_t^\alpha \overline{e}_i^T)(t) P \overline{e}_i(t) + \sum_{i=1}^{N} r_i \underline{e}_i^T(t) P(D_t^\alpha \underline{e}_i(t)) \\
& + \sum_{i=1}^{N} r_i (D_t^\alpha \underline{e}_i^T)(t) P \underline{e}_i(t) + \sum_{i=1}^{N} r_i \overline{e}_i^T(t) P(D_t^\alpha \overline{e}_i(t)).
\end{aligned}
\tag{35}
$$

Substituting the error dynamics into (35)

$$
\begin{aligned}
D_t^\alpha V(t) = {}& \sum_{i=1}^{N} r_i (\Omega_i \overline{e}_i + \overline{\Delta f}(x_i))^T P \overline{e}_i + \sum_{i=1}^{N} r_i \underline{e}_i^T P(\Omega_i \underline{e}_i + \underline{\Delta f}(x_i)) \\
& + \sum_{i=1}^{N} r_i (\Omega_i \underline{e}_i + \underline{\Delta f}(x_i))^T P \underline{e}_i + \sum_{i=1}^{N} r_i \overline{e}_i^T P(\Omega_i \overline{e}_i + \overline{\Delta f}(x_i)) \\
\leq {}& \sum_{i=1}^{N} r_i (\Omega_i \overline{e}_i + F_1 \overline{e}_i + F_2 \underline{e}_i)^T P \overline{e}_i + \sum_{i=1}^{N} r_i \overline{e}_i^T P(\Omega_i \overline{e}_i + F_1 \overline{e}_i + F_2 \underline{e}_i) \\
& + \sum_{i=1}^{N} r_i \underline{e}_i^T P(\Omega_i \overline{e}_i + F_3 \overline{e}_i + N_F \underline{e}_i) + \sum_{i=1}^{N} r_i (\Omega_i \overline{e}_i + F_3 \overline{e}_i + F_4 \underline{e}_i)^T P \underline{e}_i \\
= {}& ((R \otimes (A - LC) - \gamma R\mathcal{L} \otimes M)\overline{e} + R \otimes F_1 \overline{e} + R \otimes F_2 \underline{e})^T (I \otimes P)\overline{e} \\
& + \overline{e}^T (I \otimes P)((R \otimes (A - LC) - \gamma R\mathcal{L} \otimes M)\overline{e} + R \otimes F_1 \overline{e} + R \otimes F_2 \underline{e} \\
& + ((R \otimes (A - LC) - \gamma R\mathcal{L} \otimes M)\underline{e} + R \otimes N_3 \overline{e} + R \otimes F_4 \underline{e})^T (I \otimes P)\overline{e} \\
& + \overline{e}^T (I \otimes P)((R \otimes (A - LC) - \gamma R\mathcal{L} \otimes M)\overline{e} + R \otimes F_3 \overline{e} + R \otimes N_4 \underline{e}) \\
= {}& \overline{e}^T (R \otimes (He(PA - PLC + PF_1)) - \gamma (R\mathcal{L} + \mathcal{L}^T R) \otimes PM)\overline{e} \\
& + \underline{e}^T (R \otimes (He(PA - PLC + PF_4)) - \gamma (R\mathcal{L} + \mathcal{L}^T R) \otimes PM)\underline{e} \\
& + \overline{e}^T (PF_2 + F_3^T P)\underline{e} + \underline{e}^T (PF_3 + F_2^T P)\overline{e},
\end{aligned}
\tag{36}
$$

where $\Omega_i = A - LC - \gamma M \sum_{j=1}^{N} \mathcal{L}_{ij}$, $\overline{\Delta f}(x_i) = \tilde{f}(\overline{x}_i, \underline{x}_i) - \tilde{f}(x_i, x_i)$ and $\underline{\Delta f}(x_i) = \tilde{f}(x_i, x_i) - \tilde{f}(\underline{x}_i, \overline{x}_i)$.

According to Lemma 3, $R\mathcal{L} + \mathcal{L}^T R$ could be simplified as

$$2a(\mathcal{L})\bar{e}^T R\bar{e} \leq \bar{e}^T (R\mathcal{L} + \mathcal{L}^T R)\bar{e}, \tag{37}$$

$$2a(\mathcal{L})\underline{e}^T R\underline{e} \leq \underline{e}^T (R\mathcal{L} + \mathcal{L}^T R)\underline{e}. \tag{38}$$

Taking $M = P^{-1}$, (37) and (38) into account, it follows from (36) that

$$\begin{aligned}
D_t^\alpha V(t) \leq {} &\bar{e}^T (R \otimes (He(PA - PLC + PF_1)) - 2\gamma a(\mathcal{L})R \otimes I_n)\bar{e} \\
&+ \underline{e}^T (R \otimes (He(PA - PLC + PF_4)) - 2\gamma a(\mathcal{L})R \otimes I_n \underline{e} \\
&+ \bar{e}^T (PF_2 + F_3^T P)\underline{e} + \underline{e}^T (PF_3 + F_2^T P)\bar{e}.
\end{aligned} \tag{39}$$

Considering $\gamma > \frac{\tau}{a(\mathcal{L})}$, we have

$$\begin{aligned}
D_t^\alpha V(t) \leq {} &\bar{e}^T (R \otimes (He(PA - PLC + PF_1)) - 2\tau R \otimes I_n \bar{e} \\
&+ \underline{e}^T (R \otimes (He(PA - PLC + PF_4)) - 2\tau R \otimes I_n \underline{e} \\
&+ \bar{e}^T (PF_2 + F_3^T P)\underline{e} + \underline{e}^T (PF_3 + F_2^T P)\bar{e} \\
= {} &\epsilon^T(t) R \otimes \Pi\epsilon(t),
\end{aligned}$$

where $\epsilon(t) = [\bar{e}^T(t), \underline{e}^T(t)]^T$ and

$$\Pi = \begin{bmatrix} He(PA - PLC + PF_1) - 2\tau I_n & PF_2 + F_3^T P \\ PF_3 + F_2^T P & He(PA - PLC + PF_4) - 2\tau I_n \end{bmatrix}.$$

To satisfy the LMI toolbox, $U = PL$ is applied in Π, which leads to

$$\hat{\Pi} = \begin{bmatrix} He(PA - UC + PF_1) - 2\tau I_n & PF_2 + F_3^T P \\ PF_3 + F_2^T P & He(PA - UC + PF_4) - 2\tau I_n \end{bmatrix}.$$

Then, matrix $\hat{\Pi} \prec 0$ is equal to $D_t^\alpha V(t) < 0$, which implies that $\lim_{t\to\infty} \bar{e}(t) = 0$ and $\lim_{t\to\infty} \underline{e}(t) = 0$. The boundedness of the error dynamics could be guaranteed. $\quad\square$

Remark 3. *Constant τ is simple without an actual effect. However, it is a parameter connected with γ. If we just compute γ, the LMI toolbox just gives one feasible solution. Nevertheless, if we compute τ, $\gamma > \frac{\tau}{a(\mathcal{L})}$ would have more regions to select.*

On the basis of Theorem 2, an algorithm was constructed to design distributed interval observers for fractional-order MASs.

4. Simulation

Considering a fractional-order MAS with nonlinearity, the state-space model is similar to (7), where

$$A = \begin{bmatrix} -0.5 & 2 \\ 0 & -1 \end{bmatrix}, \quad C = \begin{bmatrix} 0 & 1 \end{bmatrix}, \quad f(x_i) = sin(x_i).$$

For nonlinear function $f(x_i)$, corresponding function $\tilde{f}(\bar{x}_i, \underline{x}_i)$ is defined as

$$\tilde{f}(\bar{x}_i, \underline{x}_i) = sin(\bar{x}_i) + \bar{x}_i - \underline{x}_i.$$

For functions $a(x_i) = sin(x_i) + x_i$ and $b(x_i) = x_i$, it is obvious that $a(x_i)$ and $b(x_i)$ are increasing functions. $a(x_i) = sin(x_i) + x_i$ and $b(x_i) = x_i$ are substituted into (13); then,

$$sin(\overline{x}_i) + \overline{x}_i - \underline{x}_i - sin(x_i)$$
$$= (sin(\overline{x}_i) - sin(x_i)) + (\overline{x}_i - x_i) + (x_i - \underline{x}_i)$$
$$\leq 2(\overline{x}_i - x_i) + (x_i - \underline{x}_i)$$
$$= F_1\overline{e}_i + F_2\underline{e}_i.$$

We chose $F_1 = 2I$ and $F_2 = I$. Similarly, we have $F_3 = I$ and $F_4 = 2I$. Laplacian matrix $\mathcal{L}$ is

$$\mathcal{L} = \begin{bmatrix} 3 & -1 & -1 & -1 \\ -1 & 2 & -1 & 0 \\ -1 & -1 & 2 & 0 \\ -1 & 0 & 0 & 1 \end{bmatrix}.$$

By using Lemma 3, we obtained generalized algebraic connectivity $a(\mathcal{L}) = 1$. By solving the LMI in Theorem 2, the observer gains can be computed:

$$L = \begin{bmatrix} -1.1771 \\ -0.2282 \end{bmatrix}, \quad M = \begin{bmatrix} 0.1906 & 0.0757 \\ 0.0757 & 0.1470 \end{bmatrix}, \tau = 2.9151.$$

Then we chose $\gamma = 3$. The initial value of the original system state is defined as $x(0) = [1\ 2\ 3\ 5\ 2\ -1\ -1\ 4]^T$. The initial value of the distributed interval observer is defined as $\overline{x}(0) = [6\ 7\ 8\ 10\ 7\ 4\ 4\ 9]^T$ and $\underline{x}(0) = [-4\ -3\ -2\ 0\ -4\ -6\ -6\ -1]^T$.

By performing Steps 5–8 in Algorithm 1, we can obtain Figures 1–6. Then, Figures 1 and 2 show the original system state of the four agents. Figures 3 and 4 display the bounds from a distributed interval observer. v_{ij} means the upper bound of the jth state of the ith agent, while u_{ij} means the lower bound of the jth state of the ith agent. From Figures 3 and 4, the bounds of the distributed observer trap the state of the original system. Define that $e_{ij}^+ = \overline{x}_{ij} - x_{ij}$ and $e_{ij}^+ = x_{ij} - \underline{x}_{ij}$. For Figures 5 and 6, the error between the observer and the original system could be reduced to a bounded value, which implies that the distributed interval observer is feasible.

Algorithm 1 Distributed interval estimation for fractional-order MASs.

Step 1: Given the constant matrix $\mathcal{L}$, compute the generalized algebraic connectivity $a(\mathcal{L})$;

Step 2: For given matrices $A \in R^{n \times n}, C \in R^{m \times n}$, compute the observer gains L and M and the constant τ from Theorem 2;

Step 3: Select appropriate constant $\gamma > \frac{\tau}{a(\mathcal{L})}$;

Step 4: Ensure that the matrix Ω is a Metzler matrix;

Step 5: For the total time $t = 10$, choosing the step size $h = 0.001$, the step $N = T/h$ can be calculated.

Step 6: Based on Step 5, construct two loops. The first loop is for the original fractional-order system and the other loop is for the distributed interval observer.

Step 7: Construct an array with $N + 1$ volumes to store the output of the original system in step 6.

Step 8: Establish a distributed interval observer based on the output in Step 7 and the other loop proposed in Step 6.

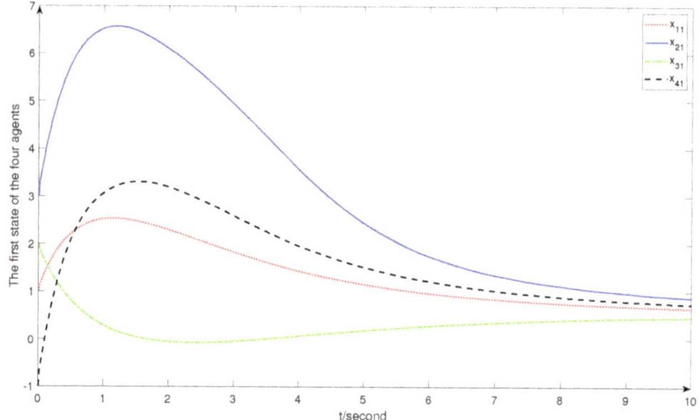

Figure 1. The first state of the four-agents.

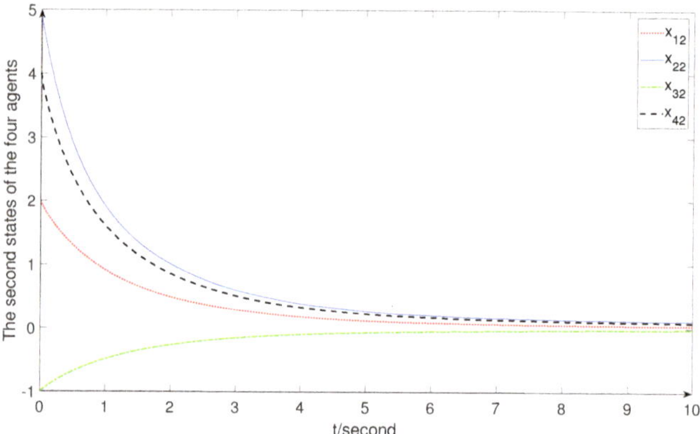

Figure 2. The second state of the four-agents.

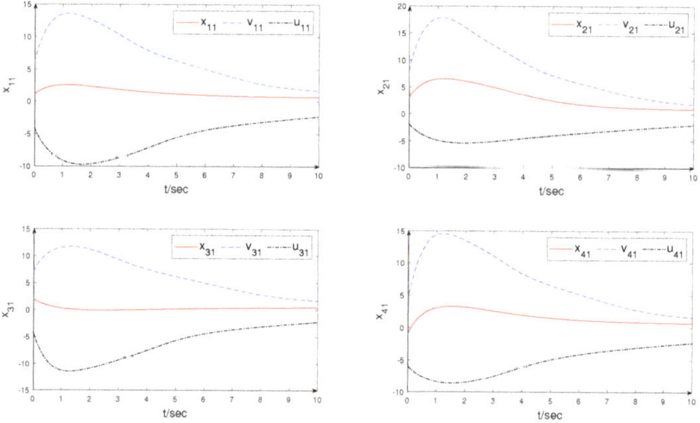

Figure 3. The bounds of the first state for the four-agents.

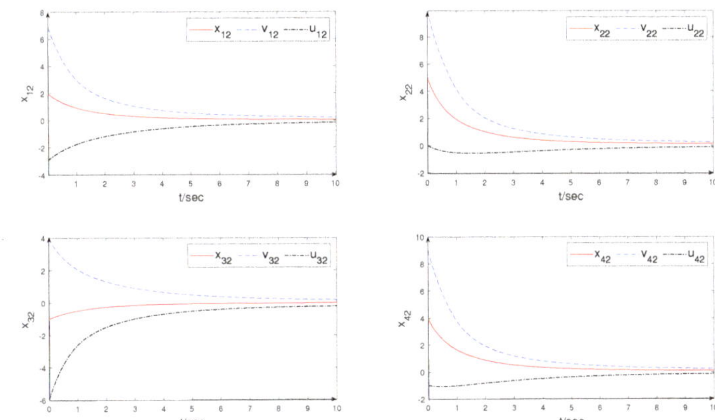

Figure 4. The bounds of the second state for the four-agents.

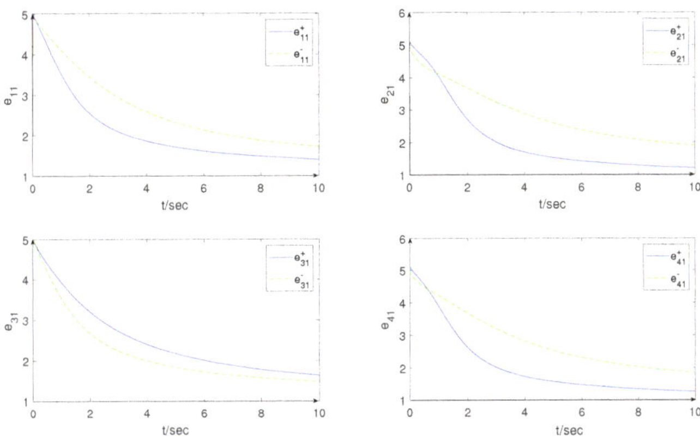

Figure 5. The error of the first state for the four-agents.

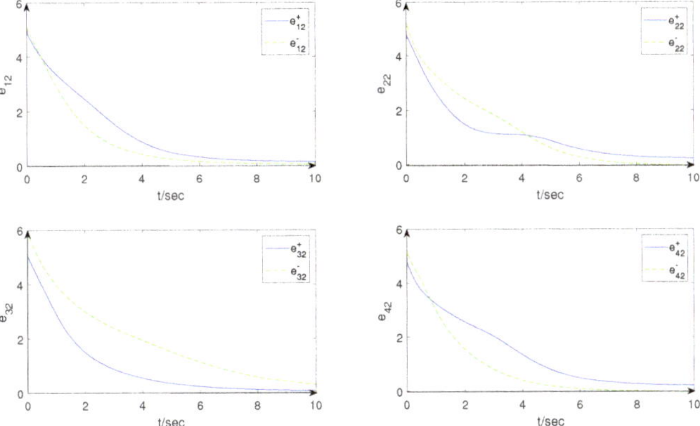

Figure 6. The error of the second state for the four-agents.

5. Conclusions

In this paper, a distributed interval observer design methodology for linear fractional-order MASs with nonlinearity was proposed. A Lyapunov method that is useful for observer and controller design was first introduced for general fractional-order systems. For MASs, graph theory was applied to fractional-order systems, and the strict LMI and an effective algorithm were presented for observer design. Lastly, an example was given to demonstrate the effectiveness of the proposed method. In the future, by using the H_∞ technique, we aim to focus on research regarding the consensus control or formation control of fractional-order MASs.

Author Contributions: Conceptualization, H.Z. and J.H.; methodology, J.H.; software, H.Z.; validation, H.Z., J.H. and S.H.; formal analysis, H.Z.; investigation, S.H.; resources, H.Z.; data curation, H.Z.; writing—original draft preparation, H.Z.; writing—review and editing, H.Z.; visualization, H.Z.; supervision, J.H.; project administration, J.H.; funding acquisition, J.H. All authors have read and agreed to the published version of the manuscript.

Funding: This research were funded by the Natural Science Foundation of Jiangsu province of China under grant BK2021-1309, and in part by the Open Fund for Jiangsu Key Laboratory of Advanced Manufacturing Technology under grant HGAMTL-2101 and by Undergraduate Training Program For Innovation and Entrepreneurship, Soochow University(202210285152Y).

Institutional Review Board Statement: Not applicable.

Informed Consent Statement: Not applicable.

Data Availability Statement: Not applicable.

Acknowledgments: we would like to express our great appreciation to the editors and reviewers.

Conflicts of Interest: The authors declare no conflict of interest.

References

1. Hartley, T.; Lorenzo, C. Fractional-Order system identification based on continuous order-distributions. *Signal Process.* **2003**, *83*, 2287–2300. [CrossRef]
2. Li, M.; Li, D.; Wang, J.; Zhao, C. Active disturbance rejection control for fractional-order system. *ISA Trans.* **2013**, *52*, 365–374. [CrossRef] [PubMed]
3. Lu, B.; Zhang, Y.; Reeves, D.M.; Sun, H.; Zheng, C. Application of tempered-stable time fractional-derivative model to upscale subdiffusion for pollutant transport in field-scale discrete fracture networks. *Mathematics* **2018**, *6*, 5. [CrossRef]
4. Turab A.; Mitrović Z.; Savić A. Existence of solutions for a class of nonlinear boundary value problems on the hexasilinane graph. *Adv. Differ. Equ.* 2021, *to be published.* [CrossRef]
5. Rezapour, S.; Souid, M.S.; Bouazza, Z.; Hussain, A.; Etemad, S. On the fractional variable order thermostat model: Existence theory on cones via piece-wise constant functions. *J. Funct. Space* 2022, *to be published.* [CrossRef]
6. Zhang, K.; Jiang, B.; Yan, X.; Shen, J. Interval sliding mode observer based incipient sensor fault detection with application to a traction device in China railway high-speed. *IEEE Trans. Veh Technol.* **2019**, *68*, 2585–2597. [CrossRef]
7. Huang, J.; Ma, X.; Che, H.; Han, Z. Further result on interval observer design for discrete-time switched systems and application to circuit systems. *IEEE Trans. Circuits Syst. II-Express Briefs* **2019**, *67*, 2542–2546. [CrossRef]
8. Gouzé, J.; Rapaport, A.; Hadj-Sadok, M. Interval observers for uncertain biological systems. *Ecol. Modell.* **2000**, *133*, 45–56. [CrossRef]
9. Mazenc, F.; Bernard, O. Interval observers for linear time-invariant systems with disturbances. *Automatica* **2011**, *47*, 140–147. [CrossRef]
10. Raïssi, T.; Efimov, D.; Zolghadri, A. Interval state estimation for a class of nonlinear systems. *IEEE Trans. Autom. Control* **2011**, *57*, 260–265. [CrossRef]
11. Dinh, T.N.; Marouani, G.; Raïssi, T.; Wang, Z.; Messaoud, H. Optimal interval observers for discrete-time linear switched systems. *Int. J. Control* **2020**, *93*, 2613–2621. [CrossRef]
12. Huang, J.; Che, H.; Raïssi, T.; Wang, Z. Functional interval observer for discrete-time switched descriptor systems. *IEEE Trans. Autom. Control* **2021**, *67*, 2497–2504. [CrossRef]
13. Xu, F.; Yang, S.; Wang, X. A novel set-theoretic interval observer for discrete linear time-invariant systems. *IEEE Trans. Autom. Control* **2021**, *66*, 773–780. [CrossRef]
14. Cai, H.; Lewis, F.L.; Hu, G.; Huang, J. The adaptive distributed observer approach to the cooperative output regulation of linear multi-agent systems. *Automatica* **2017**, *75*, 299–305. [CrossRef]

5. Han, W.; Trentelman, H.; Wang, Z.; Shen, Y. A simple approach to distributed observer design for linear systems. *IEEE Trans. Autom. Control* **2018**, *64*, 329–336. [CrossRef]
6. Huang, J.; Yang, M.; Zhang, Y.; Zhang, M. Consensus control of multi-agent systems with P-one-sided Lipschitz. *ISA Trans.* **2022**, *125*, 42–49. [CrossRef]
7. Li, C.; Yu, X.; Liu, Z.W.; Huang, T. Asynchronous impulsive containment control in switched multi-agent systems. *Inf. Sci.* **2016**, *370*, 667–679. [CrossRef]
8. Hua, C.; You, X.; Guan, X. Leader-Following consensus for a class of high-order nonlinear multi-agent systems. *Automatica* **2016**, *73*, 138–144. [CrossRef]
9. Chen, D.; Liu, G. A networked predictive controller for linear multi-agent systems with communication time delays. *J. Frankl. Inst.* **2020**, *357*, 9442–9466. [CrossRef]
20. Hong, Y.; Chen, G.; Bushnell, L. Distributed observers design for leader-following control of multi-agent networks. *Automatica* **2008**, *44*, 846–850. [CrossRef]
21. Liu, T.; Huang, J. A distributed observer for a class of nonlinear systems and its application to a leader-following consensus problem. *IEEE Trans. Autom. Control* **2018**, *44*, 1221–1227. [CrossRef]
22. Yu, Z.; Jiang, H.; Hu, C. Leader-Following consensus of fractional-order multi-agent systems under fixed topology. *Neurocomputing* **2015**, *149*, 613–620. [CrossRef]
23. Li, Z.; Gao, L.; Chen, W.; Xu, Y. Distributed adaptive cooperative tracking of uncertain nonlinear fractional-order multi-agent systems. *IEEE-CAA J. Automatic.* **2016**, *47*, 222–234. [CrossRef]
24. Zhu, W.; Li, W.; Zhou, P.; Yang, C. Consensus of fractional-order multi-agent systems with linear models via observer-type protocol. *Neurocomputing* **2017**, *230*, 60–65. [CrossRef]
25. Gong, Y.; Wen, G.; Peng, Z.; Huang, T.; Chen, Y. Observer-Based time-varying formation control of fractional-order multi-agent systems with general linear dynamics. *IEEE Trans. Circuits Syst. II-Express Briefs* **2019**, *67*, 82–86. [CrossRef]
26. Wen, G.; Zhang, Y.; Peng, Z.; Yu, Y.; Rahmani, A. Observer-Based output consensus of leader-following fractional-order heterogeneous nonlinear multi-agent systems. *Int. J. Control* **2020**, *93*, 2516–2524. [CrossRef]
27. Afaghi, A.; Ghaemi, S.; Ghiasi, A.; Badamchizadeh, M. Adaptive fuzzy observer-based cooperative control of unknown fractional-order multi-agent systems with uncertain dynamics. *Soft Comput.* **2020**, *24*, 3737–3752. [CrossRef]
28. Danca, M.; Kuznetsov, N. Matlab code for Lyapunov exponents of fractional-order systems. *Int. J. Bifurc. Chaos* **2018**, *28*, 1850067. [CrossRef]
29. Tian, X.; Yang, Z. Adaptive stabilization of a fractional-order system with unknown disturbance and nonlinear input via a backstepping control technique. *Symmetry* **2019**, *12*, 55. [CrossRef]
30. Huong, D. Design of functional interval observers for non-linear fractional-order systems. *Asian J. Control* **2020**, *22*, 1127–1137. [CrossRef]
31. Podlubny, I. Fractional derivatives and integrals. In *Fractional Differential Equations: An Introduction to Fractional Derivatives, Fractional Differential Equations, to Methods of Their Solution and Some of Their Applications*; Elsevier: San Diego, CA, USA, 1998; pp. 41–80.
32. Tavazoei, M.; Haeri, M. A note on the stability of fractional order systems. *Math. Comput. Simul.* **2009**, *79*, 1566–1576. [CrossRef]
33. Kamal, S.; Sharma, R.K.; Dinh, T.N.; Ms, H.; Bandyopadhyay, B. Sliding mode control of uncertain fractional-rder systems: A reaching phase free approach. *Asian J. Control* **2021**, *23*, 199–208. [CrossRef]
34. Ren, W.; Beard, R. Consensus seeking in multiagent systems under dynamically changing interaction topologies. *IEEE Trans. Autom. Control* **2005**, *50*, 655–661. [CrossRef]
35. Yu, W.; Chen, G.; Cao, M.; Kurths, J. Second-Order consensus for multiagent systems with directed topologies and nonlinear dynamics. *IEEE Trans. Syst. Man Cybern. Syst.* **2009**, *40*, 881–891.
36. Moisan, M.; Bernard, O. Robust interval observers for global Lipschitz uncertain chaotic systems. *Syst. Control Lett.* **2010**, *59*, 687–694. [CrossRef]
37. Aguila-Camacho, N.; Duarte-Mermoud, M.; Gallegos, J. Lyapunov functions for fractional order systems. *Commun. Nonlinear Sci. Numer. Simul.* **2014**, *19*, 2951–2957. [CrossRef]

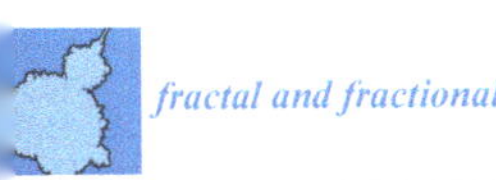
fractal and fractional

MDPI

Article

Design and Application of an Interval Estimator for Nonlinear Discrete-Time SEIR Epidemic Models

Awais Khan [1,2,3], Xiaoshan Bai [1,3], Muhammad Ilyas [4], Arshad Rauf [5], Wei Xie [6], Peiguang Yan [2] and Bo Zhang [1,3,*]

1   College of Mechatronics and Control Engineering, Shenzhen University, Shenzhen 518060, China; awaiskhan@szu.edu.cn (A.K.); baixiaoshan@szu.edu.cn (X.B.)
2   College of Physics and Optoelectronic Engineering, Shenzhen University, Shenzhen 518060, China; yanpg@szu.edu.cn
3   Shenzhen City Joint Laboratory of Autonomous Unmanned Systems and Intelligent Manipulation, Shenzhen University, Shenzhen 518060, China
4   Department of Electrical Engineering, Balochistan University of Engineering and Technology Khuzdar, Khuzdar 89100, Pakistan; m.ilyas@buetk.edu.pk
5   College of Automation Engineering, Nanjing University of Aeronautics and Astronautics, Nanjing 210016, China; arshad@nuaa.edu.cn
6   College of Automation Science and Technology, South China University of Technology, Guangzhou 510641, China; weixie@scut.edu.cn
*   Correspondence: zhangbo@szu.edu.cn

Citation: Khan, A.; Bai, X.; Ilyas, M.; Rauf, A.; Xie, W.; Yan, P.; Zhang, B. Design and Application of an Interval Estimator for Nonlinear Discrete-Time SEIR Epidemic Models. *Fractal Fract.* **2022**, *6*, 213. https://doi.org/10.3390/fractalfract6040213

Academic Editors: Thach Ngoc Dinh, Shyam Kamal, Rajesh Kumar Pandey and Norbert Herencsar

Received: 24 January 2022
Accepted: 7 April 2022
Published: 9 April 2022

Publisher's Note: MDPI stays neutral with regard to jurisdictional claims in published maps and institutional affiliations.

Abstract: This paper designs an interval estimator for a fourth-order nonlinear susceptible-exposed-infected-recovered (SEIR) model with disturbances using noisy counts of susceptible people provided by Public Health Services (PHS). Infectious diseases are considered the main cause of deaths among the top ten worldwide, as per the World Health Organization (WHO). Therefore, tracking and estimating the evolution of these diseases are important to make intervention strategies. We study a real case in which some uncertain variables such as model disturbances, uncertain input and output measurement noise are not exactly available but belong to an interval. Moreover, the uncertain transmission bound rate from the susceptible towards the exposed stage is not available for measurement. We designed an interval estimator using an observability matrix that generates a tight interval vector for the actual states of the SEIR model in a guaranteed way without computing the observer gain. As the developed approach is not dependent on observer gain, our method provides more freedom. The convergence of the width to a known value in finite time is investigated for the estimated state vector to prove the stability of the estimation error, significantly improving the accuracy for the proposed approach. Finally, simulation results demonstrate the satisfying performance of the proposed algorithm.

Keywords: interval analysis; interval estimator; finite-time convergence; bounded uncertainties; infectious diseases; SEIR epidemic model

1. Introduction

There were around 30.2–45.1 million people living with HIV with 680,000 casualties in 2020, whereas an epidemic like seasonal influenza causes 3–5 million serious illness cases with 250,000–500,000 casualties each year worldwide according to the WHO [1,2]. The surveillance of infectious diseases plays a vital role in analyzing these epidemics, for instance, origin, spread and dynamics. PHS relies upon surveillance statistics collected by agencies such as the Chinese Center for Disease Control and Prevention (China CDC) for infected people to estimate the activity level of such diseases, intervention strategy preparation, and recommendations of design policies.

Mathematical modeling of epidemics plays a major role in organizing public health responses and developing early outbreak detection systems [3–7]. The first modern mathematical epidemic model, i.e., susceptible-infectious-recovered (SIR), was proposed by

Kermack et al. in 1927 for cholera (London 1865) and plague epidemics (Bombay 1906 London 1665–1666) [8]. According to the SIR model, a fixed number of the population can be divided into three compartments at any time: susceptible individuals (not yet infected but can be infected in future), infectious individuals (those who have an infection and can infect others), and recovered individuals (who are recovered from the infection and are immune now). The number of people for each compartment represents the states of a SIR epidemic model. The total number of individuals who are assumed to be mixed homogenously remains the same, which means the probability of each individual coming in contact with others is equal [3].

However, the generic SIR epidemic model must be expanded to include a fourth compartment in case of many infectious diseases, for instance, influenza-like illness, tuberculosis, and HIV/AIDS [3,9,10]. The state of the fourth compartment corresponds to the latency period of disease, i.e., someone who is infected but still unable to infect others. This modified model is called the SEIR epidemic model [11]. Several estimation techniques have been developed to track and estimate the states of these models [3,12,13]. To design these estimators to converge to actual states, one needs to know the exact values of the uncertain quantities. However, designing such estimators for SEIR models in a real scenario is challenging, especially when the uncertain parameters are not exactly known but are defined by an interval or polytope. Interval estimator techniques can solve such issues [14–22]. Based on the monotone system theory (MST), interval estimators are designed to estimate the real states at any time instant and generate a set of acceptable values known as the interval in [20,23–28]. The ability to deal with large and unknown uncertainties in the system is one of the key advantages of interval state estimator design [29–31]. However, getting cooperative/nonnegative systems are not always possible, and solving this issue for interval estimator design is still an open area of research.

This article proposes an interval estimation-based method to track and estimate the four states of the SEIR epidemic model subject to uncertain parameters. Diaby et al., 2015 [32] proposed the first interval estimator for the continuous-time epidemic model. The results obtained by Diaby et al. were adequate but not ideal because the observer gain was manually set. Instead, we consider the discrete-time SEIR model and use an efficient method based on the observability matrix to design the interval estimator without observer gain. Finite-time convergence for the interval vectors' width is derived to verify the boundedness of the estimation error that significantly improves the accuracy of the designed method.

It is worth mentioning that the observer gain used in the conventional interval observers' design determines the magnitude of the upper bound of the interval estimation errors (for instance, see [27,28]). As a result, interval observers that converge faster may result in a state enclosure that is too conservative at steady state. The noted issue is solved in this study since we do not require an observer gain to run the proposed state estimator. In contrast, set-membership state estimators address the optimization problem at each iteration, and the problem of finite convergence time is ignored. Therefore, the proposed result on finite time convergence is intriguing. However, it is a little more demanding in terms of computation time. Furthermore, compared with the Kalman filter-type estimators, the proposed interval state estimator requires less information on state disturbances and measurement noise to generate guaranteed enclosures of the real state vector. This knowledge is advantageous when dealing with real-world situations when state disturbances and measurement noise are poorly known. More specifically, compared with the existing results in the literature, the contributions are fourfold:

1. We solve the interval estimation problem for the fourth-order SEIR epidemic model subject to disturbances and uncertainties. The estimation procedure is designed based on the observability matrix to relax the strong cooperativity assumption for designing traditional interval observers. Finite-time convergence and tight initialization problems are analyzed separately to improve the performance of the developed method;
2. In contrast to the existing interval observer design methods [19,33], we considered a nonlinear model with unknown input affecting the output with a highly uncertain

state matrix A. The bounds on the uncertain input are constructed before designing the interval estimator;

3. We introduce a novel interval state estimation method using an observability matrix and past input-output values without designing an observer gain that can alleviate some limitations of traditional interval observer design. For example, the system being cooperative/non-negative [20,34–39] and the probable inflation existence of the estimation error at the steady-state are avoided.

4. We consider the fourth compartment of the SEIR model by following the incubation stage compared with the interval estimator designed for the SIR model in [40]. Moreover, ref. [40] considered continuous-time dynamics, whereas we concentrate on the discrete time, which has grown in prominence through past years [41,42]. In addition, ref. [40] assumed that exact values of $\wp SI$ (new infectives per day) and the upper and lower bound of $\wp$ (transmission rate) are available. In our case, only noisy values of susceptible people S and probable bounds on uncertain $\wp SI$ are available, while the bounds on $\wp$ are not given by PHS. Hence, our method is more applicable in reality as $\wp$ is highly uncertain and cannot be obtained directly from biological consideration compared with [43,44]. Furthermore, its bounds are usually unavailable for such models [14].

The remainder of this work continues with notations and interval analysis in Section 2. The problem statement is described in Section 3, and the key findings are shown in Section 4. Two numerical examples are given in Section 5, where a comparison with previous results in [27,28] is demonstrated. Finally, concluding remarks are presented in Section 6.

2. Preliminaries Results

First, we reviewed some basic notations on interval estimation necessary to design the proposed state estimator.

2.1. Notations

A set of real numbers is symbolized by $\mathbb{R}$ with $\mathbb{R}_+ = \{\iota \in \mathbb{R} : \iota \geq 0\}$, whereas $\mathbb{Z}$ denotes a set of integers with $\mathbb{Z}_+ = \mathbb{Z} \cap \mathbb{R}_+$. The identity matrix of dimension n is denoted by $\mathbb{I}_n$. $\lambda_{\max}(\Delta)$ is the largest and $\lambda_{\min}(\Delta)$ is the smallest eigenvalue for a square matrix $\Delta \in \mathbb{R}^{n \times n}$. Let the L_2-induced matrix norm be $\|\Delta\|_2 = \sqrt{\lambda_{\max}(\Delta^T \Delta)}$, where the infinity norm is $\|\Delta\|_\infty = \max\limits_{1 \leq i \leq n} \Sigma_{j=1}^n |a_{ij}|$. Δ is non-negative ($\Delta > 0$) if $a_{ij} \geq 0$, whereas it is Schur stable if $|\lambda_i| < 1$ for all $i, j = 1, \ldots, n$. The relations $\Delta_1 \leq \Delta_2$ and $\nabla_1 \leq \nabla_2$ are understood elementwise for two matrices $\Delta_1, \Delta_2 \in \mathbb{R}^{n \times n}$ or vectors $\nabla_1, \nabla_2 \in \mathbb{R}^n$. For known $A \in \mathbb{R}^{m \times n}$, we define $A^+ = \{0, A\}$ and $A^- = \{0, -A\}$ with $|A| = A^+ + A^-$.

2.2. Interval Analysis

Uncertain parameters are defined by intervals that contain real values of unknown variables in a guaranteed way.

Definition 1. *An interval vector $[x]$ is determined by [45]*

$$[x] = [\underline{x}, \overline{x}] = \{a | \underline{x} \leq a \leq \overline{x}, \ \underline{x}, \overline{x} \in \mathbb{R}^n\}.$$

Lemma 1. *Let $x \in \mathbb{R}^n$ be an interval vector for some $\underline{x}, \overline{x} \in \mathbb{R}^n$ and $A \in \mathbb{R}^{m \times n}$. Then [29]*

$$A^+ \underline{x} - A^- \overline{x} \leq Ax \leq A^+ \overline{x} - A^- \underline{x}.$$

Definition 2. *The theory of monotone systems states that the solutions to the below system for given $x(0) \geq 0$ constructed by a matrix $A \in \mathbb{R}_+^n$ are non-negative,*

$$x(k+1) = Ax(k) + w(k),$$
$$x \in \mathbb{R}^n, \ w : \mathbb{Z}_+ \to \mathbb{R}_+^n, \ k \in \mathbb{Z}_+, \ k \geq 0$$

and the system is referred to as cooperative or non-negative [46].

3. Problem Statement

The SEIR discrete-time model demonstrated by Figure 1 and obtained using Euler discretization of the classical continuous-time model is given as follows [42,46]:

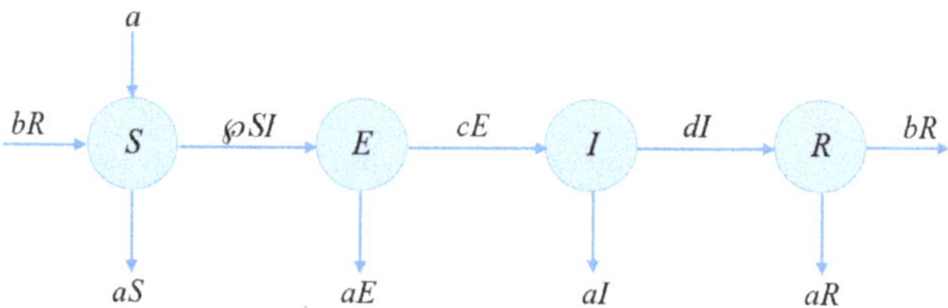

Figure 1. Block diagram for SEIR model.

$$
\begin{aligned}
S(k+1) &= (1-a(k))S(k) + b(k)R(k) - \wp(k)S(k)I(k) + a(k),\\
E(k+1) &= (1-a(k)-c(k))E(k) + \wp(k)S(k)I(k),\\
I(k+1) &= c(k)E(k) + (1-a(k)-d(k))I(k),\\
R(k+1) &= d(k)I(k) + (1-a(k)-b(k))R(k),
\end{aligned}
\tag{1}
$$

where $S(k)$, $E(k)$, $I(k)$, $and R(k)$ represent state variables corresponding to the portion of population in each compartment of the model. The time-varying non-negative parameters a stands for the natural birth-death rate, whereas b, c, d denote the uncertain transition rates from one disease state to the other. The exact values of non-negative parameters a, b, c, d are unknown. We only know the lower bound and upper bound values, i.e., $a \in [\underline{a},\ \bar{a}]$, $b \in [\underline{b},\ \bar{b}]$, $c \in [\underline{c},\ \bar{c}]$ and $d \in [\underline{d},\ \bar{d}]$ with given $\underline{a},\bar{a},\underline{b},\bar{b},\underline{c},\bar{c},\underline{d},\bar{d} \in \mathbb{R}_+$. The time-varying parameter $\wp(k)$ is extremely uncertain, and no bounds on $\wp(k)$ are available for measurements. The initial values for $x(k) \in \mathbb{R}^4$ are unknown but bounded with known bounds $\underline{x}(0), \bar{x}(0) \in \mathbb{R}^4$ such that $\underline{x}(0) \leq x(0) \leq \bar{x}(0)$. At at any given time instant k, the death rate is exactly equal to birth rate $a(k)$ in all the compartments. In fact, by summing up (1), one gets directly that the total population is constant, thus satisfying

$$
N(k+1) = N(k) = N_0 \quad \forall k \in \mathbb{Z}_{0+},
$$

for

$$
N(k) = S(k) + E(k) + I(k) + R(k), \quad \forall k \in \mathbb{Z}_{0+}.
$$

This results in

$$
S(k+1) + E(k+1) + I(k+1) + R(k+1) = (1-a(k))N_0 + a(k), \quad \forall k \in \mathbb{Z}_{0+}.
$$

Hence, if the total population is initially in unity, then (1) remains as a normalized model for all samples with the total population remaining in unity through time; therefore

$$
S(k+1) + E(k+1) + I(k+1) + R(k+1) = 1 - a(k) + a(k) = 1, \quad \forall k \in \mathbb{Z}_{0+}.
\tag{2}
$$

The transmission of disease arises because of the interactions among susceptible and infectious individuals as described by (1). The disease is transferred to $\wp(k)$ individuals through infectious individuals at each time instant. However, a new case only arises with probability $S(k)$ when contact is directly made with the susceptible individual. Therefore, in compartment S, a fraction $\wp(k)I(k)$ of people shift to exposed but non-infectious compartment E at time k. Similarly, a fraction c and d of individuals in compartments E and I migrate to the infectious I and recovered R compartments, respectively. It should be noted that the recovered compartment is composed of people not yet immune.

The output measured data consists of noisy counts of susceptible individuals obtained from different government sources such as census bureaus by PHS and are represented by the following output system model [3]:

$$y(k) = S(k) + v(k) \tag{3}$$

where $v(k) \in L_\infty$ stands for unknown measurement noise with known bounds $\underline{v}(k)$, $\overline{v}(k) \in L_\infty$ such that $\underline{v}(k) \leq v(k) \leq \overline{v}(k)$, $\forall k \geq 0$. The unknown measurement noise consists of the uncertain number of susceptible people who did not visit the health care unit for diagnosis. Therefore, Equations (1) and (3) are rewritten as follows:

$$\begin{aligned}
x(k+1) &= A(k)x(k) + E\Im(k) + w(k), \\
y(k) &= Cx(k) + v(k),
\end{aligned} \tag{4}$$

where $x(k) = [S(k)\ E(k)\ I(k)\ R(k)]^T$ and $\Im(k) = \wp(k)S(k)I(k)$ represent the unknown state vector to be determined and uncertain input, i.e., the newly confirmed infected people from the susceptible individuals at each time instant in the known population, respectively. The time-varying unknown matrix $A(k)$ and constant matrices E and C in (4) are given by

$$A(k) = \begin{bmatrix}
1-a & 0 & 0 & b \\
0 & 1-a-c & 0 & 0 \\
0 & c & 1-a-d & 0 \\
0 & 0 & d & 1-a-b
\end{bmatrix},$$

$$E = \begin{bmatrix} -1 \\ 1 \\ 0 \\ 0 \end{bmatrix}, \quad w(k) = \begin{bmatrix} a(k) \\ 0 \\ 0 \\ 0 \end{bmatrix}, \quad C = \begin{bmatrix} 1 \\ 0 \\ 0 \\ 0 \end{bmatrix}^T.$$

The uncertain unknown bounded matrix $w(k)$ for $\underline{w}, \overline{w} \in L_\infty$ is defined as

$$\underline{w} = \begin{bmatrix} \underline{a}(k) \\ 0 \\ 0 \\ 0 \end{bmatrix}, \quad \overline{w} = \begin{bmatrix} \overline{a}(k) \\ 0 \\ 0 \\ 0 \end{bmatrix},$$

such that $\underline{w} \leq w \leq \overline{w}$.

4. Interval Estimator Design for SEIR Model

We will design the interval state estimator in this section for the SEIR model (4). In the presence of uncertain parameters, the primary goal of this research is to construct an interval estimator for the SEIR model such that the unknown state signals always satisfy the following inequality:

$$\underline{x}(k) \leq x(k) \leq \overline{x}(k), \quad \forall k \geq 0, \tag{5}$$

where $\underline{x}(k), \overline{x}(k)$ represent the highest and lowest values for the interval state bounds provided that $x(0) \in [\underline{x}(0), \overline{x}(0)]$. The proposed interval estimator can help to make a deciding rule for pandemic detection. The following definition and assumption are required to design the proposed interval state estimator for the given SEIR model.

Proposition 1. *As the state $x(k_0) = x(0)$ is determined uniquely for all $u(\tau)$ and $y(\tau)$, $\tau \in [k_0, k_1]$, the SEIR epidemic model described by (4) is observable over $[k_0, k_1]$.*

Assumption 1. *There are known bounds $\underline{w}, \overline{w} \in \mathbb{R}^4$, $\underline{v}, \overline{v} \in \mathbb{R}$ such that $w(k) \in [\underline{w}, \overline{w}]$, $v(k) \in [\underline{v}, \overline{v}]$.*

The given proposition and assumption are necessary for designing the proposed interval estimator. The bounds given by Assumption 1 determine the uncertainty of initial values, input disturbance and noise.

4.1. Interval State Estimator Design

The observability matrix $\bigcirc \in \mathbb{R}^{4\times4}$ for (4) is given by

$$\bigcirc = \begin{bmatrix} C \\ CA(k) \\ CA(k+1)A(k) \\ CA(k+2)A(k+1)A(k) \end{bmatrix}$$

Then, the SEIR model (4) can be written in the input/output form in the absence of uncertain quantities and exogenous signals as follows:

$$x(k+4) = A(k+3)A(k+2)A(k+1)A(k)x(k),$$

$$\begin{bmatrix} y(k) \\ y(k+1) \\ y(k+2) \\ y(k+3) \end{bmatrix} = \Psi(k:k+3) = \bigcirc x(k). \tag{6}$$

As a result, using available input-output values, (6) can be written as follows:

$$x(k) = \bigcirc^{-1}\Psi(k:k+3) \tag{7}$$

Hence, the states of the SEIR model (4) can be obtained using the available input/output values for $k - 3 \geq 0$ by

$$\hat{x}(k) = \Delta_y(k)\Psi(k-3:k), \tag{8}$$

with

$$\Delta_y(k) = A(k-1)A(k-2)A(k-3)\bigcirc^{-1}.$$

Then, the equation of our interval state estimator for the SEIR model (4) that generates certain bounds on the real states for $k - 3 \geq 0$ subject to exogenous signals takes the following form:

$$\overline{x}(k) = \hat{x}(k) + \overline{D} + \overline{\Lambda}(k) + \overline{V},$$
$$\underline{x}(k) = \hat{x}(k) + \underline{D} + \underline{\Lambda}(k) + \underline{V}, \tag{9}$$

where $\overline{D}, \underline{D} \in \mathbb{R}^{4\times1}, \overline{\Lambda}(k), \underline{\Lambda}(k) \in \mathbb{R}^{4\times1}$ and $\overline{V}, \underline{V} \in \mathbb{R}^{4\times1}$ denote the upper and lower limits on uncertain birth and death rate, uncertain input, and measurement noise, respectively. We will define these terms one by one using Lemma 1 as follows.

4.1.1. Bounds on the Uncertain Birth and Death Rate

The bounds of the unknown birth and death rate are given by

$$\overline{D} = \phi^+(k)\overline{w}_{n-1} + \phi^-(k)\underline{w}_{n-1}$$
$$\underline{D} = \phi^+(k)\underline{w}_{n-1} + \phi^-(k)\overline{w}_{n-1} \tag{10}$$

where $\overline{w}_{n-1} \in \mathbb{R}^{12\times1}$ and $\underline{w}_{n-1} \in \mathbb{R}^{12\times1}$ denote $n-1$ bound concatenation on the uncertain birth and death rate defined by

$$\overline{w}_{n-1} = \begin{bmatrix} \overline{w} & \overline{w} & \overline{w} \end{bmatrix}^T, \quad \underline{w}_{n-1} = \begin{bmatrix} \underline{w} & \underline{w} & \underline{w} \end{bmatrix}^T,$$

with

$$\overline{w} = \begin{bmatrix} \overline{a}(k) & 0 & 0 & 0 \end{bmatrix}^T, \quad \underline{w} = \begin{bmatrix} \underline{a}(k) & 0 & 0 & 0 \end{bmatrix}^T.$$

Furthermore,

$$\phi^+(k) = \max\{0,\ \phi(k)\},\ \phi^+(k) = \max\{0,\ -\phi(k)\},\ \phi(k) = \Sigma_A - \Delta_y(k)\Sigma_{CA},$$

where

$$\Sigma_A = \begin{bmatrix} \zeta_{11} & 0 & \zeta_{13} & \zeta_{14} & \zeta_{15} & 0 & 0 & \zeta_{18} \\ 0 & \zeta_{22} & 0 & 0 & 0 & \zeta_{26} & 0 & 0 \\ 0 & \zeta_{32} & \zeta_{33} & 0 & 0 & \zeta_{36} & \zeta_{37} & 0 \\ 0 & \zeta_{42} & \zeta_{43} & \zeta_{44} & 0 & 0 & \zeta_{47} & \zeta_{48} \end{bmatrix} \mathbb{I}_4, \tag{11a}$$

and

$$\zeta_{11} = (1 - a(k_1))(1 - a(k_2)),$$
$$\zeta_{13} = b(k_1)d(k_2),$$
$$\zeta_{14} = b(k_2)(1 - a(k_1)) + b(k_1)(1 - a(k_2) - b(k_2)),$$
$$\zeta_{15} = (1 - a(k_2)),$$
$$\zeta_{18} = b(k_2),$$
$$\zeta_{22} = (1 - a(k_1) - c(k_1))(1 - a(k_2) - c(k_2)),$$
$$\zeta_{26} = (1 - a(k_2) - c(k_2)),$$
$$\zeta_{32} = c(k_2)(1 - a(k_1) - c(k_1)) + c(k_1)(1 - a(k_2) - c(k_2)),$$
$$\zeta_{33} = (1 - a(k_1) - d(k_1))(1 - a(k_2) - d(k_2)),$$
$$\zeta_{36} = c(k_2),$$
$$\zeta_{37} = (1 - a(k_2) - d(k_2)),$$
$$\zeta_{42} = d(k_1)c(k_2),$$
$$\zeta_{43} = d(k_2)(1 - a(k_1) - d(k_1)) + d(k_1)(1 - a(k_2) - d(k_2)),$$
$$\zeta_{44} = ((1 - a(k_1) - b(k_1))((1 - a(k_2) - b(k_2))),$$
$$\zeta_{47} = d(k_2),$$
$$\zeta_{48} = (1 - a(k_2) - b(k_2)),$$

$$\Sigma_{CA} = \begin{bmatrix} 0_{4\times1} & 0_{4\times1} & 0_{4\times1} \\ C & 0_{4\times1} & 0_{4\times1} \\ CA(k_1) & C & 0_{4\times1} \\ CA(k_1)A(k_2) & CA(k_1) & C \end{bmatrix},$$

$$= \begin{bmatrix} 0 & 0 & 0 & 0 \\ 1 & 0 & 0 & 0 \\ 1 - a(k_1) & 0 & 0 & b(k_1) \\ \{1 - a(k_1)\}\{1 - a(k_2)\} & 0 & \begin{matrix} b(k_1) \\ d(k_1) \end{matrix} & \begin{matrix} -b(k_1)b(k_2)+ \\ b(k_1)\{1 - a(k_1)\}+ \\ b(k_2)\{1 - a(k_1)\} \end{matrix} \\ 0 & 0\ 0 & 0 & 0\ 0\ 0\ 0 \\ 0 & 0\ 0 & 0 & 0\ 0\ 0\ 0 \\ 1 & 0\ 0 & 0 & 0\ 0\ 0\ 0 \\ 1 - a(k_1) & 0\ 0 & b(k_1) & 1\ 0\ 0\ 0 \end{bmatrix}, \tag{11b}$$

with $k_1 = k - 1$, $k_2 = k - 2$.

4.1.2. Bounds on Uncertain Input

The bounds on the uncertain input are represented as

$$\overline{\Lambda}(k) = (\phi E_{n-1})^+\overline{\mathfrak{I}}(k) - (\phi E_{n-1})^-\underline{\mathfrak{I}}(k),$$
$$\underline{\Lambda}(k) = (\phi E_{n-1})^+\underline{\mathfrak{I}}(k) - (\phi E_{n-1})^-\overline{\mathfrak{I}}(k). \tag{12}$$

Using (3) and (4), one can write

$$y(k+1) = Cx(k+1) + v(k+1),$$
$$= CA(k)x(k) + CE\Im(k) + Cw(k) + v(k+1),$$
$$-CE\Im(k) = CA(k)x(k) + Cw(k) - y(k+1) + v(k+1).$$

Moreover, from (1), (3) and (4), we derive that

$$CE = -1, \; Cw(k) = a(k).$$

This results in

$$\Im(k) = CA(k)x(k) + a(k) - y(k+1) + v(k+1),$$

with

$$\overline{\Im}(k) = (CA)^+\overline{x}(k) - (CA)^-\underline{x}(k) + |-y(k+1)| + \overline{v} + \overline{a}$$
$$\underline{\Im}(k) = (CA)^+\underline{x}(k) - (CA)^-\overline{x}(k) + |-y(k+1)| - \overline{v} + \underline{a} \; ,$$

such that $[\Im(k)] = [\underline{\Im}(k), \; \overline{\Im}(k)]$ and $E_{n-1} = [E \quad E \quad E]^T$.

4.1.3. Bounds on Measurement Noise Vector

The bounds for the measurement noise vector are described by

$$\overline{V} = \varphi_v^+(k)\overline{v}_n + \varphi_v^-(k)\underline{v}_n,$$
$$\underline{V} = \varphi_v^+(k)\underline{v}_n + \varphi_v^-(k)\overline{v}_n, \tag{13}$$

where $\overline{v}_n, \; \underline{v}_n \in \mathbb{R}^{4\times1}$, respectively, denote the n concatenation of $\overline{v} \in \mathbb{R}$ and $\underline{v} \in \mathbb{R}$, given by

$$\overline{v}_n = \begin{bmatrix} \overline{v} \\ \overline{v} \\ \overline{v} \\ \overline{v} \end{bmatrix}, \quad \underline{v}_n = \begin{bmatrix} \underline{v} \\ \underline{v} \\ \underline{v} \\ \underline{v} \end{bmatrix},$$

and

$$\varphi_v^+(k) = \max\{0, \; \varphi_v(k)\}, \; \phi_v^-(k) = \max\{0, \; -\varphi_v(k)\}$$

with

$$\varphi_v(k) = -\Delta_y(k).$$

Theorem 1. *When Assumption 1 is satisfied for the given SEIR model (4), the interval state estimator given by (9) yields the following relations:*

$$\underline{x}(k) \le x(k) \le \overline{x}(k), \quad \forall k \ge 3, \tag{14}$$

provided that $\underline{x}(0) \le x(0) \le \overline{x}(0)$.

Proof of Theorem 1. The solution to the SEIR model (4) at any time instant k for $x(0) \in [\underline{x}(0), \overline{x}(0)]$ and $w(k) \in [\underline{w}, \overline{w}]$ can be obtained as

$$x(k) = \prod_{\ell=1}^{k} A(k-\ell)x(0) + E\Im(k-1) + w(k-1) + \sum_{m=0}^{k-2}\left\{ \prod_{\ell=1}^{k-m-1} A(k-\ell) \right\}\{E\Im(m) + w(m)\}. \tag{15}$$

The given SEIR model is a 4th order system i.e., $n = 4$. Therefore, the states $x(k)$ can be determined at any time k using previous state values at $k - 3$ as follows:

$$x(k) = A(k-1)A(k-2)A(k-3)x(k-3) + A(k-1)A(k-2)\{E\Im(k-3) + w(k-3)\}$$
$$+A(k-1)\{E\Im(k-2) + w(k-2)\} + E\Im(k-1) + w(k-1). \tag{16}$$

Now using theory of interval analysis [39] for all $x(k-3) \in [x(k-3)]$, $\Im(\vartheta) \in [\Im(\vartheta)]$ and $w(\vartheta) \in [w(\vartheta)]$ with $\vartheta \in \{k-3,\ k-2,\ k-1\}$, the state vector $x(k) \in [x(k)]$ is given by

$$[x(k)] = A(k-1)A(k-2)A(k-3)[x(k-3)] + \Sigma_A\{E_{n-1}[\Im(k)] + [w_{n-1}]\}, \tag{17}$$

As a result, utilizing the past input/output values and the observability matrix, the following set inversion formula is obtained to get the state enclosure $[x(k-3)]$:

$$[x(k-3)] = \bigcirc^{-1}\{[\Psi(k-3:k)] - \Sigma_{CA}E_{n-1}[\Im(k)]\} + \bigcirc^{-1}\Sigma_{CA}[w_{n-1}], \tag{18}$$

where

$$[\Psi(k-3:k)] = \begin{bmatrix} y(k-3) \\ y(k-2) \\ y(k-1) \\ y(k) \end{bmatrix} - [v_n].$$

Consequently, by combing (17) and (18), one gets

$$\begin{aligned} [x(k)] &= A(k-1)A(k-2)A(k-3)\bigcirc^{-1}([\Psi(k-3:k)] - \Sigma_{CA}\{E_{n-1}[\Im(k)] + [w_{n-1}]\}) \\ &\quad + \Sigma_A\{E_{n-1}[\Im(k)] + [w_{n-1}]\}, \end{aligned} \tag{19}$$

$$\begin{aligned} [x(k)] &= \Delta_y(k)\{[\Psi(k-3:k)] - [v_n]\} - \Delta_y(k)\Sigma_{CA}\{E_{n-1}[\Im(k)] + [w_{n-1}]\} \\ &\quad + \Sigma_A\{E_{n-1}[\Im(k)] + [w_{n-1}]\}, \end{aligned} \tag{20}$$

$$[x(k)] = \Delta_y(k)[\Psi(k-3:k)] - \Delta_y(k)[v_n] + (\Sigma_A - \Delta_y(k)\Sigma_{CA})\{E_{n-1}[\Im(k)] + [w_{n-1}]\}, \tag{21}$$

$$[x(k)] = \Delta_y(k)[\Psi(k-3:k)] + \phi(k)E_{n-1}[\Im(k)] + \phi(k)[w_{n-1}] + \varphi_v(k)[v_n], \tag{22}$$

$$[x(k)] = \hat{x}(k) + [\Lambda] + [D] + [V], \tag{23}$$

where $[x(k)] = [\underline{x},\ \overline{x}]$, $[\Lambda] = [\underline{\Lambda},\ \overline{\Lambda}]$, $[D] = [\underline{D},\ \overline{D}]$ and $[V] = [\underline{V},\ \overline{V}]$. This completes the proof of Theorem 1. $\square$

4.2. Interval Prediction for $k < 3$

It is worth mentioning that to use the developed interval state estimator (9) for the given SEIR model (4), the initial $n - 1 = 3$ values of input-output should be accessible. However, in many real life scenarios, these values are not always available for measurements, such as in the given case, where only the initial values are given with some known bounds. Therefore, we proposed the following recursive system as an interval predictor that provides a bound on the system's states for $k = 0, 1, 2$.

Proposition 2. *The following interval predictor generates $[x(k)]$ such that for $k = 0, 1, 2$, we have $x(k) \in [x(k)]$ as*

$$\begin{aligned} [x(k)] &= \prod_{\ell=1}^{k} A(k-\ell)[x(0)] + [\partial(k-1)], \\ [\partial(k)] &= A(k)[\partial(k-1)] + E[\Im] + [w], \\ [\partial(0)] &= E[\Im] + [w]. \end{aligned} \tag{24}$$

Proof. To prove Proposition 2, we use mathematical induction. As for the initial case $k = 0$, the input and output are available. Therefore, we consider the case $k = 1$. Hence, the first cycle of the SEIR model produces the following equations:

$$\begin{aligned} x(1) &= A(0)x(0) + E\Im(0) + w(0), \\ [x(1)] &\in A(0)[x(0)] + E[\Im] + [w], \\ [x(1)] &\in A(0)[x(0)] + [\partial(0)], \end{aligned} \tag{25}$$

which implies the correctness of (24) for $k = 1$. Next, we demonstrate that (24) is true for $k = 2$. Once again, considering (4) for $k = 2$, we get

$$x(2) = \prod_{\ell=1}^{2} A(2-\ell)x(0) + E\Im(1) + w(1) + \sum_{m=0}^{1}\left\{\prod_{\ell=1}^{2} A(2-\ell)\right\}\{E\Im(m) + w(m)\},$$
$$[x(2)] \in \prod_{\ell=1}^{2} A(2-\ell)[x(0)] + A(1)[\partial(0)] + E[\Im] + [w], \tag{26}$$

$$[x(2)] \in \prod_{\ell=1}^{2} A(2-\ell)[x(0)] + [\partial(1)]. \tag{27}$$

Thus, by simple mathematical induction, one can easily show that (27) is true for $k = 2$ as well. This completes the proof of Proposition 2. $\square$

4.3. Finite-Time Convergence

The finite-time convergence of the interval width $\mho[x(k)]$ is one of the primary issues concerning the tight initialization and stability of the interval estimator. Therefore, this section proves that $\mho[x(k)]$ converges to a known upper-bounded value in finite time provided by the uncertain quantities. Hence, we introduce the following Lemma to compute the upper bound on $\mho[x(k)]$.

Lemma 2. *The following inequality determines the upper bound on the width of the interval vector provided by (9) and (24):*

$$\mho[x(k)] \leq \|\phi(k)E_{n-1}\|_{\infty}\mho[\Im] + \|\phi(k)\|_{\infty}\mho[w] + \|\varphi_v(k)\|_{\infty}\mho[v_n], \quad \forall k \geq 3. \tag{28}$$

Proof. Firstly, the recursive system (24) is employed as an interval predictor during the initialization phase for $k = 0, 1, 2$ to provide tight bounds on the interval vector of the SEIR model (4). As a result, the upper limit on the width of interval vector provided by (24) is given by

$$\mho[x(k)] \leq \left\|\prod_{\ell=1}^{k} A(k-\ell)\right\|_{\infty} \mho[x(0)] + \mho[\partial(k-1)]. \tag{29}$$

Secondly, the proposed interval state estimator (9) is used for $k \geq 3$. Then, Equation (23) implies

$$\mho[x(k)] \leq \|\phi(k)E_{n-1}\|_{\infty}\mho[\Im] + \|\phi(k)\|_{\infty}\mho[w_{n-1}] + \|\varphi_v(k)\|_{\infty}\mho[v_n], \tag{30}$$

whereas

$$\|\phi(k)E_{n-1}\|_{\infty}\mho[\Im] + \|\phi(k)\|_{\infty}\mho[w_{n-1}] + \|\varphi_v(k)\|_{\infty}\mho[v_n]$$
$$\leq \|\phi(k)E_{n-1}\|_{\infty}\mho[\Im] + \|\phi(k)\|_{\infty}\mho[w] + \|\varphi_v(k)\|_{\infty}\mho[v_n]. \tag{31}$$

Based on (30) and (31), one can easily determine that

$$\mho[x(k)] \leq \|\phi(k)E_{n-1}\|_{\infty}\mho[\Im] + \|\phi(k)\|_{\infty}\mho[w] + \|\varphi_v(k)\|_{\infty}\mho[v_n]. \tag{32}$$

This completes the proof of Lemma 2. $\square$

Remark 1. *It is worth noting that we do not need observer gain to design the interval estimator. Therefore, the impact of gain that leads to pessimistic state enclosures for the traditional-type interval observer design method [27,28] is avoided. However, it is more demanding in term of computation time. In addition, unlike Kalman filter-type state estimators, the exact values of exogenous signals are not necessarily known and hence represent an advantage while dealing with practical applications.*

5. Simulation Results

In this section, two simulated examples are used to demonstrate the efficiency of the designed interval state estimator compared with the traditional interval observer design methods [27,28]. Example 1 is a general numerical type, whereas Example 2 is based on the 2014 West African Ebola virus outbreak [47].

5.1. Example 1

Consider the SEIR epidemic model (4) with the following parameters defining the time-varying matrix $A(k)$ as

$$a(k) = a(0) + 0.05\sin(0.5\pi k)$$
$$b(k) = b(0) + 0.02\sin(0.5\pi k)$$
$$c(k) = c(0) + 0.1\sin^2(0.25\pi k)$$
$$d(k) = d(0) + 0.1\sin(0.25\pi k)$$

with $a(0) = 0.4/\text{month}$, $b(0) = 0.124/\text{month}$, $c(0) = 0.2/\text{month}$, $d(0) = 0.45/\text{month}$ and degree of seasonality $\eta = 0.4$.

The uncertain input parameter is $\wp(k) = 0.5(1 + \eta\cos(0.25k))$, while the bounded disturbance and output measurement noise are: $w(k) \in [\underline{w}(k)\ \overline{w}(k)]$ for $\underline{w}(k) = [0.35\ 0\ 0\ 0]^T$, $\overline{w}(k) = [0.45\ 0\ 0\ 0]^T$ and $v(k) = V\cos(0.25\pi k)$ with $V = -0.00001$.

The two matrices Σ_A and Σ_{CA} are, respectively, obtained using (12a) and (12b), and bounds on the uncertain quantities are calculated by (10), (11) and (13). The observability matrix for $\kappa_j = (k+j)$; $j = 0, 1, 2$ is computed as follows:

$$\bigcirc^{-1} = \begin{bmatrix} 1 & 0 & 0 & 0 \\[4pt] 1-a_{\kappa_0} & 0 & 0 & b_{(\kappa_0)} \\[4pt] \dfrac{(1-a_{\kappa_1})}{(1-a_{\kappa_0})} & 0 & b_{\kappa_1}d_{\kappa_0} & \begin{array}{l}(1-a_{(\kappa_1)})b_{(\kappa_0)}\\ +b_{(\kappa_1)}(1-a_{(\kappa_0)}-b_{(\kappa_0)})\end{array} \\[12pt] \begin{array}{l}(1-a_{\kappa_2})\\(1-a_{\kappa_1})\\(1-a_{\kappa_0})\end{array} & b_{\kappa_1}c_{\kappa_0}d_{\kappa_2} & \begin{array}{l}b_{\kappa_1}d_{\kappa_2}(1-a_{\kappa_0}-d_{\kappa_0})\\ +d_{\kappa_0}\{b_{\kappa_1}(1-a_{\kappa_2})\\ +b_{\kappa_2}(1-a_{\kappa_1}-b_{\kappa_1})\end{array} & \begin{array}{l}b_{\kappa_0}(1-a_{\kappa_2})(1-a_{\kappa_1})\\ +(1-a_{\kappa_0}-b_{\kappa_0})\\ \{b_{\kappa_1}(1-a_{\kappa_2})+b_{\kappa_2}(1-a_{\kappa_1}-b_{\kappa_1})\}\end{array} \end{bmatrix}$$

It should be noted that our model is of order four, i.e., $n = 4$. Therefore, we need to know the first two state intervals to implement the given interval estimator (9) for $k \geq 3$. Hence, the interval predictor (24) is used for $k = 1, 2$ provided that the initial values $\underline{x}(0) \leq x(0) \leq \overline{x}(0)$ are satisfied. It is worth mentioning that the given model does not have to be non-negative for the proposed interval estimator to operate.

Simulation experiments of the proposed method and the one in [28] are conducted to show the efficiency of the given approach. As perceived, the actual states are confined inside the two boundaries generated by (24) and (9). Figure 2 depicts the evolution of the actual states x_s, $s = 1, 2, 3, 4$, the estimated bounds by the proposed method (solid pink lines), and the estimated bounds by MST (blue dashed line) [28]. The figure shows that the developed approach estimates tighter bounds than those calculated using the method described in [28]. Furthermore, in regards to the design perspective, the observer gain matrix in [28] needs to be Schur and non-negative, while we do not need an observer gain to design the interval estimator.

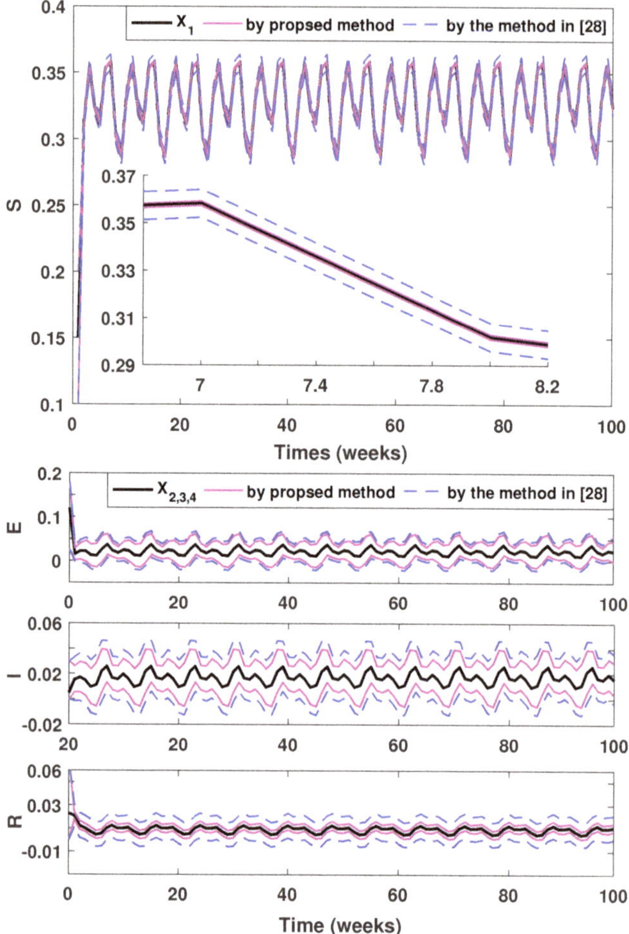

Figure 2. Interval estimations by proposed method vs. given in [28] for each state variable x_1, x_2, x_3, x_4 corresponding to S, E, I, R.

Secondly, the comparison of the interval state estimation errors e_S, e_E, e_I and e_R reflected in Figures 3 and 4 further clarify that the estimated bounds generated by the proposed method are more accurate and precise compared with [28]. Finally, Figure 5 shows the convergence of the interval widths given by (28). After three steps, the interval widths converge to their final values, proving the finite-time convergence performance of the proposed technique. Thus, it is concluded that the proposed method has a better performance.

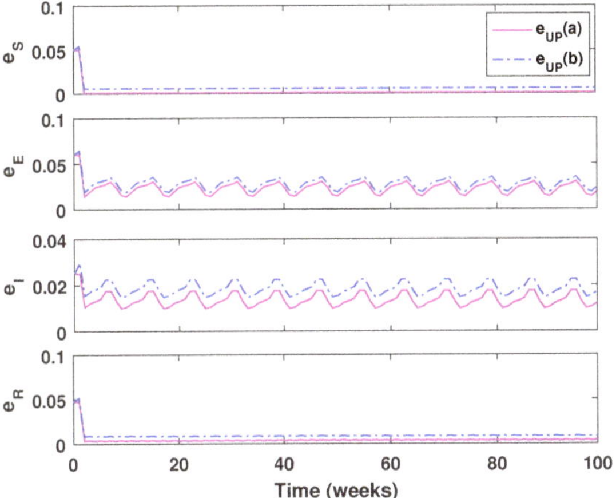

Figure 3. Upper-bound error: (a) proposed method; (b) method given in [28].

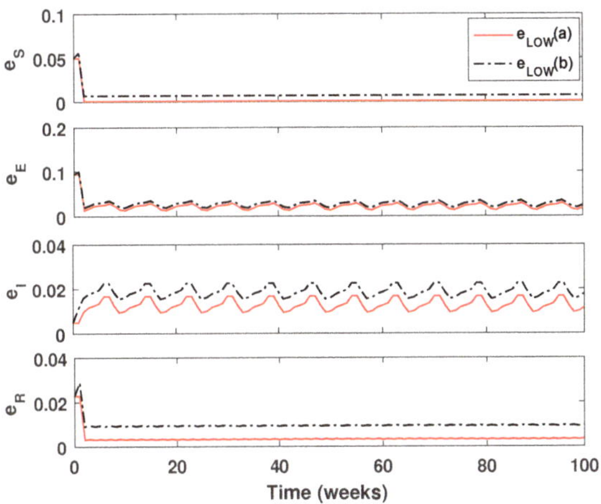

Figure 4. Lower-bound error: (a) proposed method; (b) method given in [28].

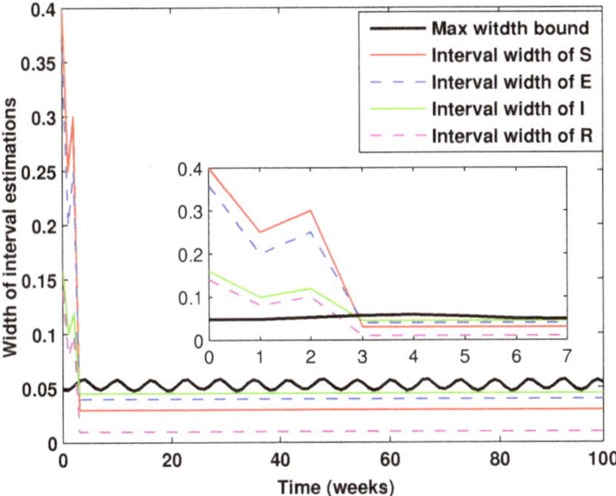

Figure 5. Finite-time convergence of S, E, I and R after 3rd iteration.

5.2. Example 2: Ebola Outbreak in West Africa

The 2014 West Africa Ebola outbreak [47] is considered using the following parameters to demonstrate the efficiency of the suggested technique having a period of one day as

$$a = 0.0099/day, \ b = 0.00128/day, \ c = 0.1887/day,$$
$$d = 0.1/day, \ \wp = 0.4/day.$$

Using these parameters, we get matrix A, as follows:

$$A = \begin{bmatrix} 0.9901 & 0 & 0 & 0.00128 \\ 0 & 0.8014 & 0 & 0 \\ 0 & 0.1887 & 0.8901 & 0 \\ 0 & 0 & 0.1 & 0.9888 \end{bmatrix}.$$

The output measurement noise $v(k)$ is uniformly distributed with given bounds
$$-V \le v(k) \le V; \quad V = 0.001.$$

We assumed that $\Im(k) = \wp S(y - v)$ is unknown but bounded with the following constraints:

$$\Im(k) - \tau = \underline{\Im}(k) \le \Im(k) \le \overline{\Im}(k) = \Im(k) + \tau; \quad \tau = 0.0001.$$

The observability matrix is obtained using A and C as

$$O = \begin{bmatrix} 1 & 0 & 1 & 0 \\ 0.9991 & 0.1887 & 0.8991 & 0.0013 \\ 0.9982 & 0.3226 & 0.8085 & 0.0026 \\ 0.9973 & 0.4140 & 0.7272 & 0.0038 \end{bmatrix}.$$

The disturbance matrix to obtain bounds on the uncertain input, and uncertain birth and death rate is computed as
$\phi(k) = \Sigma_A - \Delta_y(k)\Sigma_{CA}$ with $\Delta_y(k) = A^3 O^{-1}$ and

$$\Sigma_A = \begin{bmatrix} 0.9696 & 0 & 0.0001 & 0.0024 & 0.96 & 0 & 0 & 0 & 1 & 0 & 0 & 0 \\ 0 & 0.5016 & 0 & 0 & 0 & 0.66 & 0.66 & 0 & 0 & 1 & 0 & 0 \\ 0 & 0.4348 & 0.7638 & 0 & 0 & 0.3 & 0.8493 & 0 & 0 & 0 & 1 & 0 \\ 0 & 0.0332 & 0.2057 & 0.9672 & 0 & 0 & 0.1107 & 0.9588 & 0 & 0 & 0 & 1 \end{bmatrix},$$

$$\Sigma_{CA} = \begin{bmatrix} 0 & 0 & 0 & 0 & 0 & 0 & 0 & 0 & 0 & 0 & 0 & 0 \\ 1 & 0 & 0 & 0 & 0 & 0 & 0 & 0 & 0 & 0 & 0 & 0 \\ 1.01 & 0 & 0 & 0.0012 & 1 & 0 & 0 & 0 & 0 & 0 & 0 & 0 \\ 0.9696 & 0 & 0.0001 & 0.0024 & 1.01 & 0 & 0 & 0.0012 & 1 & 0 & 1 & 0 \end{bmatrix}.$$

Similarly, the measurement noise matrix is calculated as $\varphi_v(k) = -\Delta_y(k)$. The initial unknown bounded states are part of the interval $\underline{x}(0) \le x(0) \le \overline{x}(0)$ with $\underline{x}(0) = [0.88\ 0.06\ 0.049\ 0]$ and $\overline{x}(0) = [0.93\ 0.08\ 0.052\ 0.05]$.

The bounds on the uncertain input, uncertain birth-death rate and measurement noise are obtained using (10), (11) and (13), respectively. For the proposed SEIR model (4), we have $n = 4$; therefore, interval predictor (24) is used for $k = 1, 2$ whereas (9) is used for $k > 2$ to obtain guaranteed bounds on $x(k)$ provided that $\underline{x}(0) \le x(0) \le \overline{x}(0)$.

The simulation results of the proposed method and the one in [27] are depicted in Figure 6 to compare the observers' dynamics. As shown in Figure 6, the bounds generated by the developed method are tighter than those resulting from the work of Degue et al. [27]. In addition, the proposed method is easy to implement compared with [27] as it does not need observer gain and nonnegativity of the system dynamics to design the interval estimator. Moreover, Figures 7 and 8 show that the interval estimation errors $e_i^- = \underline{x}_r - x_r$,

$e_i^+ = \overline{x}_r - x_r$ for $r = 1, 2, 3, 4$; $i = S, E, I, R$, respectively, provided by our work are much smaller compared with [27].

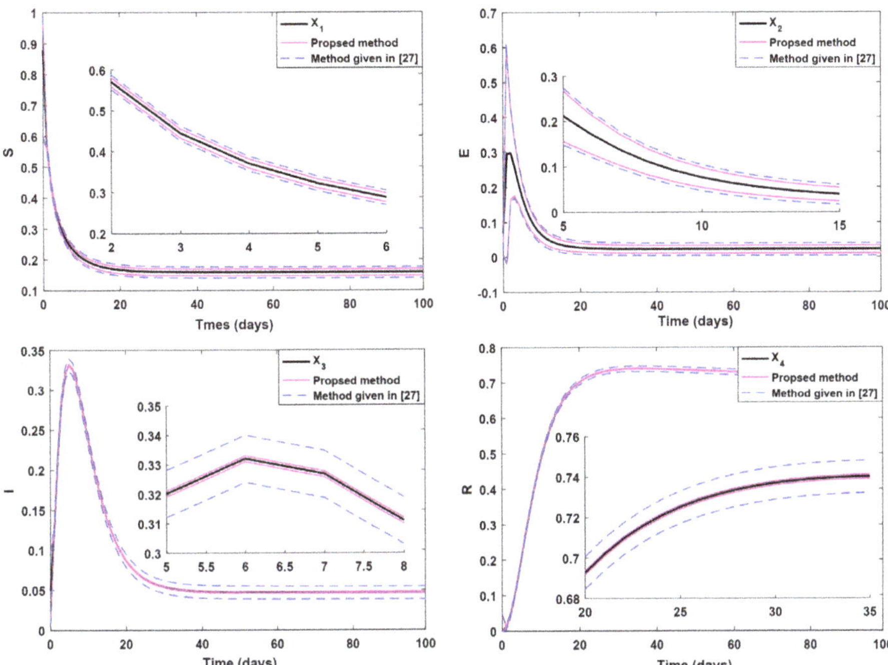

Figure 6. Interval estimations by proposed method vs. the method given in [27] for each state variable x_1, x_2, x_3, x_4 corresponding to S, E, I, R.

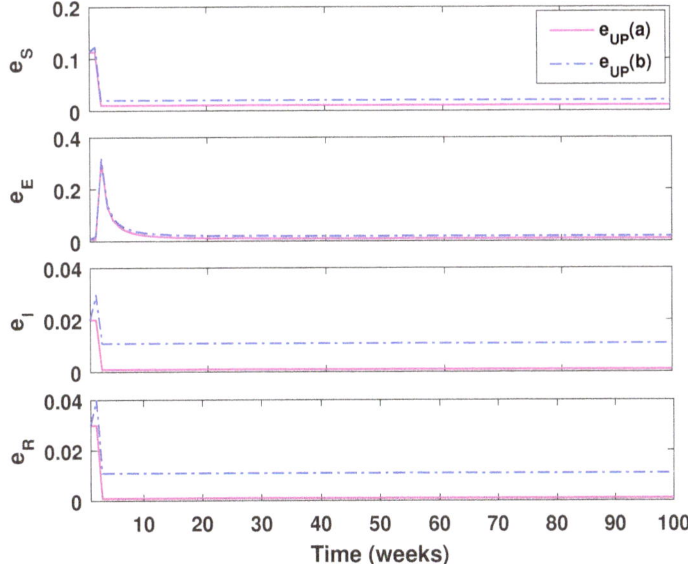

Figure 7. Upper-bound error: (a) proposed method; (b) method given in [27].

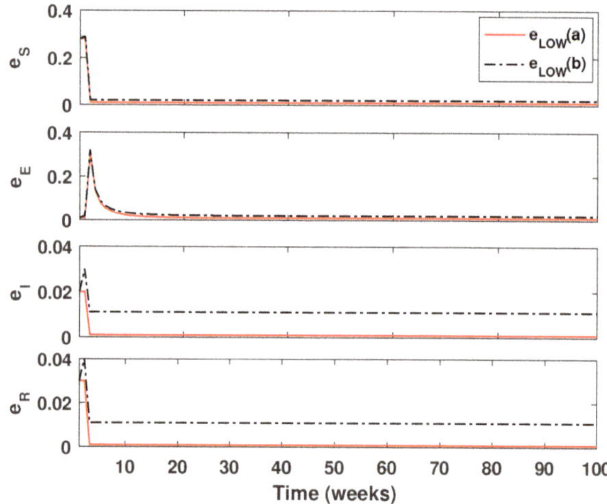

Figure 8. Lower-bound error: (a) proposed method; (b) method given in [27].

6. Conclusions

We developed a new strategy to design an interval state estimator for a fourth-order nonlinear discrete-time SEIR epidemic model subject to uncertain input, disturbances and measurement noise. The proposed method only requires the bounded values instead of exact values for state disturbance, unknown input, and parameters. In addition, no bounds on the time-varying transmission rate (susceptible to the infected stage) are required. The MST is widely used to design such an observer, but obtaining a non-negative model is not always feasible. Therefore, the proposed interval state estimator relaxes such restrictions by estimating the four compartment states using the observability matrix instead of point-wise estimation. The finite-time convergence of the interval width for the proposed approach is investigated to demonstrate its stability and performance. In addition, the interval widths' upper bound is estimated a priori. Finally, two numerical simulations are conducted to test the performance of the designed method. It is concluded that the proposed interval estimator generates more accurate boundaries and performs better. However, the proposed technique can currently be applied in linear and nonlinear discrete-time models. The interval estimator design approach for continuous-time systems will be investigated in the future.

Author Contributions: Conceptualization, A.K.; methodology, A.K.; software, A.K.; validation, A.K., X.B.; formal analysis, A.K., B.Z.; investigation, A.K., B.Z., P.Y.; writing—original draft preparation, A.K., X.B. and M.I.; writing—review and editing, A.K., A.R. and W.X.; supervision, B.Z. All authors have read and agreed to the published version of the manuscript.

Funding: This work was supported in part by the National Natural Science Foundation of China (NSFC) under the grant 62173234 and 62003217.

Institutional Review Board Statement: Not applicable.

Informed Consent Statement: Not applicable.

Data Availability Statement: Not applicable.

Conflicts of Interest: The authors declare no conflict of interest.

References

1. World Health Organization (WHO). *HIV/AIDS*; Global Health Observatory (GHO) Data; WHO: Geneva, Switzerland, 2020. Available online: https://www.who.int/data/gho/data/themes/hiv-aids (accessed on 6 April 2022).
2. World Health Organization. *Seasonal Influenza*; Global Health Observatory (GHO) Data; WHO: Geneva, Switzerland, 2020. Available online: https://www.who.int/news-room/fact-sheets/detail/influenza-(seasonal) (accessed on 6 April 2022).

1. Dukic, V.; Lopes, H.F.; Polson, N.G. Tracking epidemics with Google flu trends data and a state-space SEIR model. *J. Am. Stat. Assoc.* **2012**, *107*, 1410–1426. [CrossRef]
2. Fallas-Monge, J.J.; Chavarria-Molina, J.; Alpizar-Brenes, G. Combinatorial metaheuristics applied to infectious disease models. In Proceedings of the 2016 IEEE 36th Central American and Panama Convention (CONCAPAN XXXVI), San Jose, Costa Rica, 9–11 November 2016; pp. 1–6.
3. Grassly, N.C.; Fraser, C. Mathematical models of infectious disease transmission. *Nat. Rev. Microbiol.* **2008**, *6*, 477–487. [CrossRef] [PubMed]
4. Kaplan, E.H.; Craft, D.L.; Wein, L.M. Emergency response to a smallpox attack: The case for mass vaccination. *Proc. Natl. Acad. Sci. USA* **2002**, *99*, 10935–10940. [CrossRef] [PubMed]
5. Keeling, M.J.; Rohani, P. *Modeling Infectious Diseases in Humans and Animals*; Princeton University Press: Princeton, NJ, USA, 2011.
6. Kermack, W.O.; McKendrick, A.G. A contribution to the mathematical theory of epidemics. *Proc. R. Soc. Lond. Ser. A Contain. Pap. A Math. Phys. Character* **1927**, *115*, 700–772.
7. Feng, Z.; Huang, W.; Castillo-Chavez, C. On the role of variable latent periods in mathematical models for tuberculosis. *J. Dyn. Differ. Equs.* **2001**, *13*, 425–452. [CrossRef]
8. Castillo-Chavez, C.; Cooke, K.; Huang, W.; Levin, S.A. On the role of long incubation periods in the dynamics of acquired immunodeficiency syndrome (AIDS). *J. Math. Biol.* **1989**, *27*, 373–398. [CrossRef]
9. Hethcote, H.W. The mathematics of infectious diseases. *SIAM Rev.* **2000**, *42*, 599–653. [CrossRef]
10. Alonso-Quesada, S.; la Sen, M.D.; Agarwal, R.; Ibeas, A. An observer-based vaccination control law for an SEIR epidemic model based on feedback linearization techniques for nonlinear systems. *Adv. Differ. Equs.* **2012**, *161*, 1–32. [CrossRef]
11. Alonso-Quesada, S.; la Sen, M.D.; Ibeas, A. On the discretization and control of an SEIR epidemic model with a periodic impulsive vaccination. *Commun. Nonlinear Sci. Numer. Simul.* **2017**, *42*, 247–274. [CrossRef]
12. Degue, K.H.; Efimov, D.; Iggidr, A. Interval estimation of sequestered infected erythrocytes in malaria patients. In Proceedings of the 2016 European Control Conference (ECC), Aalborg, Denmark, 29 June–1 July 2016; pp. 1141–1145.
13. Efimov, D.; Perruquetti, W.; Raissi, T.; Zolghadri, A. On interval observer design for time-invariant discrete-time systems. In Proceedings of the 2013 European Control Conference (ECC), Zurich, Switzerland, 17–19 July 2013; pp. 2651–2656.
14. Efimov, D.; Polyakov, A.; Fridman, E.; Perruquetti, W.; Richard, J.-P. Delay-dependent positivity: Application to interval observers. In Proceedings of the 2015 European Control Conference (ECC), Linz, Austria, 15–17 July 2015; pp. 2074–2078.
15. Efimov, D.; Raissi, T.; Zolghadri, A. Control of nonlinear and LPV systems: Interval observer-based framework. *IEEE Trans. Autom. Control* **2013**, *58*, 773–778. [CrossRef]
16. Gouze, J.-L.; Rapaport, A.; Hadj-Sadok, M.Z. Interval observers for uncertain biological systems. *Ecol. Model.* **2000**, *133*, 45–56. [CrossRef]
17. Gucik-Derigny, D.; Raissi, T.; Zolghadri, A. A note on interval observer design for unknown input estimation. *Int. J. Control* **2016**, *89*, 25–37. [CrossRef]
18. Mazenc, F.; Dinh, T.N.; Niculescu, S.I. Interval observers for discrete-time systems. *Int. J. Robust Nonlinear Control* **2014**, *24*, 2867–2890. [CrossRef]
19. Khan, A.; Xie, W.; Zhang, L.; Liu, L.-W. Design and applications of interval observers for uncertain dynamical systems. *IET Circuits Devices Syst.* **2020**, *14*, 721–740. [CrossRef]
20. Khan, A.; Xie, W.; Bo, Z.; Liu, L.-W. A survey of interval observers design methods and implementation for uncertain systems. *J. Frankl. Inst.* **2021**, *358*, 3077–3126. [CrossRef]
21. Degue, K.H.; Efimov, D.; Richard, J.-P. Stabilization of linear impulsive systems under dwell-time constraints: Interval observer-based framework. *Eur. J. Control* **2018**, *42*, 1–14. [CrossRef]
22. Moisan, M.; Bernard, O. Interval observers for non monotone systems. Application to bioprocess models. In Proceedings of the IFAC Proceedings Volumes, Puebla, Mexico, 14–25 November 2005; Volume 38, pp. 43–48.
23. Rotondo, D.; Cristofaro, A.; Johansen, T.A.; Nejjari, F.; Puig, V. State estimation and decoupling of unknown inputs in uncertain LPV systems using interval observers. *Int. J. Control* **2018**, *91*, 1944–1961. [CrossRef]
24. Yousfi, B.; Raissi, T.; Amairi, M.; Gucik-Derigny, D.; Aoun, M. Robust state estimation for singularly perturbed systems. *Int. J. Control* **2017**, *90*, 566–579. [CrossRef]
25. Degue, K.H.; Ny, J.L. Estimation and outbreak detection with interval observers for uncertain discrete-time SEIR epidemic models. *Int. J. Control* **2020**, *93*, 2707–2718. [CrossRef]
26. Degue, K.H.; Ny, J.L. An interval observer for discrete-time SEIR epidemic models. In Proceedings of the 2018 Annual American Control Conference (ACC), Milwaukee, WI, USA, 27–29 June 2018; pp. 5934–5939.
27. Efimov, D.; Raissi, T. Design of interval observers for uncertain dynamical systems. *Autom. Remote Control* **2016**, *77*, 191–225. [CrossRef]
28. Mazenc, F.; Bernard, O. Asymptotically stable interval observers for planar systems with complex poles. *IEEE Trans. Autom. Control* **2009**, *55*, 523–527. [CrossRef]
29. Mazenc, F.; Bernard, O. Interval observers for linear time-invariant systems with disturbances. *Automatica* **2011**, *47*, 140–147. [CrossRef]
30. Diaby, M.; Iggidr, A.; Sy, M. Observer design for a schistosomiasis model. *Math. Biosci.* **2015**, *269*, 17–29. [CrossRef] [PubMed]

33. Robinson, E.I.; Marzat, J.; Raissi, T. Interval observer design for unknown input estimation of linear time-invariant discrete-time systems. *IFAC-PapersOnLine* **2017**, *50*, 4021–4026. [CrossRef]

34. Efimov, D.; Raissi, T.; Chebotarev, S.; Zolghadri, A. Interval state observer for nonlinear time varying systems. *Automatica* **2013**, *49*, 200–205. [CrossRef]

35. Chebotarev, S.; Efimov, D.; Raissi, T.; Zolghadri, A. Interval observers for continuous-time LPV systems with L1/L2 performance. *Automatica* **2015**, *58*, 82–89. [CrossRef]

36. Guo, S.; Zhu, F. Interval observer design for discrete-time switched system. *IFAC-PapersOnLine* **2017**, *50*, 5073–5078. [CrossRef]

37. Khan, A.; Xie, W.; Zhang, L. Interval state estimation for linear time-varying (LTV) discrete-time systems subject to component faults and uncertainties. *Arch. Control Sci.* **2019**, *29*, 289–305.

38. Liu, L.-W.; Xie, W.; Khan, A.; Zhang, L.-W. Finite-time functional interval observer for linear systems with uncertainties. *IET Control Theory Appl.* **2020**, *14*, 2868–2878. [CrossRef]

39. Yi, Z.; Xie, W.; Khan, A.; Xu, B. Fault detection and diagnosis for a class of linear time-varying (LTV) discrete-time uncertain systems using interval observers. In Proceedings of the 2020 39th Chinese Control Conference (CCC), Shenyang, China, 27–29 July 2020; pp. 4124–4128.

40. Bliman, P.-A.; Barros, B.D.A. Interval observers for SIR epidemic models subject to uncertain seasonality. In Proceedings of the International Symposium on Positive Systems, Fukuoka, Japan, 4–5 November 2016; pp. 31–39.

41. Hu, Z.; Teng, Z.; Jiang, H. Stability analysis in a class of discrete SIRS epidemic models. *Nonlinear Anal. Real World Appl.* **2012**, *13*, 2017–2033. [CrossRef]

42. Mickens, R.E. Numerical integration of population models satisfying conservation laws: NSFD methods. *J. Biol. Dyn.* **2007**, *1*, 427–436. [CrossRef] [PubMed]

43. Bichara, D.; Cozic, N.; Iggidr, A. On the estimation of sequestered infected erythrocytes in Plasmodium falciparum malaria patients. *Math. Biosci. Eng.* **2014**, *11*, 741–759. [CrossRef]

44. Hooker, G.; Ellner, S.P.; Roditi, L.D.V.; Earn, D.J. Parameterizing state–space models for infectious disease dynamics by generalized profiling: Measles in Ontario. *J. R. Soc. Interface* **2011**, *8*, 961–974. [CrossRef] [PubMed]

45. Moore, R.E. *Interval Analysis*; Prentice-Hall: Hoboken, NJ, USA, 1966.

46. Ibeas, A.; Sen, M.D.; Alonso-Quesada, S.; Zamani, I. Stability analysis and observer design for discrete-time SEIR epidemic models. *Adv. Differ. Equs.* **2015**, *2015*, 122. [CrossRef]

47. Rachah, A. Analysis, simulation and optimal control of a SEIR model for Ebola virus with demographic effects. *Commun. Fac. Sci. Univ. Ank. Ser. A1 Math. Stat.* **2018**, *67*, 179–197.

MDPI

St. Alban-Anlage 66

4052 Basel

Switzerland

Tel. +41 61 683 77 34

Fax +41 61 302 89 18

www.mdpi.com

Fractal and Fractional Editorial Office

E-mail: fractalfract@mdpi.com

www.mdpi.com/journal/fractalfract